IET TRANSPORTATION SERIES 30

ICT Solutions and Digitalisation in Ports and Shipping

Other related titles:

Volume 1	**Clean Mobility and Intelligent Transport Systems** M. Fiorini and J.-C. Lin (Editors)
Volume 2	**Energy Systems for Electric and Hybrid Vehicles** K.T. Chau (Editor)
Volume 5	**Sliding Mode Control of Vehicle Dynamics** A. Ferrara (Editor)
Volume 6	**Low Carbon Mobility for Future Cities: Principles and applications** H. Dia (Editor)
Volume 7	**Evaluation of Intelligent Road Transportation Systems: Methods and results** M. Lu (Editor)
Volume 8	**Road Pricing: Technologies, economics and acceptability** J. Walker (Editor)
Volume 9	**Autonomous Decentralized Systems and Their Applications in Transport and Infrastructure** K. Mori (Editor)
Volume 11	**Navigation and Control of Autonomous Marine Vehicles** S. Sharma and B. Subudhi (Editors)
Volume 12	**EMC and Functional Safety of Automotive Electronics** K. Borgeest
Volume 16	**ICT for Electric Vehicle Integration with the Smart Grid** N. Kishor and J. Fraile-Ardanuy (Editors)
Volume 17	**Smart Sensing for Traffic Monitoring** N. Ozaki (Editor)
Volume 18	**Collection and Delivery of Traffic and Travel Information** P. Burton and A. Stevens (Editors)
Volume 25	**Cooperative Intelligent Transport Systems: Towards high-level automated driving** M. Lu (Editor)
Volume 26	**Traffic Information and Control** R. Li and Z. He (Editors)
Volume 38	**The Electric Car** M.H. Westbrook
Volume 45	**Propulsion Systems for Hybrid Vehicles** J. Miller
Volume 79	**Vehicle-to-Grid: Linking electric vehicles to the smart grid** J. Lu and J. Hossain (Editors)

ICT Solutions and Digitalisation in Ports and Shipping

Edited by
Michele Fiorini and Natalie Gupta

The Institution of Engineering and Technology

Published by The Institution of Engineering and Technology, London, United Kingdom

The Institution of Engineering and Technology is registered as a Charity in England & Wales (no. 211014) and Scotland (no. SC038698).

First published 2021

The Institution of Engineering and Technology
Michael Faraday House
Six Hills Way, Stevenage
Herts, SG1 2AY, United Kingdom

www.theiet.org

British Library Cataloguing in Publication Data
A catalogue record for this product is available from the British Library

ISBN 978-1-83953-086-9 (hardback)
ISBN 978-1-83953-087-6 (PDF)

Typeset in India by MPS Limited
Printed in the UK by CPI Group (UK) Ltd, Croydon

Contents

About the editors

Michele Fiorini, MBA, Ph.D., CEng, FIET acts as a principal in large international projects and global organisations where he leads (or is a member of) cross-cultural teams providing guidance, problem-solving attitude and engineering management as well as educational mentoring. His work focuses on intelligent and cognitive systems and engineering management for research and innovation strengthening cooperation between industry and academia. He has been the Chair of the Council at the Institution of Engineering and Technology (London, United Kingdom, 2017–18) and a session chair at the Euro-Asia Economic Forum (Xi′an, China, 2013). He is a project engineering manager for Leonardo s.p.a. in Rome, Italy, and an industrial member of the e-Navigation Committee at the International Association of Marine Aids to Navigation and Lighthouse Authorities (IALA-AISM) in Paris, France.

Michele is serving as a member of the advisory board of the MBA programme at the Gdańsk University of Technology in Poland and a judge for the E&T Innovation Awards, intelligent systems category in London, United Kingdom.

Natalie Gupta is an independent consultant focusing on ports and trade facilitation. She works through her own consultancy Port Processes Ltd and is the Chair of Independent Port Consultants (IPC), a network of port consultants working internationally. Natalie works on three main streams of projects, namely feasibility studies for ports and the maritime sector, trade facilitation in the context of ports and trade corridors, and port operations-related assignments to do with performance and management. Natalie also has experience of working on bunkering in the sector and is involved in projects concerning the greenification of the sector in various country contexts.

List of contributors

Angelica M. Baylon, Ph.D., MBGPH, MSBM, Ph.D. MBA, MS Chem., BS Chem., AFNI, MIMarEST, FDR, DFRIMM, SFRIEdr, RIIVPF, DFRIEdr, RIRF is a Director at the Maritime Academy of Asia and the Pacific (MAAP) in the Philippines. She has more than 25 years of teaching experience and honorary and visiting professor positions around the world. At age 30, she earned Professor IV rank at the Philippine Women University (PWU) Graduate School. She was awarded Honorary Professor by the Australia-Asia Research and Educational Foundation (AAREF). She is regularly invited and sent locally and globally as a resource speaker, trainer, presenter and facilitator in Europe and Africa (Italy, Poland, Sweden, Germany, Turkey, France, Russia, Denmark, United Kingdom, Romania, Bulgaria, Egypt and Ghana Africa); the Americas (Canada and USA – Tennessee, Arizona, Nevada, California, Florida, Maryland, New Jersey, New York, Washington, DC and Guam); and the Asia-Pacific and Oceania (Thailand, Malaysia, Korea, Indonesia, Myanmar, Japan, China, Hong Kong, Nepal, Singapore, India, Taiwan, Bangladesh, Vietnam, Doha Qatar, Pakistan and Australia). She is a member of various International Editorial Board for numerous publications in Australia (AUMII), China (Bit Conference), Poland (TRANSNAV), Romania (JMTE), Turkey (JEMS), USA (AUSN) and the United Kingdom (KIE). She is PAEPI-Global Chairman Emeritus.

Valentin Carlan is a PostDoc researcher at the department of Transport and Regional Economics at the University of Antwerp. His research interests include innovation assessment models, cost and benefits quantification methods for transport-related innovations, and applications of cost-effectiveness analysis for IT introduction. He obtained his Ph.D. title in Applied Economics (2019) from the University of Antwerp by defending his Ph.D. thesis titled 'Maritime supply chain innovation: costs, benefits and cost-effectiveness of ICT introduction'. During the past years he had carried research within the BNP Paribas Fortis research Chair for Transport Logistics and Ports (2014–16), and on projects like Clusters 2.0, Truck Guidance System (together with IMEC), developing a Universal Middleware Framework for Data Streaming or the SPEED Interreg project. Since 2019 he has been the cofounder of a spin-off of the University of Antwerp, DigiTrans. DigiTrans provides tailor-made digital assistants for planning operations and AI algorithms for data parsing from emails, digital documents or external sources.

Jillian Carson-Jackson, B.Ed., FNI, FRIN has experience of more than three decades in the maritime environment, promoting education, training, diversity and inclusion. With practical experience, both afloat and ashore, she has management experience in different government areas, including aids to navigation (AtoN), vessel traffic services (VTS), regulation of maritime pilots, vessel tracking and maritime technology related fields. Through a varied career, Jillian has worked at the operational, technical, policy and legislative level. This includes work with the International Maritime Organization (IMO), International Telecommunication Union (ITU), International Electrotechnical Commission (IEC) and the International Association of Marine Aids to Navigation and Lighthouse Authorities (IALA) in addition to her work with the Canadian Coast Guard and the Australian Maritime Safety Authority. She currently manages her own maritime consultancy, providing a focus on technical advice, maritime education and training, and port risk assessment.

Jillian is currently the Chair of the IALA ENAV Committee WG 2 (emerging digital technologies) and the IALA VTS Committee WG3 (Training and Personnel). She has recently been appointed as a director of GlobalMET and is currently the President of the Nautical Institute.

Konstantinos Chatzikokolakis received the B.Sc., M.Sc. and Ph.D. degrees from the Department of Informatics and Telecommunications, National and Kapodistrian University of Athens, in 2008, 2012 and 2016, respectively. He participated in several European projects, including METIS 2020, BigDataOcean and SELIS. His research interests include datamining, machine learning, big data, and distributed and cloud computing.

Ioannis (Yannes) Filippopoulos is a 'technology architect', a 'team leader', an information systems and a digital transformation specialist with a passion exceptional aptitude for working with computers combined with an astounding academic background in computer and software engineering (Ph.D.) and business management (MBA). He is also teaching and researching of innovative systems and platforms, and he is specialised in Geographical Information Systems. For more than 8 years, he has been specialising in Internet of Things (IoT), big data, artificial intelligence (AI), blockchain platforms in shipping industry as CIO at a top-performing maritime company (Angelicoussis Group). Today, he is expanding the research activities to other industries as an Academic Director of Informatics and Engineering programmes and an Assistant Professor at Hellenic American University, and as an Adjunct Professor and Postdoctoral Researcher at University of Thessaly.

Jiequn Guo currently serves as a Dean of Ningbo China Institute for Supply Chain Innovation (NISCI), which is a member of MIT Global SCALE Network. Jay was a Deputy Director of the Monetary Policy and Financial Stability Research Center at Tsinghua University, China. Prior to that role, he was managing director at the Zais Group, a hedge fund based in Redbank, NJ and served as a General Manager for Zais Asia-Pacific Group. Jay received his Ph.D. in Economics from Indiana University.

Craig Matthew Hancock is currently a Senior Lecturer in Surveying at Loughborough University. He has a Ph.D. in Space Geodesy from Newcastle University and has published almost 100 papers on GNSS, Laser Scanning, Geospatial Engineering and BIM.

Georgios Kapogiannis is currently the M.Sc. Director of Geospatial Engineering with BIM at the University of Nottingham Ningbo China, an Asst. Professor in Building Information Modelling at the same University and is a Faculty Visiting Member of Ningbo China Institute for Supply Chain Innovation (NISCI), which is a member of MIT Global SCALE Network. He has won four prizes in Innovation in Construction in China and Worldwide. His Ph.D. is in Construction and Project Management from the University of Salford (UK) with EU scholarship. He has been a member of the peer review college of the Economic and Social Research Council (ESRC) since 2015 (after invitation) and served as a Chairman for the Chartered Institute of Building (CIOB) for Coventry and Warwickshire branch between 2013 and 2015.

Jeptoo Kipkech is a Doctoral Researcher in Maritime Cybersecurity in the Systems Security Group at the Institute of Future Transport and Cities (IFTC) at Coventry University. She specialises in cybersecurity in the maritime domain. She has a Master's of Science in Cybersecurity and Management from the University of Warwick and has worked previously as a Cybersecurity Consultant, IT Manager and Software Developer.

Aristotelis Komianos is an Associate Fellow of the Nautical Institute. He is a former Naval Officer with seagoing experience up to the duties of Commanding Officer, and a Class 1 Master Mariner. He holds a Master of Science in Shipping and an MBA in Human Resources Management. His research interests include innovation in shipping, environmental protection, maritime safety and cybersecurity. His latest book is *Towards the Autonomous Ship-Operational, Regulatory, and Quality Challenges*.

Kristen Kuhn is a Researcher in Maritime Cybersecurity in the Systems Security Group at the Institute of Future Transport and Cities (IFTC) at Coventry University, United Kingdom. She specialises in cybersecurity decision-making and the maritime industry. She is currently working on the 'Cyber Readiness for Boards' (CR4B) project, a 2-year initiative funded by Lloyds Register and the UK National Cyber Security Centre. The CR4B project investigates how corporate boards assess cyber-risk and make investment decisions about cybersecurity.

Alexandra Middleton is an Assistant Professor at Oulu Business School, University of Oulu (Finland). Alexandra's areas of expertise include Arctic sustainable business development, logistics, human capital, Arctic innovations and connectivity solutions.

Eirik Næsje is a Senior Vice President of Vessel Insight at Kongsberg Digital. Næsje has a Master of Marketing from BI Norwegian Business School and broad experience with SaaS, technology, IT, data-driven market analysis and customer satisfaction. He has 13 years of international experience from different management positions in a number of Software-as-a-Service companies, including Cxense, where he was both General Manager for EMEA and VP of Marketing, and Meltwater where he held the position of an Area Director of Norway/Denmark.

Jan-Hendrik Oltmann is a senior strategic adviser to the German Federal Waterways and Shipping Administration and works at the Aids-to-Navigation (AtoN)/Vessel Traffic Services (VTS) department of that Administration. He received his Master Degree (Dipl.-Ing.) in Electrical Engineering from Darmstadt Technical University, Germany, in 1992. Since he joined the Administration in 1993, he worked in diverse fields of technology application comprising terrestrial radio navigation services for the maritime domain, the application of transponder technology to VTS as well as ICT system architecture frameworks for shore-based AtoN/VTS systems. In that occupation, he had an active role in the developments of the global automatic identification system (AIS), the overarching system architecture framework for IMO's e-navigation strategy and internationally harmonised data structures for the previous fields. He managed several domestic and international projects, chaired several international groups and was a speaker at international conferences.

Peter Pokorny is currently a consultant in the field of maritime and radiocommunications. Previously, he was a Manager of Maritime Services (Australia) with Kordia Pty Ltd and was a Principal Advisor, Maritime Communications at the Australian Maritime Safety Authority. Peter was active in International Maritime and Radiocommunications radio fora and chaired the joint Australia/New Zealand Standards Committee (Radiocommunications equipment – Maritime and Safety of Life).

Peter spent 5 years at sea as a deck/navigation officer on international trading vessels, then he went ashore to pursue radio-electronic engineering in 1982. Highlights include wintering as a radiocommunications technical officer on Australian Antarctic scientific bases (Davis Station and Macquarie Island) and an Antarctic marine science voyage. Peter has been a Senior Technical Officer at Perth Radio (VIP) maritime communications station, a lecturer on the Global Maritime Distress and Safety System, Celestial Navigation, Visual Communications and Electronics at the Australian Maritime College. Peter holds a Bachelor of Applied Science (Nautical Science), Advanced Diploma of Applied Science (Chief Mate/Shipmaster), Deck Watchkeeper Certificate, Advanced Diploma of Industrial Data Communications, Networking and IT, Diploma in Electronic Engineering and GMDSS Radio Electronic Certificate.

Satya Prasad Sahu ('SP' to colleagues) is a Senior Trade Facilitation Specialist with the World Bank's Global Practice on Trade and Regional Integration. He specialises in the use of ICT in trade facilitation and works extensively on projects and analytical activities concerning the digitisation of international trade, transport and regulatory procedures.

Periklis Saragiotis is a Senior Private Sector Specialist with the World Bank's Trade and Regional Integration Global Practice. He leads the design and implementation of World Bank programmes aiming at best practice national trade facilitation policies and reforming action plans. He specialises on seaport and maritime transport and leads advisory and lending projects built around the adoption of port-centric trade facilitation reforms leading to trade competitiveness and private sector development.

Jens Schröder-Fürstenberg studied Navigation at the University of Rostock. He sailed on bulk carrier and container feeder vessels before moving ashore to work with TEAMLINES GmbH in Hamburg at ship operation and the ship-handling department for many years.

Since 1999, Jens has worked for the German Maritime and Hydrographic and Agency (BSH). He started as a Sailing Direction Editor and is now the Head of the Nautical Information Service Division. He is the Chairman of the IHO Standardization of Nautical Publication Working Group.

Since 2003, he has been working as an ECDIS basics instructor at the Maritime Education and Training Centre Warnemünde of University of Applied Sciences Wismar. He is married and has one daughter.

Siraj Ahmed Shaikh is Professor of Systems Security and Director of Research at the Institute of Future Transport and Cities (IFTC) at Coventry University. He is also a Co-Founder and Chief Scientist at CyberOwl, which addresses risk analytics for maritime and CNI sectors. He is currently involved in the 'Cyber Readiness for Boards' initiative, funded jointly by the National Cyber Security Centre (NCSC) and Lloyds Register Foundation (LRF) to assess the cyber-risk perception of boards in the maritime sector.

Giannis Spiliopoulos received the Diploma and M.Sc. degrees from the Electronics and Computer Engineering Department, Technical University of Crete, in 2008 and 2010, respectively. He has been a member of Technical Chamber of Greece, since 2009. He has worked as a Software Engineer and an Analyst in the private sector, for more than 6 years. His research interests include statistical analysis of spatial data and data-driven extraction of spatiotemporal patterns using unsupervised machine learning methods.

Marios Vodas received the B.Sc. and M.Sc. degrees in informatics from the University of Piraeus, Greece, in 2011 and 2013, respectively. He is currently pursuing a Ph.D. degree with the Department of Informatics, University of Piraeus. He is also a Researcher with MarineTraffic. His research interests include spatiotemporal indexing and mining on big data.

Yung Ho Yu graduated from Korea Maritime and Ocean University (KMOU) in 1974. After doing Korean Navy service, he worked on a commercial ship as a chief engineer. He joined the Korea Register of Shipping as a ship technical surveyor then received his Masters and Ph.D. degree on Control Engineering at KMOU. He was a professor of IT Division for 25 years and the head of Advanced IT and Ship Convergence Centre in KMOU. After retiring from KMOU, he worked for MEIPA (Korea Marine Electronic Industry Promotion Association) as the head of researcher of OSICT (Ocean and Ship ICT Convergence Centre). He is a member of IALA eNav and ARM Committee, IEC TC80 WG10, MT5, WG15, WG16, WG17, ISO TC8 WG10, NMEA OneNet, NMEA 2000 and RTCM.

He is the Chairman of Korea Smart and Autonomous Ship Forum and PG607 of TTA (Telecommunication Technology Association) for domestic standard activity. He is very much interested in secured ship network standard and technology, e-navigation and smart ship platform, MSP (Maritime Service Portfolio), etc.

Dimitris Zissis is currently an Associate Professor with the University of the Aegean. His published scientific work includes more than 70 publications that have received more than 2,300 citations to date. His research interests and areas of expertise include several aspects of architecting and developing complex information systems (IS), including distributed and cloud-based big data deployments. He is a member of the editorial boards of Future Generation Computer Systems (FGCS) (Elsevier) and the International Journal of Internet of Things and Cyber-Assurance (Inderscience) and a PC Member for numerous conferences, including CLOUDCOM, GECON, SerCO and others. He is an IEEE Senior Member and a member of the IEEE Computer Society, IEEE Oceanic Engineering and the IEEE Intelligent Transportation Systems Societies and the Young Researchers Committee of the World Federation on Soft Computing. His professional experience includes senior consulting and researcher positions in a number of private and public institutions.

Foreword

The authors of this comprehensive book, 'ICT Solutions and Digitalisation in Ports and Shipping', edited by **Michele Fiorini** and **Natalie Gupta** (IET 2020/2021), came from a wide variety of maritime institutions composed of eminent engineers, researchers and scientists in the global maritime arena.

Dr Michele Fiorini, Dr Natalie Gupta and other fellow writers have cumulative global experiences and expertise that, when combined, had resulted in an influential book on 'ICT Solutions and Digitalization in Ports and Shipping'. Indeed, an 'A rate' because my good friend Dr Michele Fiorini, the Chair of the Institute of Engineering and Technology (IET) Italy, is one of the prime movers of this book project. In 2009, we initially met at the International Conference on Marine Navigation and Safety of Sea Transportation (TRANSNAV) organised by the Faculty of Navigation of the Gdynia Maritime University in Gdynia, Poland. Our professional cooperation continues, constantly presenting papers in TRANSNAV and other publications regularly. I have known Dr Michele Fiorini for more than 10 years as a dynamic respectable scholar, highly technical person who can transform difficult topics into easy ones to make both maritime students and professionals understand instructions.

The book is designed for researchers, chief technology officers, system architects, system and principal engineers, consultants and students who are or will be involved in information and communications technology (ICT) in maritime transport and ports, including software design, testing, performance analysis and validation. Chairpersons, chief executives, commercial directors of ports, managing and operations directors somewhat involved in ports, port operations, port environments and traffic monitoring and management systems, authorities and business persons who aim to have a complete overview of ICTs and trade initiatives in seaports and port-to-port traffic would also find this book very useful. No previous experience in ICT, systems engineering or port-logistics is required to appreciate the book. However, a generic knowledge of communication engineering and logistics would be beneficial.

The challenges in communications and information technology are both problematic and fascinating. People are working on them with enthusiasm, tenacity and dedication to develop new analysis methods and provide unique solutions to keep up with the ever-changing threats. In this modern age of global interconnectivity and interdependence, it is necessary to offer ICT practitioners, professionals and students some state-of-the-art knowledge on the frontiers. The book provides an overview and analysis of the technologies suitable for different types of port operations and traffic monitoring and management systems. It may also fill a need and provide solutions through various options and alternative examples of many issues and concerns. I admire and salute the initiators and contributors for their joint effort and conceptualising this unique and vital book through the various well-thought chapters.

The broad topics covered by this book indicate the breadth of perspectives and insights, the scope of experiences and practices, and the functional integration of theory and concrete examples brought together by the global authors who are experts on the subject of ICT in shipping, ports and international logistics. The book is composed on the basis of high-quality, scholarly research that is structured in 13 chapters. The topics are comprehensive from global trade facilitation regulatory framework, maritime ports and cybersecurity, e-navigation and shore-based monitoring systems, maritime transportation along the Northern Sea Route, SMART shipping beyond e-navigation, maritime communications, a data-driven methodology for navigational Patterns of Life discovery, real-time information in ports with ITS technology, digital supply chain and port information modelling, hinterland users of ICT innovation in the MarSC: achieving cost-effective tools by integrating horizontal data, decarbonisation technologies in shipping and the question of transition fuels and shore-based navigation: from vessel traffic services to e-navigation Maritime Service Portfolio.

It can be surmised that the authors from all over the world had contributed in examining the application of ICT on the movement and clearance of freight at seaports globally and evaluating the feasibility, use and impact of different types of ICTs on port and maritime navigation port-to-port.

The shipping and ports industry has been comparably slow in engaging with the concept of digitalisation relative to other modes of freight and passenger transfer. This is partly due to the focus on freight rather than passengers (except ferries and cruise terminals) and market and ownership factors. The importance of ICTs in the industry is, however, rapidly being recognised by many. Very advanced examples already exist in the number one ports by volumes such as Singapore, Hamburg, Rotterdam and others. However, only 3% of the world's ports are massive, with over 50 million tonnes of cargo throughput per year. Furthermore, over 90% are small or very small, with less than 5 million tonnes of cargo throughput annually. The implementations of ICT solutions and digitalisation result are the determinants of investment decisions on the development of these emerging ports.

Students, teachers, port operators, shipping managers, policy-makers, project planners, professionals, researchers and scientists are encouraged to read this exciting book on 'ICT Solutions and Digitalisation in Ports and Shipping' and be inspired to think, rethink and contribute their share for sustainable development in ports and shipping. This book is a good step in that direction.

Prof. **Angelica M. Baylon**, Ph.D., MBGPH, MSBM, Ph.D. MBA, MS Chem, BS Chem, AFNI, MIMarEST, FDR, DFRIMM, SFRIEdr, RIIVPF, DFRIEdr, RIRF

Chairman Emeritus of **PAEPI-Global**

Fellow TRANSNAV, **GMU, and NI Poland**

Hon Sec, **IMarEST Ph Branch, NI Ph Branch** and **Institute of Maritime Management, RI Singapore**

ERO Director, **Maritime Academy of Asia and the Pacific – Philippines**

Preface

The society of the future requires discernment and adaptation. The economic crisis, triggered by the COVID-19 pandemic, is a major disruptor to many sectors but also an opportunity to evolve towards more purpose-driven sustainable actions. In 2020, the economy contracted at a rate comparable to that experienced by large countries during the World War II. Polls at the EU level show that three out of four people consider the protection of environment a priority for governments and expect changes in industries production cycles and lifestyles to cope with public health. In the United States it is estimated that 20% of office working hours are already permanently shifted to homes.

The VUCA (volatile, uncertain, complex and ambiguous) time we are living in is forcing decision-makers to demonstrate flexibility and adaptability from traditional activities to the more innovative online ones. Adapt or die seems to be the more appropriate message to tradition business-owners nowadays. We are entering a decade of disruption. The maritime sector is traditionally very conservative; however, it is not excluded from the dynamics. The introduction of the electronic means concept to collect, integrate and display maritime information aboard and ashore started by an International Maritime Organization (IMO) Expert Group on e-navigation in 2006 as an ambitious project is currently more actual than ever. This book aims to be a portfolio of scientific and operational outputs from experts and leading workers in the ports logistics and maritime shipping from around the world.

The basic idea started to be conceived from various conversations during events, working groups and online meeting during the last 3 years, including the 'Cyber Security in Transport' seminar organised by the IET Transport systems Technical Professional Network (TPN) held at the IET Birmingham: Austin Court, in December 2018 chaired by Dr John Walker. Experts in the field suggested the development of an edited book particularly dealing with the role played by information and communication technologies (ICT) in ports. This evolved into a book that touches on ICT technologies in the sector but not only.

The resulting material is organised as follows:

- Chapter 1 introduces modern trade automation and digital systems as core trade facilitation programmes. Information and communications technology (ICT) applied in modern trade control and facilitation tools constitutes a remarkable enabler for the implementation of the World Trade Organization – Trade Facilitation Agreement.
- Chapter 2 is about cybersecurity threats and vulnerabilities and how they are being addressed in the port industry. It introduces the modern port in Section 2.1 and explores its importance in Section 2.2. It then presents

cybersecurity in Section 2.3 and draws a connection between ports and cybersecurity in Section 2.4, highlighting ports as a cyber–physical environment. To illustrate the cyberthreat landscape, five known attacks in ports are explored in Section 2.5. This chapter then examines control mechanisms in place for cyber risk management for ports in Section 2.6 and reviews current cybersecurity guidelines and standards in Section 2.7. These concepts are encapsulated in the summary presented in Section 2.8, which mentions digital trends and the future of maritime ports and cybersecurity.

- Chapter 3 describes the current composition of information needed to navigate a ship safely based either on paper documents or on electronic systems. It provides an overview of the achievements and the ongoing work of the main involved parties, the IMO and the International Hydrographic Organization (IHO). The IMO as the main legal body has introduced the e-navigation concept and associated services. In parallel, the IHO has developed the S-100 standard as the Universal Hydrographic Datamodel which in turn has been accepted as the underlying basis of IMO's e-navigation.
- Chapter 4 provides a comprehensive review of shipping activity along the Northern Sea Route (NSR) by type of shipping, type of cargo, including container shipping. Additionally, a case study of how global demand for liquefied natural gas is affecting the development of the NSR is highlighted. Finally, the future of the NSR development by looking at sustainability requirements in the Arctic and worldwide that shape the future of the NSR is discussed. Smart shipping is likely to affect global shipping; thus several initiatives for the development of smart shipping for the NSR are discussed too. In conclusion, the chapter provides an outlook for the development of shipping along the NSR in the future.
- Chapter 5 discusses the purpose of smart and autonomous ships as well as the technologies and services necessary for efficiency in the context of the recent investments in this field by the South Korean Government and compliance requirements with IMO environmental regulation.
- Chapter 6 is looking at the developments around maritime communications in a digital age; the existing and evolving roles and responsibilities for maritime communications; what is changing in the industry, and why; an overview of existing and evolving technologies for maritime use; and a glimpse into the future.
- Chapter 7 is a review with some critical considerations from the author about modern radiocommunication systems and technologies used on maritime domain.
- Chapter 8 presents a discussion on a data-driven methodology suitable for discovering maritime traffic patterns and revealing 'roads of sea'. The proposed solution exploits the MapReduce paradigm to perform parallel distributed processing of large vessel tracking datasets, collected through the automatic identification system, and analyses vessels' navigational patterns in a computationally efficient and accurate way.

- Chapter 9 evolves around comprehensive literature on the integration of ITS technologies in the maritime industry describing how ITS technologies can be used as decision support systems in the hands of seafarers as well as operators to maximise efficiency, which in turn, supports the financials of the companies.
- Chapter 10 presents how digital transformation could inspire innovation within the supply chain in ports. It consists of three core themes: digital supply chain transformation, port information modelling and innovation and integrated supply chain management.
- Chapter 11 presents findings from primary data collection activities aimed to establish the cost-effectiveness and decision-making criteria for ICT innovation in ports specifically geared towards improving the port–hinterland interface processes and procedures.
- Chapter 12 discusses the advent of alternative fuels in the shipping industry and the technologies concerning this dynamic. The question of transition fuels in the context of existing regulatory initiatives to mitigate climate change and decarbonise the sector by mid-century provides the basis of this chapter.
- Chapter 13 introduces the shore-based navigation concept presenting the improvement of maritime services from the three traditional vessel traffic services (VTSs) to the sixteen e-navigation Maritime Service Portfolio.

  The chapter includes three appendixes presenting respectively:

  Appendix A: a system engineering approach to VTSs /fleet operations centres/ coastal surveillance systems (VTS/FOC/CSS) architecture;
  Appendix B: an example of VTS/FOC/CSS system's functions; and
  Appendix C: an overview on maritime autonomous surface ship.

The overall aim of this book is to identify, discuss, and in so doing, provide insights for port authorities, VTS operators and administrations, ship owners and investors or those acting on their behalf, on ICT available in support of trade facilitation initiatives in seaports and from port to port. The book aims to provide an overview of the processes and technologies suitable for different types of port operations, port environments and traffic monitoring and management systems and authorities. It seeks to bring together a review of the application of different types of ICT and their application in order to assess possible gaps in the current supply- and demand-side market for this type of technological solution.

Among the reasons for editing this book, one is to provide some conceptual clarity on the applicability of the different dimensions of ICT solutions and digitalisation in maritime domain which are broad concepts that include many different technologies (both infrastructure and processes/information/software related).

Many topics touched upon in the chapters are currently subject of further investigation and research: they are open arguments. The presented concepts and operative implementations, often just speculative, should be seen as preliminary or following empirical approaches currently under discussion at international level for possible further investigation, review and suggestion. The book does not pretend to be exhaustive, so criticisms and comments are most welcome.

Enjoy reading the book.

Michele Fiorini and Natalie Gupta
Scientific Editors
Rome and London, November 2020

Acknowledgements

We are indebted to many colleagues, too numerous to mention all, for having inspired, supported – with many conversation, criticisms and comments – and encouraged us on a number of occasions. International cooperation forms the foundation of this project. Technical discussion, opportunities to consult with experts, informative conversations have been invaluable for enriching and shaping the final outputs. Cooperation between academia and industry are always challenging, we are very thankful to Mathilde Vik Magnussen, Vice President Communications & Marketing, Kongsberg Digital, for helpful cooperation in realising the interview with Eirik Næsje, Senior Vice President of Vessel Insight at Kongsberg Digital (cf. Introduction).

Our appreciation goes to all authors of the chapters for their valuable contribution to such an ambitious book project. They provided timely, authoritative and readable contributions on their areas of expertise and experiences. With some of them, our professional cooperation runs from years and brings to friendship which is definitely an added value to the profession. We are confident that this book is a noteworthy exhibit of the chapter authors' professions.

The Editors would also like to acknowledge the professional skills of the IET editorial team, the publisher – academic books, Paul Deards, the senior commission editor, Christoph von Friedeburg and assistant editor Olivia Wilkins. We thank them very much for their excellent support in maintaining the high quality of this book.

Michele Fiorini and Natalie Gupta
Scientific Editors
Rome and London, November 2020

Introduction

This edited volume provides commentary on the nature of the changing dynamics in the sector and pin-points some of the key challenges that are also being faced. Enabling the maritime sector to become more technologically advanced and digital is increasingly being seen as one of the main priorities of the international maritime community. During the preparation of the book, we have had the privilege to discuss it with Eirik Næsje, Senior Vice President of Vessel Insight at Kongsberg Digital. His view and insights were to us the best introduction to this edited volume.

I.1 How far has the maritime industry come in terms of digitalisation?

We view digitalisation as a journey and believe we are still in the early stages. The end state for digital transformation with everything connected and running automatically and semi-autonomously is still only a theoretical end state and a lot needs to be done before we get there. The first step is ensuring vessel connectivity on the operator side and enough software on the vendor side to ensure data captured from vessels can be used to increase efficiency, safety and sustainability at sea.

In the last years, digitalisation of the industry has been proceeding at an increasing pace, and, while in the beginning it was mostly pursued by larger and shipping companies, we see that increasingly also smaller shipping companies are starting to explore maritime software solutions to optimise their businesses. It is still somewhat slow on a broad scale; however, something that can probably be credited to several factors, but in our opinion the three main reasons are the lack of quality data, high entry costs and lack of 'real world' use-cases with healthy ROIs. That is about to change. With large parts of the older inefficient tonnage being scrapped during the last years, ship operators are now inevitably left with the task to optimise the operations of their existing fleets, and digital tools will play a key role in this. Also, the number of newbuilds equipped with additional sensors is slowly but steadily increasing – also across the less 'sophisticated' shipping segments. This is indicating that ship owners have come to acknowledge the fact that they need to get ready for the digital age ahead of them.

In addition, we see that the solutions and services available to maritime players are getting more mature, and it can be expected that the focus will broaden in the short term from fuel and energy management solutions – which have dominated maritime digitalisation for a long time – towards other less well-established areas, such as document handling, repair and maintenance and port call optimisations.

We expect cybersecurity to receive its deserved attention with the new IMO guidelines effective from 1 January 2021. Connecting vessels in order to get access to vessel data and sharing it with value-added applications obviously also poses a risk, making it crucial to understand the exposure and take mitigating actions. The ones that will be successful will find the right balance between protecting their systems and opening them up in order to explore the business potential of digitalisation.

Furthermore, we also see that 'real-world' use cases are starting to emerge, for instance Golden Ocean significant savings on operational fuel usage, as solutions are getting more mature and the cost of getting access to quality data is going down [1]. Finally, some will say, as conferences and articles on maritime digitalisation have tended to dive into topics and technologies already years ahead of the daily world of most ship owners in the industry. Combining that with a strengthened focus on more sustainable, efficient and greener operations, we feel confident that 2021 will be the year where the early adopters will start to reap the benefits of their investments.

I.2 What are the drivers?

There are many drivers for digitalisation in the maritime industry. First, digital solutions can help reduce operational expenses (OPEX) such as fuel, port charges. In an industry that is highly competitive and where margins are low, the pressure to reduce OPEX is high and various digital solutions can help with operators' specific needs. Second, new technologies are increasing the viability of making the business more digital. Connectivity is becoming cheaper, processing power is increasing, it is possible to print spare parts in 3D printers and electric and hybrid solutions are becoming more common. Third, new regulations require vastly more reporting. Emission monitoring, EU-MRV (monitor, reporting and verification) reporting and ensuring IMO compliance can be done digitally and, to a large extent, automatically. This will reduce man hours and increase accuracy in reporting. Finally, new expectations are emerging when it comes to transparency and the sharing of information between operators and clients, operators and vendors and operators and governments. Digitalisation will make it easier to increase transparency and share what should be transparent, while ensuring that business information is held securely in the operators' hands.

I.3 What are the most obvious benefits of digitalisation (i.e. for fleet owners and operators)?

The most obvious short-term benefits of digitalisation are tied to OPEX reductions such as energy and fuel savings, smarter maintenance, smoother logistics and automatic reports to reduce workload. Potential cost savings on existing OPEX on fuel management, the largest bulk cost, is 20% according to a report by a leading consulting company. In a low margin industry, this is substantial. As an example, KONGSBERG's Vessel Performance application, available in the Kognifai Maritime Ecosystem, takes data from the ship's control system and a multitude of sensors on the vessel, logged at frequent intervals and processed in real time [2].

Data sources include the engines, propulsion system, automation, deck machinery along with other equipment. Selected information is displayed graphically to focus the crew's attention immediately to key elements. For example, fuel consumption levels are displayed against a baseline of historical data, showing whether consumption is higher than need be for the conditions. From this insight, the crew could start optimising operations, for instance, by changing the number of running auxiliary engines in order to bring the operating engines into the load range with the lowest specific fuel consumption or to monitor the correct level of lube oil on the engines. The on-board interface is available on any connected device using built-in remote access to meet ever-changing customer requirements and governmental regulations.

Smarter maintenance means monitoring equipment health digitally and planning maintenance according to need instead of pre-scheduled visits. Digital tools that enable the sharing of data between value chain actors allows for smoother and more efficient logistics.

Longer term benefits are tied to end-to-end digitalisation of the whole value chain, but we believe that is still some years away.

I.4 What are the first steps towards fleet owners and operators should take?

The first step that vessel owners and operators should take is to make their fleet digital ready. That means ensuring that contextualised data are extracted from the vessel in real time and sent to the cloud for further analysis in software and applications. With massive amounts of data collected every minute, changes in one point will correlate with changes in other points. Contextualised data means data points that are mapped in a system so that the source of each data point and how it correlates with other data are mapped. Without contextualising data analyses will not make sense, so that is the first thing that needs to be in place. The amount of data depends on the number of sensors. Obviously as we are looking at more high frequent data, like vibration and acoustic sensors with sampling frequencies well into the 100s Hz, transfer of all the samples does not make sense and thus will need to be processed onboard the vessel.

I.5 What does the maritime software vendor landscape look like and have there been changes in the recent years?

For the last 3 years, Kongsberg Digital has mapped the maritime software landscape. A surge in the first year was followed by a decrease in the growth in 2019 and 2020. We have seen just a small increase in the number of new vendors entering the arena in the last couple of years, which was partly offset by a number of consolidations across existing vendors such as BMT Smart being acquired by Greensteam, and ChartCo and Marine Press merging to form OneOcean. It seems that the market has finally reached a saturation in the area of fuel and energy monitoring solutions, where not many new entrants seem to be willing to challenge the well-established existing vendors.

Also, we observed a continuing large growth in supply-chain-related vendors, as the interest of investors seems to have shifted from purely maritime startups in a narrow sense, to focus on intermodal supply-chain-related startups, which are often closely linked to the container shipping segments and ports. As port plays an important role in the logistics, value chain is also an important OPEX cost driver for vessels. There are several use cases which could have significantly positive effects for ship owners/operators like just in time arrival, electronic documents exchange and real-time depth monitoring to assure optimum loading of cargo. In this space, we see that large vendors like Maersk and IBM go up against small startups. Furthermore, some forward leaning ports like Rotterdam and Singapore are investing into the development of their own solutions.

I.6 Conclusion points from the editors

Drawing from the previous points highlighted in the interview from Kongsberg, and the chapters presented in this volume, there are a number of definite pathways for researchers and practitioners alike to consider as a way forward for the research agenda on technological advancement in the industry. One clear pathway is digitising the exchange of information further to increase transparency in the sector, particularly in the operations sphere. This will not only be useful endeavour for the private sector, but increasingly so also a requirement by regulators. This can take various forms and be seen to be functional for various activities. One example, which is increasingly becoming significant for the sector, is the improved exchange of information to track, trace and predict sustainability-related indicators and performance (e.g. emissions, fuel consumption, pollution, predictive maintenance). Another important and emerging area of work that clearly comes out as core from the chapters of this book is enabling more efficient marine–landside interface solutions in the context of an industry that is becoming more and more focused on the development of autonomous technological solutions. This emerges in the context of vessel traffic services requirements as much as autonomous vessels. Finally, a third area of work that will clearly dominate research in the sector going forward is the greenification of the industry and the role of technology in pursuing the green agenda. This takes various forms and requires a multidisciplinary approach. The book touches on this through a discussion on alternative fuels and other engineering solutions that not only have the potential to increase efficiency in day-to-day port operations but also mitigate emission from operations. This, together with adaptation and resilience, is a core topic that will be increasingly discussed in the context of the evolution of technological advancement of the sector.

References

[1] "Golden energy offshore and Kongsberg collaborate to reduce operational fuel consumption by 20%", available online at https://www.kongsberg.com/no/maritime/about-us/news-and-media/news-archive/2019/latest/ (accessed 07.11.2020).

[2] "Vessel Performance by Kongsberg", available online at https://www.kongsberg.com/digital/kognifaiecosystem/kognifai-marketplace/maritime/vessel-performance/ (accessed 07.11.2020).

Chapter 1
The global trade facilitation regulatory framework and the role of ICT

Periklis Saragiotis[1] *and Satya Prasad Sahu*[1]

1.1 Introduction

The case for trade facilitation is even stronger in today's world. Exponential increase of global trade volumes puts tremendous pressure on border inspection agencies, both in terms of operational performance and funding sustainability of personnel and control equipment. Constantly increasing global containerized trade leads to port congestion and delays in the movement of goods. Container volumes, measured in 20-foot equivalent units (TEUs) recorded an average annual growth rate of 5.8 per cent over the past two decades [1]. In some parts of the world, trade facilitation policies are implemented together with measures to secure the supply chain, especially in regions that experience high risks of criminal or terrorist activities. The role of trade facilitation under this prism becomes critical because, it ensures that compliant traders are rewarded with faster clearance, while allowing border agencies to efficiently detect illicit or dangerous cargo. In this context, border control agencies should implement mechanism to exchange information at every level within and across borders, apart from building partnerships with the private sector [2].

However, the most vital argument in favor of trade facilitation is related to improved firm competitiveness in the global markets. Trade facilitation reforms lead to clear financial benefits for export-oriented companies, especially in less-developed countries. Africa stands to gains more than other regions by undertaking procedural improvements covering regulatory controls at-the-border and before-the-border. Such measures will boost manufacturing exports [3]. Reduction of trade costs and time spent at the borders increases levels of predictability and leads to improved supply chain efficiency and increased national competitiveness. In an era of global value chains (GVCs), trade facilitation is particularly important for the global agribusiness and light manufacturing, since procedural inefficiencies and complex procedures negatively impact not only the cost of imported production components but also the exports of the final product [4].

[1]World Bank Group, Washington, DC, USA

Trade automation and digital systems are at the core of trade facilitation programs. Information and communication technology (ICT) applied in modern trade control and facilitation tools constitutes a remarkable enabler for the implementation of the World Trade Organization's Trade Facilitation Agreement articles. For instance, it is hard to imagine a successful risk management policy being implemented without the support of a Customs Information System (CIS), such as the Automated System for Customs Data (ASYCUDA)* that allows the storing and analysis of trade data at a declaration level. Similarly, exchange of data between trade stakeholders, including government inspection agencies and the private sector, cannot be achieved without fully automated Electronic Single Window (eSW) platforms. Today, the abundance of public and commercial datasets, combined with the availability of affordable technological solutions, places ICT in the center of this agenda and encourages governments to deploy hard and soft ICT infrastructure to digitalize the flow of international trade [5]. However, such policies should not be designed in isolation and in the absence of supporting legal and regulatory framework to enable their proper operationalization.

1.2 The international trade ICT framework

Modern supply chain business is geared to respond to consumers' needs and operates in a continuous sourcing environment in which the ordering and delivery of goods are just a few clicks away. But behind this ultrasimple "click and get" world, there exists a hidden complex network of transactions that occur in the background, carried out by numerous participants in the logistics chain. One may relate this scenario to the e-commerce transactions that have become commonplace but even traditional commerce now works in a similar fashion. Participants in the logistics chain connect with each other via multiple automated networks and highly sophisticated enterprise or cross-enterprise platforms to fulfill consumers' orders. Decisions on production, shipping, storage and routing are taken with data inputs in real time and those decisions ultimately result in on-time delivery at least cost.

When goods have to be shipped between countries, border control agencies, such as Customs, get involved. This adds another layer of sophistication and complexity to the supply chain network. The rest of this section discusses the role of ICT-based solutions in facilitating international trade and how border agencies use ICT to support contemporary supply chain networks.

Increasingly, businesses are automating the entire lifecycle of the international trade procedures in order to reduce transaction costs. The use of ICT in trade facilitation must be seen in the wider context of international trade procedures. Information and documentation are exchanged between multiple parties as part of the process of buying and selling goods across borders. A lot of paperwork is generated when goods are ordered, paid for and are physically shipped to the customer. A series of transactions is recorded before, during and after the actual

*https://asycuda.org/en/

movement of goods, which broadly falls into three types of business processes—buy, ship and pay [6].

The "Buy" process refers to commercial aspects of the trade transaction, which starts with procurement, the trade agreement, ordering, invoicing and initiation of order fulfillment. The "Ship" processes cover the transport transactions, including the contracting for transport services, insurance services, the movement of freight and associated means of transport and documentation to support regulatory and compliance formalities, especially when goods move across national borders. The "Pay" processes cover the trade finance agreement, trade delivery and trade settlement processes, including remittances and payment for the goods and associated services. Information flows drive "Buy," "Ship" and "Pay" processes, and all parties involved end up generating and consuming large volumes of data and documents. There are rules and regulations that apply to the "Buy" and "Pay" processes but cross-border regulatory agencies play a major role in controlling the "Ship" process around the physical transportation of goods crossing international borders.

The UNCEFACT Buy-Ship-Pay model amply demonstrates that all the three processes utilize a common set of business data to fulfill the trade transaction end to end. The e-commerce environment best illustrates the concept of an end-to-end model, where buy, ship and pay data persist in the same ICT platform, supporting the entire process from the time a customer orders goods online until it is delivered to the customer. Even in traditional commerce, the involvement of ICT-based networks is inescapable, and data are shared between supply chain participants using multiple platforms. This has been possible because of the adoption of a holistic view of ICT infrastructure incorporating commercial, compliance and regulatory functions adequately reflects its wide scope throughout the ship, buy and pay stages.

Therefore, ICT for trade facilitation has to be viewed as an integral component of the broader trade transaction framework. In this context, it has been proved that the adoption of automated solutions provides border management authorities with effective means to facilitate trade; however, the sole execution of trade automation should not be considered as the silver bullet for reducing time and cost for international traders [7]. Rather, it should be regarded as one of the many tools to support the implementation of sound border management practices.

The use of ICT has a positive impact on trade facilitation in terms of reduction in trade costs and faster throughput of import and export goods, but ICT alone may not be sufficient. Globally, it has been shown that the levels of corruption and hidden costs have reduced as a result of automation [8]. Further, it has also been observed that the use of ICT in the absence of legal, regulatory and procedural reforms to modernize Customs practices might not yield the desired results [9]. Although ICT is a necessary but not sufficient condition for the full and effective implementation of trade facilitation reforms, it can accelerate the adoption of most measures contained in the WTO TFA [10]. On a macroeconomic level, it is also recognized that the use of ICT has a significant positive effect on bilateral trade in emerging economies [11].

In this context, the historical evolution of trade facilitation ICT needs also to be considered. Customs and border agencies' automation is now being recognized as a powerful method for facilitating trade and increasing the efficiency of border operations. However, their evolution process has always followed the corporate ICT revolution that transformed supply chain management. Advanced industrial nations adopted mainframe solutions in the early 1970s and electronic data interchange (EDI) in the mid-1980s, while several middle-income countries followed suit in the following decade. Today, corporate ICT interconnectivity is based on big data analytics, use of artificial intelligence (AI) and the full utilization of the Internet of Things, thus offering significant possibilities and unlimited potential for the provision of high quality, cost-efficient customized services. However, this evolution did not happen overnight. It was rather a result of a long progression of developments, principally due to the development of (a) global standards for point-to-point connectivity, and (b) sophisticated technology in hardware and software solutions.

(a) In the early days of ICT, businesses used automation largely for gains in internal efficiencies. As it became clear that there were further advantages in exchanging information between established business partners, bilateral exchanges became possible through point-to-point connectivity, where information was shared on the basis of mutually agreed schemes. As ICT matured and the business value of sharing information became apparent, electronic data and telecommunication global standards were developed by international bodies such as ISO-IEC, ITU and the UN. National standard bodies such as the ANSI and industry consortia stepped in to create popular standards such as EDIFACT, ANSI X.12 and Odette [12]. This happened at the same time when data communication businesses started providing value-added network services to help deliver standardized data across business partners and under standard agreements. The world of EDI emerged, and the process of exchanging structured electronic data and messages was industrialized on a global scale.
(b) The emergence of the Internet and the rapid growth of technology led to the swift development of numerous ICT-based corporate applications. The processing power of computing hardware increased exponentially[†] along with the corresponding reduction in the cost and size of equipment. The growth of high-speed data communication using optical fibers and wireless media meant that the challenge of data connectivity is all but overcome. The cheap and ubiquitous availability of mobile-networked computers has brought the world to the threshold of the digital revolution where potentially all aspects of international trade can be conducted in the electronic medium, with increasing velocity and reducing costs. Thus, multiple architectures emerged to correspond with the evolutionary stages of technology deployment.

[†]This is referred to as Moore's law, named after the inventor, which said that the number of transistors per silicon chip, and therefore computing power doubles every year.

Today, EDI platforms have emerged as a globally recognized tool to interconnect traders' enterprise systems with Customs information solutions, exchange documents or data electronically and ultimately facilitate the lodging of electronic declarations. Initially, there was reluctance expressed by least-developed countries for the adoption of automation mandates, as they implied resources and know-how which they did not possess. These have been overcome in time and now there is more or less universal and unqualified support to the adoption of ICT for trade facilitation. Higher levels of ICT penetration combined with pressure generated from the mandatory implementation of the WTO TFA provisions have improved the levels of acceptability [8]. Governments are moving at full speed in order to upgrade their IT systems and bring them to the same level of sophistication with the ones applied by the private sector.

1.3 Elements of trade ICT functional environment

While ICT is indispensable for conducting electronic international trade transactions, its use raises many legal and regulatory issues. Some of the questions that ICT implementers need to address are related to whether systems and their respective electronic methods of regulatory reporting are: (i) well founded in law; (ii) afford adequate protection to intellectual property; (iii) are protected against unauthorized access and manipulation; (iv) are configured to protect users against the loss of privacy and commercial secrecy and (v) are exposed to security risks that can comprise the availability of information assets.

Facilitation of electronic trade requires that countries establish soft infrastructure in the form of laws and regulations to ensure that electronic transactions are valid, electronic contracts are enforceable and electronic records and documents are admissible as evidence in a court of law. Laws and regulations also should recognize the methods (e.g., cryptographic methods) to help one to overcome security risks that render electronic data and records vulnerable to theft, destruction or manipulation. For the same reason, they should also include penal provisions to help one to deter criminal and injurious conduct, such as data theft, unauthorized access to and manipulation of data.

1.4 Electronic Single Windows

eSWs or as best known as national single windows (NSWs) are defined as "facilities that allow parties involved in trade and transport to lodge standardized information and documents with a single-entry point to fulfill all import, export, and transit-related regulatory requirements. If information is electronic, then individual data elements should only be submitted once" [13]. The early notions of single window (SW) were conceived in the context of an ICT environment, but the Recommendation 33 did not necessarily imply the implementation and use of high-tech ICT. Experts agree that an SW could "hardly be conceived without the adoption of relevant ICT technologies" [14]. Collaboration and business simplification are the basic drivers behind the idea of eSWs. eSWs are facilities in which traders can submit all information and documentation pertaining to import, export and transit to government agencies through a single electronic entry point. In the absence of an SW, traders are compelled to submit basically the same information over and over again to various regulatory agencies. SW works on the principle that one piece of information needs to be submitted only once. Like all "community-oriented" systems, SWs involve multiple public and private sector stakeholders, including cross-border–trade-related regulatory authorities, traders, freight forwarders, banks, terminal operators, cargo carriers and transporters.

The concept of an eSW is very relevant to international trade processes. The typical cross-border trade transaction spans across multiple organizations situated in different parts of the world. These organizations deal with different aspects of the buy–ship–pay processes and are involved in the extensive sharing of data and documentation. An average international trade transaction involves between 20 and 30 different parties, up to 40 documents and around 200 data elements, many of which are repeated multiple times in documents [15]. Traders moving goods across borders face this complexity when they are required to produce a range of data to be taken out of these documents in order to fulfill regulatory requirements. Some of these documents such as licenses, permit and certificates are issued by the government agencies themselves. In the past, when there was little automation, government agencies required traders to produce hard copies of regulatory declarations and applications. But even with the digitization of processes, the data remained with the respective government agencies, and the trader had to make repeated data submissions.

The eSW envisaged the organization of data in logical, nonredundant structures that can serve the purpose of multiple participating agencies, who will collaborate with one another by sharing transactional data and documentation among themselves under predefined business rules covering conditions of access and confidentiality. This helps one to eliminate data duplication, rekeying of data, human intervention, errors, delays, multiple versions of the truth and the attendant costs and inefficiencies.

Maritime Single Window

Under the Maritime Single Window concept, public authorities governing the flow of maritime traffic at ports should arrange to receive regulatory data and information via a single entry point thus avoiding duplication and repetitive submission of data. The International Maritime Organization (IMO) has established the Convention of Facilitation of International Maritime Traffic (FAL Convention) under which parties to the convention are obliged to prescribe standardized documentation and reporting formalities for ships that use ports. The convention includes a compendium of electronic reporting formats, which have long been in use in the maritime industry.

The IMO took a significant step in digitalizing maritime regulatory reporting in April 2019, when it made a mandatory requirement for national governments to introduce electronic information exchange between ships and ports. Countries bound by the FAL Convention will be able to use standardized digital reporting templates specified by the convention. This measure is expected to cut down costs and time for international logistics service providers via the reduction of red tape and the increase of the efficiency of maritime trade and transport.

Services offered by an SW can be broadly classified into the following three categories on the basis of the stage at which the services are rendered:

1. *Before the border* transaction data that are submitted at the earliest point will be carried forward to later points. Processes carried out "before the border" include the submission of requests or applications in the form of licenses, permits, certificates and other types of regulatory approvals (also called LPCOs). Another type of "before the border" control is performed on the basis of the principle of pre-arrival processing. Increasingly, border regulations are being oriented toward advance information, based on which regulatory agencies' decision-making prior to the arrival of cargo at the borders. SWs are prepared to process "advance cargo information" electronically. Electronic submission of preliminary information provided by the carrier, freight forwarder or importer prior to the arrival of goods is considered useful because it allows Customs and other border agencies to accelerate the release of goods at the border.
2. *At the border* electronic Customs declaration processing is a core building block of transaction data submitted at this stage. SW systems allow traders to prepare, complete and submit the Customs declarations electronically, covering the entire range of border procedures. Modern SWs similarly handle payment processing in a comprehensive manner offering different options to the trader to pay various charges, including duties, taxes and various types of fees, imposed not only by Customs but also by other government agencies. There are many situations where payments need to be secured by way of guarantees for which SWs may provide facilities to manage obligations and financial guarantees. Prior to release, the goods may have to be inspected, screened or weighed, for which the SW system may be interfaced with different types of equipment.
3. *Beyond the border* controls imply actions taken by the trader and the authorities outside the paradigm of border controls. These are based on partnerships—between governments across borders and between governments and businesses. In SWs, programs such as the authorized economic operator (AEO) are recognized under which legislation requires traders to take responsibility and be invested in improved compliance. Traders are encouraged to invest

heavily into human and technical capabilities in ensuring compliance. Likewise, regulatory authorities are also invested in facilities that allow them to access a trader's information systems to access information at an early stage that provides one an opportunity to apply controls or take corrective action. An arrangement of partnership between the regulatory agencies and businesses helps one to provide continued assurance with regard to the level of compliance. "Beyond the border" controls include cooperation agreements between governments for sharing information pertaining to controls. eSWs are a part of the flows of information between regulatory authorities—be it in the pre-arrival processing or in sharing electronic phytosanitary certificates [16]. As long as a business process or a functional service is related to the application of regulatory controls for import, export or transit goods, SWs may be involved in delivering it.

There are several benefits anticipated from the implementation of an SW environment. These, as outlined by UN/ECE working paper, are recognized in terms of (i) reduction in administrative costs to trade, (ii) increased transparency and predictability for traders through streamlining and automating regulatory obligations related to import/export/transit clearance and border control and (iii) better national coordination with relevant authorities with an interest in cross-border trade [17]. SWs open the way for increased cooperation, coordination and collaboration of border management agencies at bilateral and regional levels, through standardized data exchanges, joint risk management and the development of trusted trader regimes leading toward traders' long-term compliance.

Evidence also suggests that countries that have single-window systems have performed better on measures related to trade facilitation than those that did not have such systems. Benefits of an SW can be seen through direct and measurable outcomes such as lower trade transaction costs and reduction in the time taken to clear goods. Indirect benefits accrue when improvements in interagency coordination allow more effective and efficient deployment of resources and time savings. These benefits accrue not only to both governments—through improved trade compliance, enhanced security and often increased revenue yield, but also to traders—through faster clearance and release, predictable application and explanation of rules, less opportunity for interpretation of laws and regulations and ultimately cutting costs through reducing delays and avoiding informal payments.

While automation is an important component, government entities and the community of trade stakeholders need to jointly decide about a wide range of aspects related to the establishment and implementation of SWs. These are summarized in the following interrelated areas corresponding to core SW implementation principles:

1. *Legal and regulatory framework* ensures that there are no legislative impediments hindering their efficient operations;
2. *Business model* defining the responsibility structure for governance, roles for agencies that will play the operational role and funding;

3. *Financial model* that defines the structure's capital and operating expenses as well as a fee/revenue model for its financial sustainability;
4. *Business process analysis* that maps existing processes (as is) and proposes a simplified and streamlined reengineered business process (to be) model;
5. *Data model* that defines the data requirements of the SW and organizes them as a standardized and harmonized model; and
6. *Change management and communications strategy* that ensures that the right level of capacity, readiness and commitment of the trade community and government departments are in place.

Today, 74 countries have notified the operation of a trade SW under Article 10.4.3 of the TFA.[‡] While countries can claim to have launched NSWs, the quality of implementations differs from one another.

1.5 Cargo Community Systems

Cargo Community Systems (CCSs) often referred to as a Port Community System (PCS) or Airport Community Systems (ACSs) have developed globally as integral interconnectivity components of international supply chains. These are ICT-based logistics platforms dedicated to the management of terminal storage and multi-modal freight flows at ports, airports and land border warehousing stations through which cargo passes.

Roughly speaking, port and ACS are community-led ICT facilities operated by a community of participating businesses and regulatory organizations for the benefit of its users. They are established for a specific purpose and typically involve functions that span across the participating organizations. Regardless of who owns or operates a community system, it is usually run by structures and mechanisms defined by a community of stakeholders. Community systems differ from enterprise ICT systems. The latter falls within the private domain of the enterprises that own and/or operate them. Enterprise systems are meant to be used by their respective users (both internal and external) and are governed by the policies set by them.

Data processed by CCS relate to the freight handled by public and private parties using a cargo handling facility or a cluster of facilities, which ports and airports are by definition. Information handled through the CCS aims at enabling a variety of decisions related to freight operations, including deliveries, loading, unloading, docking sequence, truck appointment management or cargo tracking and tracing. In addition, trade compliance agencies, such as Customs, are included in the cargo community ecosystem and may use their data to manage regulatory control activities.

- **Port Community Systems (PCSs)** represent an integrated approach to delivering digitally enabled services to users of the port. IPCSA has defined [33] PCS as a "neutral and open electronic platform enabling intelligent and secure" exchange of

[‡]WTO Notification tracker at https://tfadatabase.org/notifications/transparency.

information between public and private stakeholders in order to improve the competitive position of the sea and air ports' communities. It is a common platform that is open to all actors within the port and covers business functions associated with the operations carried out at the port. PCS acts as a trusted source of data that is available in real time to be used for executing business processes. PCSs represent the single-window approach insofar as the submission and usage of data is concerned. Transactional data once submitted helps one to drive a broad range of business processes. A PCS is modular in design and construction and comprises functional modules representing groupings of services. PCSs are characterized by the services they provide to their users. At their simplest, PCSs can be port information portals providing operational and practical information to all port users, such as service hours, vessel arrival schedules, port tariff and gate waiting times. The functional services covered by a PCS can be divided into three categories on the basis of services in support of:

- maritime and nautical operations—these are broadly related to the entry, stay and exit of vessels;
- hinterland operations—inward and outward movement of cargo using different means of transport (rail, road and waterways); and
- onshore regulatory services—other regulatory-services-based onshore covering vessel security, waste disposal, dangerous goods, veterinary and sanitary inspections, harbor and port dues collection, vessel inspection, etc.

Appendix A contains a list of potential services that can be offered by PCSs.

- **Air Community System (ACS)** acts as a single-point interface by enabling electronic communications within the user community of an airport. The community of businesses and individuals using an airport for moving cargo is very diverse. Apart from traders (importers and exporters), the private sector players include airlines, air cargo agents, freight forwarders, warehouse operators, cargo (ground) handlers, road transport operators (involved in pre-carriage and onward carriage), freight stations, insurance operators, security service providers, banks and insurance service providers. It also includes public authorities such as Customs, airport security, regulatory agencies, the public authorities concerned with the management of airport premises, air traffic control, immigration, quarantine, health and other cross-border regulatory authorities. Air CCSs are similar to PCSs.

Airport CCSs promote business collaboration through data sharing. Principally, an Air CCS relies on flight and cargo manifest data and enables the inventory keeping operations and supports the cargo accounting function. It serves upstream and downstream revenue activities by making data available for cargo handling and Customs reporting. Customs declaration processing and release of goods are facilitated through the interface with the warehouse systems. In this case, the Air CCS enables the interface between the various players. The community systems play a crucial role in implementing global programs such as IATA's e-Freight, which includes components such as e-Air Waybill and e-CSD (Cargo Security Declaration) and electronic dangerous goods declaration.

1.6 Could CCSs and NSWs work together?

CCSs are essentially data exchange systems and, in this sense, their architecture, purpose and mandate are similar to ones of NSWs as they constitute a single point of entry for data and documentation, eliminate redundant handling of data and enable rule-driven access to information and business processes. However, there are subtle differences between the CCS and SWs. An NSW is a facility that enables a trader to submit information to fulfill trade compliance formalities for import, export and transit at a single point of entry and based on information received, these agencies are expected to coordinate their respective controls, collect their respective duties, taxes, fees and charges and provide a coordinated response to the trader. Going strictly by the UN/CEFACT definition however, the SW does not cover transport regulatory formalities concerning the cross-border movement nor the flow of information associated with operations between logistics operators for efficient flow of cargoes. These information flows and collaborations are addressed through CCS.

Therefore, CCSs and NSWs represent somewhat related but complementary functions. The former deals with the transport and logistics aspect of trade flows and the latter covers the application of regulatory controls on trade. Both the facilities deal with the same underlying data arising from trade and transport transactions. Therefore, there is ample scope for CCS and SWs to form strategic alliances. CCS could act as the platform for any-to-any exchange of electronic information among public and private sector players operating at the port-of-entry ecosystem, including those in the logistics sectors, whereas the SW would focus on bidirectional data exchange between trade and regulatory agencies to fulfill reporting formalities at the border formalities.

While it is true that CCSs have traditionally grown as port-of-entry data-sharing platforms, a more advanced form of CCS could take the shape of a single national facility to support the community data exchange infrastructure. PCSs tend to be port specific or port-cluster-based, whereas NSWs tend to be "national" in scope. There is a case, therefore, to have a national approach to PCSs. These two types of infrastructure (NSW and PCS) do have natural complementariness in terms of the coverage of services. They are closely interrelated from a data and process perspective. Therefore, countries could consider PCSs and NSWs to jointly form part of a larger national strategy for trade facilitation and logistics improvement. It would help one to have a common understanding on the potential interactions between PCS and NSW systems and how such interactions will improve trade facilitation and logistics efficiencies. That would help in convincing governments and private sector stakeholders to pursue investments in this space.

1.6.1 Interoperability and systems integration

Interoperability refers to the ability of computer systems or software to exchange and utilize information. CCSs and NSWs are predicated on interoperability and various systems working together. At a wider level, interoperability is fundamental to the seamless sharing of data between ICT systems operated by transport operators, freight forwarders, enterprises and government agencies. There are several factors impinging on interoperability such as (i) a shared understanding of the processes by which parties in trade interact, (ii) simple and transparent rules and regulations and (iii) globally harmonized data standards.

A shared, unambiguous understanding of the meaning of data is the key to platform integration. CCSs and NSWs rely on harmonized and standardized data models, in order to promote a shared understanding of the meaning of the information that is exchanged between ICT systems. Businesses and governments would like to see a uniform set of data standards, and international organizations and standards consortia are working closely to support the effort in developing international data standards. The United Nations has been at the forefront on this with a stack of standards and recommendations. A United Nations Trade Data Element Directory that is a joint UN/ISO publication is the foundational data standard. UN/CEFACT has produced the "Buy-Ship-Pay reference Data Model" covering the entire trade chain. The World Customs Organization (WCO) has developed the WCO data model covering the requirements of an NSW, the Maritime Single Window (MSW) (jointly with International Maritime Organization (IMO)) as well as models for some of the licenses, permit certificates and other authorizations. Alignment of the WCO data model with existing global standards produced by UN/CEFACT and ISO is crucial. Fully harmonized semantic models were developed for cross-utilization of data originating from trade, logistics, transport, finance and supply chain domains for regulatory purposes and foster the introduction of CCSs and NSWs.

1.6.2 Technological disruption and ICT in trade facilitation

The coming era of technological changes in the field of transportation and logistics is being termed "disruptive." These changes are not about improving the current technological base through small increments, but a total transformation through paradigmatic changes, which will likely enhance trade facilitation.

The business drivers for these changes emerge from the integration of marketing with logistics. The visibility of in transit inventories coupled with high levels of predictability in supply chains allows businesses to produce reliable schedules for delivery. In turn, this enables businesses to fulfill promises made on online platforms with upfront transparency. The institutions supporting cross-border regulatory trade procedures cannot afford to be out of step with this new reality of business. They have to act with matching speed, efficiency and reliability by perhaps using the same technological capabilities and digital platforms that are transforming modern business. That is why, logistics platforms are considered foundational to national competitiveness. When supply chain information is brought on to a digital platform, it opens ways for quickening the business processes by bringing participants together for interactions in real time, thereby helping faster decision-making at crucial points. Cross-border regulatory agencies that are responsible for facilitating trade can now ride on the same information to speed up regulatory clearances.

Under the "Immediate Release Guidelines" published by the WCO, Customs and express cargo operators are encouraged to conclude agreements that may include the flow of data, a Customs declaration for obtaining the release/clearance of consignments, and to enable operators to conform to relevant Customs data requirements, including advance information and documentation. The world of express cargo and e-commerce has witnessed increased levels of digital collaboration between the commercial operators and regulatory agencies for mutual benefit. With such collaboration, the former can operate a predictable and efficient supply chain, and the latter can better achieve the goals of societal protection, regulatory compliance and security.

The express cargo model involves a limited number of players in the transport chain and cannot be easily replicated to regular maritime cargo because there are multiple layers of intermediation involved in the handling of ocean freight. Besides, the maritime cargo is not fully digitalized because of legacy reasons. The interests of freight forwarders and shipping lines do not always converge when it comes to the timely sharing of commercially sensitive logistics data. But the issues of trust can be solved through appropriate platforms and technologies.

Digitalization of freight logistics has the potential to reduce trade costs in multiple ways—by lowering the burden of document creation and data sharing, improving accuracy, reducing procedural friction and by curbing information asymmetry and increasing predictability. The advances in technology that will make digitalization imminent, pervasive and universal are listed in Table 1.1.

Technological disruption is expected to result broadly in establishing a ubiquitous, low-cost platform for commerce and logistics, driven by innovation and creating new business models and players in its wake.

1.7 Regulatory underpinnings for cross-border data sharing

While logistics and supply chain platforms have come alive thanks to a series of technological innovations, technological change is not the only factor impacting the future of trade facilitation. Appropriate international regulatory frameworks must also come into being providing universally acceptable rules for sharing, storing and using supply chain data for commercial, logistical and regulatory purposes. Regulatory regimes are also evolving more broadly to allow systematic cross-border exchange of trade data. Two categories of exchange are gaining trend. Cross-border sharing of pre-arrival information between Customs administrations has gained ground. The system-to-system exchanges of advance cargo and conveyance information are usually based on regional or bilateral agreements. These exchanges serve to facilitate trade, implement border controls and support goals ranging from risk assessment to revenue accounting. Another category of government-to-government (G2G) sharing of data is based on international conventions such as IPPC and CITES. These are carried out through technology platforms to enable cross-border exchange of vital regulatory documents and data that are crucial for the release of consignments at the border.

Table 1.1 Technological disruptions

Technological disruption and how they can impact trade and logistics facilitation		
Technology	**Description**	**Impact**
Internet of Things	The Internet of Things (IoT) is a system of smart devices that are fitted with sensors that can generate and transmit data IoT devices are assigned unique identifiers (UIDs) and can perform functions or transfer data with little or no human or machine interactions	Allows the creation of "smart" objects in the supply chain to self-report their position and condition, creating a chain of notification of supply chain events
Distributed ledger technology/ blockchain	A system of records that enables seamless sharing of data between multiple parties, in ways that provide ironclad guarantees of security, privacy, immutability and consistency	It can help one to establish platforms for "end-to-end" data sharing in the supply chain to improve efficiency in the management of the flow of information, cargo and payments. It can eliminate obstacles to sharing of data, reduce costs, ensure the timely regulatory data
Rules as Code (RAC)	It is "a methodology for creating and applying legal rules in the digital age."[§] Rules are written with machine processing in mind. It allows the conversion of legal text into electronic code that is standardized, portable and machine readable	Make rendering legal rules of related to trade agreement (rules of origin), technical barriers, prohibitions, restrictions and conditions, into machine processable code
The fifth-generation (5G) mobile network	Allows larger volumes of data to be transferred wirelessly in real time	Will make way for an unbroken visibility of freight objects; second-order benefits include analytics of a higher quality, driving up supply chain efficiencies
On-demand logistics services	Use of platforms for logistics bookings and drivers with mobile phones and GPS devices, to create on-demand freight consolidation	Technology-enabled process to implement drivers to consolidate cargo along a route in order to optimize capacity utilization and maximize value for the customers

Source: Authors.

There are not globally dominant regulatory modals using which cross-border sharing of regulatory supply-chain data can be industrialized. The WCO has also promoted the concept of "globally networked Customs (GNC)" as a fundamental building block of Customs in the twenty-first century. GNC does not envisage a

[§]https://www.lawpracticetoday.org/article/rules-code/

universal digital platform but assumes that governments will come together to agree on the sharing of data on a bilateral, regional or plurilateral framework to serve specific purposes and strategic goals. Each of these types of exchanges will be governed by "utility blocks," serving as the legal, technical and administrative blueprint for implementing a particular use case of international G2G data exchange. Since 2011, the WCO has been working toward the standardization of different types of "utility blocks." A few concrete examples of utility blocks have emerged. Under the "AEO" programs, governments enter into agreements or arrangement of mutual recognition of each other's AEOs. Under these agreements, governments are obliged to share up-to-date information on each other's AEOs. The WCO has published a utility block to describe how such exchanges can be organized. The utility block is expected to industrialize such exchanges.

A more profound and all-encompassing framework has been established in the form of a treaty of the United Nations. The Framework Agreement on Facilitation of Cross-border Paperless Trade in Asia and the Pacific has the goal to accelerate the digitalization of trade facilitation measures, by providing the necessary instruments and by gradually building the capacity of governments to engage in cross-border paperless trade. It defines paperless trade as trading carried out using electronic rather than paper documents and aims to provide the institutional framework to identify and plug in gaps in various technical and legal/regulatory layers.

Apart from technical challenges of interoperability and systems integration, successful initiatives of ICT such as CCSs and NSWs also require gradual, incremental action on multiple fronts to bring the stakeholders around to invest in the underlying strategic vision and framework. Initiatives such as NSW and PCS may involve improvements in the system's root and branch. Sustained policy support would be necessary to affect changes to long-standing practices, working methods, attitudes and behavioral aspects of government agencies and trade actors. One can detect the absence of political will in the loss of momentum of reform-oriented actions.

The development of collaborative platforms such as CCS and NSW not only requires wide ranging consultations among stakeholders but also an appropriate administrative apparatus to deliver the desired objectives. The stakeholders should be willing to provide adequate wherewithal to pursue the common vision and goals. The availability of funds for trade facilitation and ICT-related activities combined with an effective governance framework signals political will.

1.8 Conclusions

1. Trade facilitation is now a global imperative. Based on the realization that trade drives sustainable economic growth, countries around the world are pursuing policies to facilitate trade, with the aim of removing administrative difficulties faced by trade and reducing costs. Analytical studies conducted in this sphere provide a compelling case for the furtherance of simplified procedures for cross-border trade.

2. The facilitation of cross-border trade improves access for goods to regional and global markets. Better trade facilitation should result in increased volumes in international trade. Trade facilitation also creates a favorable environment for businesses, which should lead to more investments, improved volumes of trade and transit flows and increased regional integration. The gains for developing countries are much more considering than the reductions in trade costs that they can achieve on a low base with the rapid implementation of reforms. Trade facilitation is an important trade policy instrument for fostering intra-regional trade and ramping-up a country's linkages to GVCs.
3. There is now a global mandate in the form of the WTO TFA, which requires the WTO members to put in place a time-bound action plan for instituting measures that will replace outdated procedures imposed by Customs and other agencies. Several multilateral organizations have programs to lend support to countries' efforts in achieving their goals on trade facilitation.
4. ICT underpins programs for reform and modernization of trade procedures. The demands for efficient and effective border controls cannot be met without the extensive and judicious use of ICT. Businesses have come to expect digitization and paper-free processing of trade, akin to the everyday experience of widely e-commerce transactions. While these imperatives may propel countries to invest heavily into ICT, countries should ensure that the business imperatives of trade facilitation should drive automaton programs. This would ensure that the best business practices are adopted, and adequate attention is paid to the human aspects when switching over from outdated overly bureaucratic processes to modern ICT-driven ones.
5. Introduction of ICT-driven trade facilitation not only demands investments into technology but also requires a strong governance framework, an enabling consultative environment that fosters a spirit of cooperation among stakeholders, an appropriate operational model, a technology framework that supports interoperability and investment in a sustained change management process.
6. Trade facilitation comes about through extensive collaboration between border management agencies and the private sector. Collaboration is expressed in different forms, ranging from coordination of various stakeholders to running of community-based ICT platforms using which stakeholders share information and conduct business. Trade information portals help traders in making informed business decisions by providing crucial information regarding the documentation and formalities to be fulfilled by traders at borders. Community platforms such as CCSs and NSWs reflect collaboration in action, as they bring various stakeholders into a common transactional platform.
7. On NSWs and PCSs, traders exchange business-critical information at a single point, lowering the costs of clearance of goods and speeding up transaction processing. Apart from facilitating trade, community platforms also improve border agency collaboration for better enforcement and compliance. MSWs help one to enhance safe and efficient flow of maritime traffic, faster

turnaround of vessels and better connectivity with the hinterland operations. For traders, this implies reduced costs; for regulatory agencies, this means greater levels of compliance, better utilization of resources, enhanced security and compliance.

8. The impact of introduction of facilities such as NSWs and PCSs can be assessed on the basis of certain indicators of trade facilitation and logistics efficiencies. The question to pursue therefore is what these indicators are and to what extent do systems like PCS and NSW influence them. Such an assessment would help one to develop and navigate through initiatives to build PCSs and to integrate with NSWs.
9. The importance of standardization and global harmonization of data used in international trade is widely acknowledged as the basis for interoperability. In addition, political will is required in ample measure to bring a sense of partnership between the government and private sector stakeholders. This partnership is a vital part of implementing ICT solutions for trade facilitation, even more so in the case of NSWs and CCSs. Trade encompasses a wide range of interests, which can affect and be affected by ICT solutions implemented in this sphere.

Appendix A List of potential services offered by port community systems

	Service	Description
A	**Ship–port interface**	
1	**Nautical services**	
1.1	Vessel profile information/ vessel register	Shipping line provides vessel profile for all vessel details that will call on the port
1.2.	Voyage information	Shipping line provides information on scheduled voyages and obtains a stay reference number (or call reference number) that is a unique number for the call; port authorities share this information with all concerned (Customs, terminals, trading, transporting and forwarding community)
1.3	Berthing	Shipping line requests for berthing, un-berthing and re-berthing, and confirmation by port authority is transmitted to all auxiliary services, including Customs for action or information
1.4	Pilot boarding/pilot memo	Shipping line requests harbor master/port authority for pilot services
1.5	Confirm ATA	Harbor authorities confirm actual time of arrival to terminal, Customs and other shore-based service providers
1.6	Tugging and other shore-based services	Shipping line requests harbor master/port authority for tugboat and other wharf-related services

(Continues)

(*Continued*)

	Service	Description
1.7	Master data management services	Services to share master data (vessel code directories; call signs, LOCODE, MLOs carrier codes, NVOCC codes, identifier lists for forwarder, trader, Customs broker, truck registration and truck crew, master for NDG Code and IMDG Codes) between PCS/NSW and other port user organizations; this is the lifeblood for PCS–NSW interface
2	**Regulatory services**	
2.1	Conveyance report	Reporting to port authorities and Customs regarding the call of a vessel contains miscellaneous regulatory requests and permits concerning the purposes for which the vessel is calling (Issue: whether it is agreed that PCS will receive this report on behalf of Customs) Covers general declaration, weapons on board declaration; equipment carried by privately contracted security personnel (ad hoc personnel hired for protection against piracy)
2.2	Vessel clearance	Authorities grant clearance for vessel to enter or leave harbor premises
2.3	Dangerous goods on board	Shipping line notifies port authority or harbor master regarding dangerous goods on board
2.4	Sharing of dangerous goods info with terminal operators	Special handling requirements and other dangerous goods information shared with terminal handling groups and port authorities
	Customs-bonded supplies	
2.5	Bunkering	Customs clearances for ordering and loading of bunker fuel
2.6	Bonded stores/ship supplies	Customs clearances concerning loading/unloading of bonded ship stores, ship spares, fitments and all other ship supplies
2.7	Supply of repairs/fitments	By convention, ship parts and fitments supplied to ocean-going vessels are typically imported duty free and supplied as an "export" much like other bonded ship stores
	Maritime/immigration/security	
2.8	Submit vessel security report	A requirement which is mandatory as per the ISPS Code; a ship data provider submits a report to the port it seeks to call regarding the "security level" at which it has operated through the last ten ports of call, whether there was a security incident on board, whether it has a valid vessel security certificate and details of ship-to-ship operations undertaken in the last ten ports of call; now a mandatory IMO SOLAS Convention requirement

(*Continues*)

(*Continued*)

	Service	Description
2.9	Passenger list/crew list	Shipping line/cruise line submits passenger and crew lists to immigration authorities and Customs; that latter performs controls and uses data to process passenger/crew clearance and shore leave
	Financial	
2.10	Notification of port dues, including light house dues	Amounts paid or payable against regulatory services provided by the port authority/harbor master filling out details of health questions
	Health	
2.11	Declarations under IHR; maritime declaration of health	To be submitted by master of the vessel to competent health authorities on shore
2.12	Certificate of free pratique	Port authority issues a certificate of free pratique stating that a ship that is without infectious disease or plague on board. This is linked for some ports' requirement for entry of vessels for unloading and permission for crew to disembark
	Environmental services	
2.13	Ballast water discharge	Authorities require vessels to report number of ballast water tanks, each tank's volume and origin of the ballast water to be discharged (concern for invasive species and maritime pollutants in ballast water)
2.14	Waste disposal	Service by which shipping line notifies to port authority or harbor master regarding waste disposal
B	**Terminal operations**	
1	**Cargo handling within terminal**	
1.1	Cargo reporting; includes empties and freight remaining on board	Shipping lines or other ship data providers reporting to port authorities terminals and Customs regarding the cargo brought by a vessel, it is the cargo manifest and can be combined with multiple regulatory requests concerning a further disposal of cargo (Issue: whether it is agreed that PCS will stage cargo report data receive this report on behalf of Customs); this is linked closely to the vessel stowage plan and advance import list and advance export list
1.2	Share stowage position with terminal operator	Shipping line shares stowage position of containers on board vessel
1.3	Cargo status information	Real-time status information, bill of lading or container number wise regarding unloading and loading operations at the terminal
1.4	Advance import list/ advance export list	Shipping line informs terminal operator regarding the list of containers and their stow position
1.5	Discharge completion summary	Terminal operators inform shipping lines of the completion of discharge of containers

(*Continues*)

(*Continued*)

	Service	Description
2	**Regulatory services**	
2.1	Notification of verified gross mass	Terminal uses in line weigh bridges to measure the weight of the trailer truck with container and after the unloading of the container to verify/report VGM; covers both inbound, outbound and trans-shipment containers
3	**Trans-shipment logistics**	
3.1	Trans-shipment manifests; associated amendments	Shipping line submits trans-shipment manifests to terminal operator; the latter acknowledges and plans stacking of containers accordingly; transfers manifests to Customs, which treats it as a request from the shipping lines for a trans-shipment procedure
3.2	Trans-ship permits; rewarehousing	Customs updates liner bonds and grants trans-shipment permit to shipping lines and informs terminals, including conclusion of bonded movement and rewarehousing
3.3	Interterminal movements	In a special type of trans-shipment permit or due to a maneuver for other reasons, shipping line submits a request for interterminal movement to Customs and the terminals involved and receives confirmations from Customs, the "from terminal operator," the "through terminal operator" (if applicable) and the "receiving terminal operator"
C	**Interface with hinterland logistics**	
1	**Inbound logistics**	
1.1	Notify ETA of incoming vessel	Shipping line informs forwarder/importer regarding ETA; requires manifest data from the port of loading; manifest should have importer, coded/forwarder/broker; and/or notify party; parties seeking notification should be registered on PCS along with contact details
1.2	Shipping agent issues unloading instructions to terminal operator	Note that the liner (mainline operator) that filed the BAPLIE may not be the same as the shipping agent that instructs the terminal operator; shipping agent issues instructions to the terminal operator to unload containers, against which terminal will expect payment; terminal operator compares the shipping agent's request with the BAPLIE to check if the request was in respect of containers reported in the BAPLIE message
1.3	Create transport order	Consignee or forwarder arranges with the transport company (in the case of merchant haulage)/(shipping line arranges with the transport company in the case of carrier haulage)

(Continues)

(*Continued*)

	Service	Description
1.4	Share transport order created by the forwarder with terminal operator	In the case of merchant haulage shipping agent shares the transport order created by the forwarder with the terminal operator; this allows the terminal operator to take the container into account in the load planning for the targeted shit that will be berthed
1.5	Registration by transport company for container pickup	Terminal operator knows which truck is going to be loaded, with which container and at what time; allows services such as terminal gate appointment for trucks and electronic queuing of trucks at terminal gate
1.6	Load inbound hinterland means of transport	Terminal receives container release message (commercial release message from shipping line /Customs release message from Customs) to container, which is loaded on the means of transport exiting terminal (truck/rail/barge)
1.7	Container release authentication at terminal gate	Representative of terminal operator or freight company taking cargo out of terminal receive a PIN code (or another mode of authentication) to allow release of container
1.8	Register containers for pickup	The shipping company or its shipping agent informs the terminal operators with COPARN; a means by which transport operator or forwarder informs the terminal about containers being delivered or picked up shortly
1.9	Container status messages (CSM)	Shipping lines to report to Customs to update container status (including empty, LCL and full containers); there are over 12 status codes and covers end-to-end container logistics
1.10	Track and trace of containers	Allows all parties concerned to be informed about the exact location of the container
1.11	Creating electronic invoices	Electronic invoices can be generated in multiple ways—shipping lines can raise invoices on forwarders; forwarders can raise invoices to shippers; trucking company can raise invoices on shipping lines or forwarders; for container release order (commercial release) it is crucial that the shipping line creates an electronic invoice for the forwarder to collect documentary charges, terminal handling charges, bill of lading and all other container-related charges
1.12	Electronic payment	Shipping line can create an electronic invoice for the forwarder to collect documentary bill of lading and other charges (of course, e-payment can work against any of the invoicing situations presented earlier)
1.13	Share delivery order	After the confirmation is received, shipping line generates delivery order online on PCS and shares it with the forwarder and the terminal operator

(Continues)

(*Continued*)

	Service	Description
1.14	Notify delivery	Transport operator can share information about delivery and stripping of container with forwarder and shipping line
2	**Outbound logistics**	
2.1	Transport booking request	Shipper or forwarder reports a request for a transport booking on a shipping line (or in multiline portals)
2.2	Booking confirmation	Shipping line records confirmation of booking of space
2.3	Locate empty containers	Forwarder/shipper seeks empty container availability status with liners' empty container yards; possible to have multiliner services to report empty container status in respective yards
2.4	Container status messages (CSM)	Shipping lines to report to Customs to update container status (including empty, LCL and full containers); there are over 12 status codes and covers end-to-end container logistics
2.5	Shipping instructions	Forwarder/shipper provides information (or reuses—information provided at the booking stage)
2.6	Register containers for delivery	The shipping company or its shipping agent informs terminal operators; a means by which a shipping company/agent or forwarder informs the terminal about containers being delivered for exports shortly
2.7	Registration by transport company for container delivery	Terminal operator knows which truck is bringing which container at what time; allows services such as terminal gate appointment for trucks and electronic queuing of trucks at terminal gate
2.8	Report cargo arrival at terminal	This function allows trader to inform Customs office regarding the physical arrival of goods/containers at the terminal. (Check exact national regulatory requirement.) Terminal operator or PCS can receive the data from exporter and take-up the responsibility on behalf of the exporter to inform Customs
2.9	Creating electronic invoices	Electronic invoices can be generated in multiple ways—shipping lines can raise invoices on forwarders; forwarders can raise invoices to shippers; trucking company can raise invoices on shipping lines or forwarders; for container release order (commercial release), it is crucial that the shipping line creates electronic invoice for the forwarder to collect documentary charges, terminal handling charges, bill of lading and all other container-related charges
2.10	Electronic payment	Shipping line can create electronic invoice for the forwarder to collect documentary, bill of lading and other charges (of course, e-payment can work against any of the invoicing situations presented earlier)

Appendix B The global trade facilitation compliance framework

Since 2013 the global trade facilitation compliance framework is synonymous to the WTO TFA. The agreement is a product of a long discussion process to identify, categorize and systematize measures to facilitate trade and provides a robust and predictable basis for reform program design and progress evaluation. However, prior to the TFA, several international organizations had developed their own compliance frameworks in their attempt to address challenges related to their respective technical areas, as shown in Table B.1. For instance, the Electronic Communications Convention (2005) aimed at the promotion of electronic communications in international trade, while the TIR Convention (1975) paved the way for a harmonized international Customs transit system.

Table B.1 Trade and transport facilitation agreements and conventions

Organization	Conventions and Agreements	Year
International Civil Aviation Organization (ICAO)	The Convention on International Civil Aviation (Chicago Convention)	1944
	The Convention for the Unification of Certain Rules for International Carriage by Air (Montreal Convention)	1999
International Maritime Organization (IMO)	The Convention of Facilitation of International Maritime Traffic (FAL Convention)	1965
United Nations Commission on International Trade Law (UNCITRAL)	The Convention on the Use of Electronic Communications in International Contracts (Electronic Communication Convention)	2005
UN Economic Commission for Europe (UNECE) and International Road Transport Union (IRU)	The Customs Convention on the International Transport of Goods under Cover of TIR Carnets (TIR Convention)	1975
UN Economic Commission for Europe (UNECE)	The International Convention on the Harmonization of Frontier Controls of Goods	1982
United Nations Economic Commission for Asia and the Pacific (UNESCAP)	The Framework Agreement on Facilitation of Cross-border Paperless Trade in Asia and the Pacific	2018
World Customs Organization (WCO)	The International Convention on the Harmonized System (HS Convention)	1988
	The International Convention on the Simplification and Harmonization of Customs Procedures (Revised Kyoto Convention)	1999
World Trade Organization (WTO)	The Trade Facilitation Agreement	2013
	The Valuation Agreement	1994

Source: Compilation by authors.

The TFA builds on the previous work conducted by the WTO in the context of other agreements and treaties, such as the General Agreement on Tariffs and Trade (GATT), Agreements on Customs Valuation, Rules of Origin, Import Licensing and the Sanitary and Phytosanitary Measures or Technical Barriers to Trade. It constitutes a holistic attempt to incorporate previously fragmented and sporadically implemented efforts to facilitate trade under the one umbrella.

The most important one is the Revised Kyoto Convention (RKC) developed by the WCO and entered into force in 2006.[¶] It aims at facilitating trade by harmonizing and simplifying Customs procedures and practices and provides standards and recommended practices for modern Customs procedures and techniques.

In particular, the General Annex of the RKC lays down the key principles covering transparency, predictability of Customs procedures, standardization of information and documentation, use of automation where it is cost effective and efficient, simplified procedures for authorized persons, application of risk management and audit based controls, border agency co-operation and customs-business partnerships.

The TFA builds on and expands the scope of the RKC provisions. Its core contribution is the adoption of a "one-government" approach to trade facilitation and the expansion of the agenda beyond the Customs administrations. The agreement recognizes the obstacles that the rest of the border inspection agencies (health, security, standards, etc.) impose on the efficiency, predictability and transparency of trade formalities and procedures.

Formal negotiations of an international agreement centered around trade facilitation were launched in 2004 and concluded at the Bali Ministerial Conference in December 2013. The TFA entered into force in 2017, as it was ratified by two-thirds of the WTO member countries and, in its current form, it comprises 12 articles found in Section I of the agreement, guiding countries through the reform implementation process (Table B.2). Each one of the articles attempts to offer solutions that the trade industry stakeholders, especially importers, exporters and intermediate logistics service providers (i.e., Customs brokers, freight forwarders and truckers) face during the process of moving cargoes internationally.

Obstacles to trade can be broadly categorized into those that occur at and those that take place beyond border crossing points.

- Prior to reaching the border, traders encounter difficulties to receive accurate information on importation and exportation procedures, spend time and money submitting identical data several times to different border inspection agencies and face different formalities and procedures when trading across countries. The TFA includes provisions that aim at the removal of these obstacles and the subsequent reduction of transaction costs that occur prior to cargo arrival. For instance, it promotes the publication of trade information in an easily accessible manner, the implementation of SW platforms enabling traders to submit

[¶]http://www.wcoomd.org/en/topics/facilitation/instrument-and-tools/conventions/pf_revised_kyoto_conv.aspx

Table B.2 Trade facilitation Section I articles

Art	Description
1	Publication & Availability of Information
2	Opportunity to Comment, Information Before Entry into Force & Consultations
3	Advance Rulings
4	Appeal or Review Procedures
5	Other Measures to Enhance Impartiality, Non-Discrimination & Transparency
6	Disciplines on Fees & Charges Imposed on/in Connection with Importation & Exportation
7	Release & Clearance of Goods
8	Border Agency Cooperation
9	Movement of Goods Under Customs Control Intended for Import
10	Formalities Connected with Importation, Exportation & Transit
11	Freedom of Transit
12	Customs cooperation

Source: WTO (https://www.wto.org/english/docs_e/legal_e/tfa-nov14_e.htm).

all trade documents and data one time only and endorses the acceptance of electronic copies.

- Legal, regulatory and operational obstacles that traders face at the borders are even more evident. They are often characterized by the lack of cooperation and coordination among inspection agencies, absence of practical automated processes and scarce use of information technology, inefficient clearance procedures by an excessive number of border authorities, unnecessary formalities and documentation requirements and nontransparent administrative fees and charges. The adoption of the TFA strengthens border agencies' cooperation (i.e., harmonizing working days and hours, aligning procedures and formalities or sharing common facilities), fosters electronic processing, including electronic payment of fees and duties. Most importantly, it introduces selectivity in inspections via automated risk management techniques, transfers the inspection process beyond borders via the introduction of PCA systems and rewards long-term compliant traders with less formalities and physical cargo inspections and other clearly defined benefits.

Alignment to each of the articles and their measures is expected to increase the transparency, efficiency and predictability of border management agencies. For instance, the agreement includes transparency measures related to access to information and periodic public–private consultations, such as the establishment of enquiry points or providing the opportunity to the private sector to comment on laws and regulation prior to their enactment. Similarly, the application of pre-arrival processing or measures to separate cargo release from clearance increases the levels of supply chain predictability, especially in terms of time spent at the borders. Finally, measures not only aim to increase coordination between border agencies or to minimize the complexity of import, export and transit formalities but also to take an initiative to promote electronic processing, and trade automation increases the levels of procedural efficiency. In addition to these articles, the agreement includes provisions for the institutional support of trade facilitation both at the national and international levels. The establishment of a national committee on trade facilitation is suggested in order to facilitate both domestic coordination and implementation of its provisions by all inspection agencies at the borders.

The value of the TFA lies in the fact that it provides for 'a special and differential treatment', which is a flexible dispensation provided for least-developed countries to implement the TFA measures. Unlike other WTO agreements, the TFA permits countries to determine their own time-frames for implementation for each of the TFA measures.

In the previous context, countries notify Category A, B and C provisions. Each category comes with specific requirements for the country member of the WTO:

- Category A commits developed countries for immediate implementation upon its entry into force, while developing countries will only apply those substantive provisions that they have indicated they are able to do so from the date of the TFA's entry into force. Least-developed countries were given an additional year to do so.
- Category B notifications allow developing countries and least developed countries (LDCs) to list the TFA provisions to be implemented after a transitional period following the entry into force of the TFA.
- Category C notifications contain provisions that a developing country or LDC designates for implementation on a date after a transition period and require the acquisition of implementation capacity through the provision of technical assistance and capacity building.

The share between the three categories per geographic region, as it can be found in Figure B.1, shows the great need for technical support around the globe but in particular in developed countries of the sub-Saharan Africa and Asia and the Pacific regions. In this respect, several international organizations provide technical assistance via projects funded by the donor community.

Figure B.1 also shows that information technology holds a prominent place in the technical assistance requests received under Category C notifications. Requests for support sometimes originate from governments to construct or acquire new border facilities and install new or upgrade existing ICT tools. ICT has proven to be one of the most important factors in implementing the most important of trade facilitation measures [18]. They could include a combination of surveillance and facilitation solutions, i.e., X-ray scanners to complement risk management procedures and Customs computerized systems to submit electronically and process pre-arrival documents.

TFA promotes the utilization of information technology in a very discreet and subtle manner. As a matter of principle, the agreement does not directly endorse the use of ICT in fear of being used as an excuse for unjustifiable hard or soft infrastructure investments. For the same reason, it does not touch upon issues related to ICT architecture, data standards, interoperability challenges, security concerns and data management issues. However, it is difficult to imagine its modern border management operations without the use of technology.

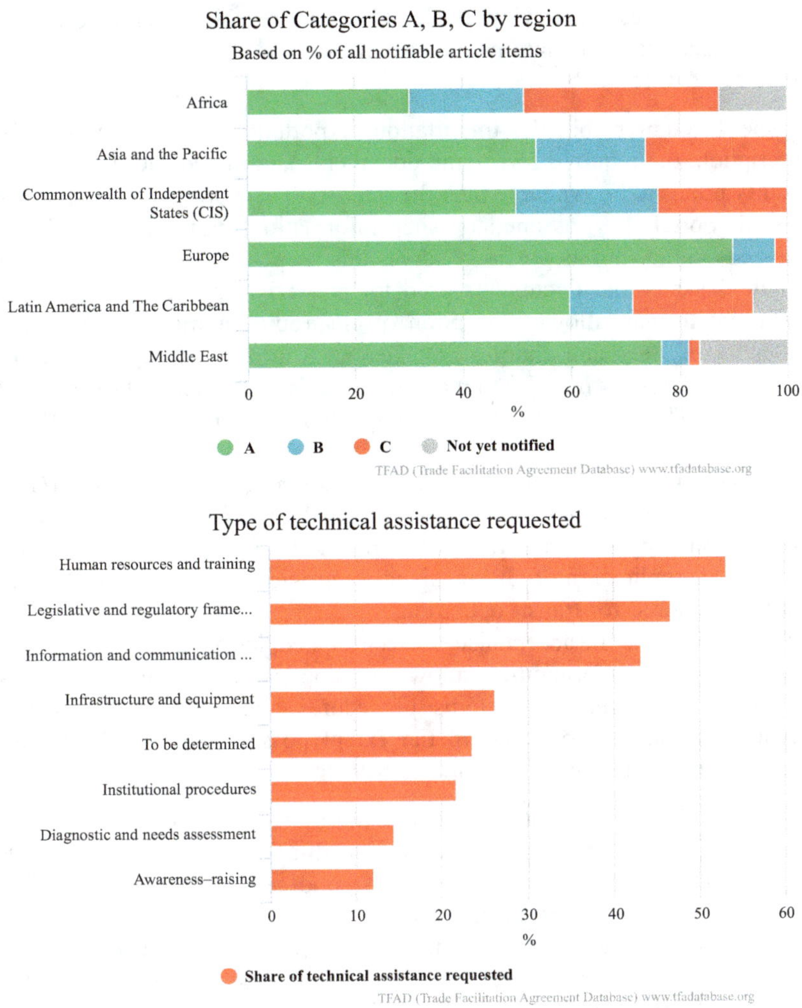

Figure B.1 Share of TFA categories by Region and Type of Assistance Requested (2020)

Out of the myriad IT tools and solutions, two are of particular importance CISs and eSWs. The former is usually found in the form of the ASYCUDA,[||] carrying the necessary technology to store huge amounts of data derived by traders' manifests and declaration forms. eSWs are facilities that allow parties involved in trade and transport to lodge standardized information and documents with a single entry point to fulfill all import, export and transit-related regulatory requirements.[**]

[||]ASYCUDA In Action: Compendium 2019, UNCTAD [1].
[**]Recommendation No. 33; United Nations Centre for Trade Facilitation and Electronic Business, Geneva, July 2005; ECE/TRADE/352; United Nations Publication—Sales No. 05.II.E.9; ISBN 92-1-116924-0.

Even though a careful review of the TFA may reveal scattered references to (electronic) data and trade-automated systems, only a few articles explicitly promote the use of ICT:

1. Article 1 requires that the importation, exportation and transit procedures' description, the required forms and documents and the contact information on enquiry points to be available over the Internet. The worldwide development of (i) trade portals [19] as one-stop shop information hubs and (ii) technology-oriented enquiry points precisely serves this purpose.
2. Article 7 states that documents used for pre-arrival processing should be in electronic format, allows the option of electronic payment for trade-related duties, taxes, fees and charges collected by Customs and allows the use of tracking technology in order to maintain high degree of control over expedited shipments. The provisions of this article cannot be implemented without the effective use of ICT.
3. Articles 9 and 11 promote the use of global positioning systems (GPS) to monitor and secure the timely move of goods, not only within the national Customs territory (internal transit) but also across borders (international transit), via the application of GPS-enabled electronic seals or other technological solutions tailored for transit operations.
4. Article 10 permits the acceptance of electronic copies of supporting documents required for trade formalities, allows the application of electronic filing or processing on common Customs procedures and uniform documentation requirements and encourages, to the extent possible and practical, the use information technology to support the SW.

The WTO TFA does not replace the mosaic of existing conventions and agreements but rather congregates core trade facilitation principles into one document and elevates them to a higher level beyond sectoral (highways, railways, ports, airports, etc.) or thematic limitations.

For instance, in the port and maritime sector, government authorities and regulatory agencies view trade facilitation through the lenses of the Convention of Facilitation of International Maritime Traffic (FAL Convention).[††] The Convention's core objective is to prevent unnecessary delays of maritime traffic. This, by definition, includes more than cargo operations and incorporates procedures relevant to the arriving vessel itself, its crew and passengers. Obviously, the main concern of FAL's standards and recommended practices is to improve the processes and procedures related to the movement of the ship as opposed to the TFA that focuses on the cargo.

The FAL Convention contains standards and recommended practices on formalities, documentary requirements and procedures that should be applied on arrival, during their stay and on departure to the ships, including their crews, passengers, baggage and cargo. Standards are viewed as internationally agreed measures that are "necessary and practicable in order to facilitate international maritime

[††] http://www.imo.org/en/OurWork/Facilitation/ConventionsCodesGuidelines/Pages/Default.aspx

traffic," while recommended practices are defined as measures, the application of which is optional. The Convention is a work in progress and since its inception in 1967, it is being continuously amended and improved. In its current form, it has achieved two major accomplishments: the first one is related to the harmonization of formalities and documentation submission requirements, while the second one refers to the digitalization of data exchanged between maritime cargo carriers and the land–port operators and regulatory authorities.

Nonetheless, TFA marks a milestone in terms of defining the concept and provides a clear reference framework for its member countries. In reality, it may take some time until all the benefits of trade facilitation take effect. Since 2013, developing countries have embarked on a comprehensive reform program to align to the TFA provisions. Their programs are implemented gradually, and their development impact will be visible over time considering their short-, medium- and long-term nature.[‡‡] According to the WBG's TFA tracking tool, the overall average alignment score for all countries is less than 50%.[§§] As governments move forward to the deep and comprehensive reform implementation, they realize that the most impactful reforms are the ones that require policy synchronization between cross-border agencies, which raises the complexity level of the implementation process.

The widespread liberalization of the global trade economic system has moved the policy discussion pendulum from tariff reductions and NTBT measures to the design and implementation of trade facilitation programs. The positive impact of trade facilitation on trade competitiveness and economic growth is stronger and more direct compared to the impact of other drastic trade liberalization measures, such as tariff reductions. A 2007 study revealed that the removal of trade facilitation barriers along international supply chains could increase global trade between 2.6% and 4.7% as opposed to a slight increase of 0.7% due to the complete elimination of tariffs [20].

Today, the positive impact of trade facilitation is well documented. Its full and effective implementation leads to substantial benefits not only for the national economies but also for the global trade logistics system, as a whole. The underlying assumption is that trade facilitation leads to a substantial decrease of trade costs that consequently reduces prices of internationally traded commodities thus increasing their export competitiveness [21]. Recent research demonstrates that full implementation of both binding and nonbinding TFA measures is associated with an average 15% in trade cost reductions in Asia–Pacific, while when other paperless and cross-border trade facilitation measures (digital trade facilitation) are included, projected trade costs reduction reached up to 26% in the same region [18]. The nature of the trade costs related to complex processes and procedures, red tape, informal payments and corruption is not easily detectable. They are usually referred to as transaction cost that imposes excruciating financial burdens and serious operational impediments to international traders.

[‡‡]The World Bank Group (WBG) via its Tracking Tool portal tracks development countries' alignment with the TFA (https://www.tfacountrytrackingtool.org/).

[§§]As of February 2020.

Excessive paperwork, overlapping processes and procedures and enhanced bureaucracy impose nontariff barriers to trade, especially to small and medium enterprises (SMEs). Their immediate impact is prolonged waiting times, high transactional costs and infusion of unpredictability and risks along supply chains. The previous factors are translated into monetary terms and increase the total cost and price of the commodity traded. Trade facilitation reforms, implemented in or outside the context of the TFA, generate efficiencies in border management operations that trickle down to the corporate world and generate measurable impact at national and global economic levels, as shown in Figure B.2.

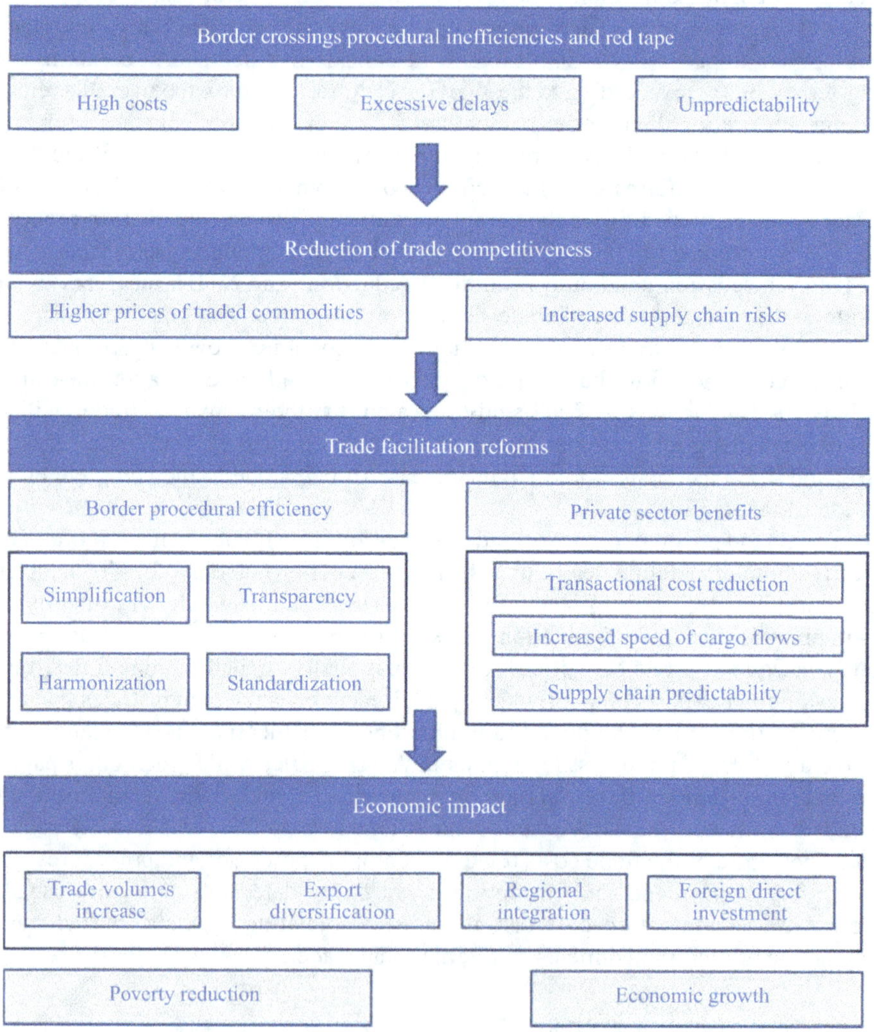

Figure B.2 Impact of trade facilitation

Trade facilitation reforms aim at the improvement of the trading business environment with emphasis placed on practical measures that increase cargo flows passing international border crossings. These measures should be carefully applied in a way not to compromise the fiscal, safety and security goals of the border control agencies. Trade facilitation reforms could be also leveraged to promote other public policies outside the narrow trade and transport domain. For instance, they could be used to pilot—in a small scale and in a controlled environment—reforms supporting broader anti-corruption practices or e-governance programs. In the same context, trade facilitation—contrary to the widespread belief that it puts revenue collection at risk—leads to cost rationalization of border management agencies and better monitoring and control of tax and duties collection.

The most direct impact of trade facilitation is related to the modernization and operational improvement of border management. This is translated to tangible improvements in the procedural simplification, transparency, harmonization and standardization. Procedural simplification refers to the simplification and harmonization of-often-overlapping multi-organizational business processes imposed to traders by public agencies with compliance and regulatory mandate. Transparency measures may be connected to publicly available trade procedural, legal and regulatory information or even allow private sectors to provide inputs in the legislative process prior to enactment. Harmonization efforts lead to the alignment of national procedures, operations and documents with international or regional rules and regulations (i.e., as part of a regional integration process). Cross-border agencies may also benefit from the standardization of documentary and information exchange formats thus enabling the alignment of processes and procedures [34].

All these measures have a dual positive effect on public sector efficiency and on the enabling environment for the trade industry. On one hand, they lead to improved operational efficiency of the border agencies that allow them to achieve their fiscal, control and security goals. They also improve the speed of decision-making process, the reduction of corruption, the effectiveness of legal and regulatory interventions and the rationalization of their human and financial resources.

On the other hand, traders will experience a notable decrease in the time their cargoes spent at the borders via the introduction of risk management tools and techniques, including AEOs and PCA systems, and the application of technology. The use of modern risk management systems significantly reduces the time traders spent at or beyond borders. The recent use of AI on risk management [22] increases the system's effectiveness both in terms of achieving high levels of revenue collection while conducting few but targeted cargo inspections. In addition, the ability to submit the declaration online and prior to the shipment arrival allows traders to process trade paperwork from their premises thus saving time and efforts vis-à-vis in-person submission.

The cost of trade will also be reduced not only due to the concurrent reduction of transaction cost, as bureaucracy and red tape retreat, but also due to the direct reduction of fees and charges that traders had to pay for compliance services—some of them non-justifiable. In developing countries, the size of trade costs is significantly higher compared to production costs. According to [23], trade costs in developing countries are equivalent to applying a 219% ad valorem tariff on international trade making exported goods less competitive in the global markets.

The importance of supply chain predictability in international trade is quite often overlooked. Clear and transparent rules and regulations increase the confidence level of expected clearance time and allow for better planning of supply chain operations and inventory levels. This has a positive impact on traders' indirect cost that may take the form of inventory or opportunity costs. Research [24] also shows that uncertainty about the conditions of crossing borders plays an important role in shaping the demography of exporters across different destinations.

Reductions of transaction costs and time as well as predictability enhancements can all be translated to monetary terms widely known as private sector savings. These savings can even be passed over to the final consumer in the form of lower traded commodity prices thus improving supply chain competitiveness or could be reinvested in the production process or supply chain operations. Lowering private sector costs can allow extrovert SMEs to link to GVCs and export to markets that otherwise they could have no access to [25]. Electronic submission of declarations and awarding long-term compliant companies with quick release benefits are some of the mechanisms that constitute trade facilitation such a valid policy tool for SME development.

At a macro level, trade facilitation has a positive impact on a number of economic areas. First and foremost, it calculated that the implementation of trade facilitation leads to an increase in trade, especially for exports. This is true for both national and global levels. The full and effective implementation of the TFA will lead to tremendous benefits for the global trade. By some estimates, full implementation of the reform under the TFA could reduce members' trade costs by a minimum of 14.3% [26]. By rapidly implementing the measures contained in the TFA, developing countries stand to gain the most. Today, it is widely recognized as an important trade policy instrument in regions with low intraregional trade and weak connections to GVCs.

In South Asia, high trade and transport costs hinder intraregional trade and lead high poverty levels. Trade facilitation reforms improve the quality of governance, transparency, documentation processing time, infrastructure and transit facilities, and ICT is one of the most important components that needs to be put in place to reduce regional trade costs. Its practical applicability and the multifaceted impact at a macro- and micro level have increased its popularity at a national and international front. Although formalities and red tape are found to be more cumbersome on the import side, the economic benefits derived by export supply chains are much higher. This is found to be true both in terms of export volume increase [27] and in terms of diversification of export-oriented firms in favor of SMEs.

Being an integral component of the regional trade agreements, cross-border simplification and harmonization reforms may strengthen intraregional trade. This is particularly important for landlocked countries where costs associated with movement to the port of export are nontrivial. In sub-Saharan Africa, research has "shown that these costs arise from complex interactions among multiple physical and policy features of the trading partners" [28]. Transit facilitation reforms, including transit guarantee systems, Customs seals, e-tracking systems and others, promote regional integration and the practical development of the trade corridors [29] thus allowing landlocked countries to access global markets.

Looking through the lenses of trade and investment policy, the swift movement of inward and outward trade is a prerequisite for the attraction of foreign direct investments [30]. International investor usually enters in an economy, where market mechanism is more efficient and international trade is less costly, quicker and easier. Thus, implementation of trade facilitation can increase FDI inflow to a particular economy.

Reduction of time and cost for traders is particularly important for the trade of time-sensitive goods and for specific economic sectors, such as agribusiness [31]. Taking a modern and holistic approach to trade facilitation which goes beyond Customs to include, inter alia, sanitary and phytosanitary (SPS) agencies, adequately addresses traders' protests, who attribute border delays to inefficient procedures of technical inspection agencies. Therefore, measures to improve interconnectivity, automate SPS entities, apply risk management and create linkages with the Customs benefit the rural poor in developing countries who export perishable products [32].

Finally, an often overlooked benefit of trade facilitation—in the context of the TFA—is the knowledge transfer and collaboration between participating countries. In fact, developed countries, through the adoption of border management best practices and modern reform ICT solutions, have paved the road and set performance standards for rest of the world. In the early stages of the TFA, developing countries benchmarked themselves to best practices formed in developed countries from the North America, West Europe or the Pacific (i.e., Australia, New Zealand). Nowadays, given the increasing number of low-income countries implementing the TFA, LDCs tend to also learn from each other via south-to-south knowledge exchange and learning.

References

[1] UNCTAD. (2019). Review of Maritime Transport.

[2] WCO. (2018). SAFE Framework of Standards.

[3] Iwanow, T. and C. Kirkpatrick. (2009). Trade facilitation and manufactured exports is Africa different? World Development, 37(6), 1039–1050. https://www.sciencedirect.com/science/article/abs/pii/S0305750X0800329X.

[4] Saslavsky, D. and B. Shepherd. (2014). Facilitating international production networks: The role of trade logistics. Journal of International Trade and Economic Development, 23(7), 979–999. https//doi.org/10.1080/09638199.2013.811534.

[5] López González, J. and M.-A. Jouanjean. (2017). Digital Trade Developing a Framework for Analysis. OECD, (205), 24. https//doi.org/https//doi.org/10.1787/524c8c83-en.

[6] UN/CEFACT, United Nations Centre for Trade Facilitation & Electronic Business. (2019). Business Requirement Specifications for the Buy-Ship-

Pay Reference Data Model. Geneva United Nations. Retrieved from https//www.unece.org/fileadmin/DAM/cefact/brs/BuyShipPay_BRS_v1.0.pdf.

[7] Yasui, T. and M. Engman. (2009). The Role of Automation in Trade Facilitation. Paris OECD.

[8] World Bank Group—Doing Business. (2017). Trading Across Borders—Technology Gains in Trade Facilitation. Washington, DC: World Bank Group.

[9] McLinden, G. and L. DeWulf. (2005). The Role of Information Technology in Customs Modernization. Washington, DC: The World Bank.

[10] World Customs Organization. (2018). Use of ICT—WTO Agreement on Trade Facilitation. Brussels: World Customs Organization.

[11] Xing, Z. (2017). The impact of Information and Communications (ICT) and E-commerce on bilateral trade flows. International Journal of Economics and Economic Policy, 15, 565–586.

[12] Pfeiffer, H. (2012). The Diffusion of Electronic Data Interchange. Berne Springer Science & Business Media—Business & Economics.

[13] United Nations Centre for Trade Facilitation & Electronic Business. (2005). Recommendation and Guidelines on Establishing a Single Window (Recommendation 33). New York & Geneva United Nations.

[14] Understanding Single Window Environment. In: *Building a Single Window Environment*. Brussels: World Customs Organization.

[15] APEC Business Advisory Council. (1996). Report to the Economic Leaders.

[16] Food and Agricultural Organization. (2017). International Standards for Phytosanitary Measures (ISPM) 12. Rome Food and Agricultural Organization.

[17] United Nations Centre for Trade Facilitation & Electronic Business (UN/CEFACT). (2005). Case Studies on Implementing a Single Window to Enhance the Efficient Exchange of Information Between Trade and Government. Geneva: United Nations.

[18] Duval, Y., C. Utoktham, and A. Kravchenko. (2018). Impact of Implementation of Digital Trade Facilitation on Trade Costs (No. 174). Retrieved from http//hdl.handle.net/10419/177771%0AStandard-Nutzungsbedingungen.

[19] Nowak, J. (2014). The Evolution of Electronic Trade Facilitation: Towards a Global Single Window Trade Portal. (January).

[20] Ferrantino, M.J. and M. Tsigas. (2013). Enabling Trade Valuing Growth Opportunities. World Economic Forum. Retrieved from http//www3.weforum.org/docs/WEF_SCT_EnablingTrade_Report_2013.pdf.

[21] Mann, C.L. (2012). Supply chain logistics, trade facilitation and international trade a macroeconomic policy view. Journal of Supply Chain Management, 48(3), 7–14. https//doi.org/10.1111/j.1745-493X.2012.03270.x.

[22] Okazaki, Y. (2017). Implications of Big Data for Customs—How It Can Support Risk Management Capabilities. Retrieved from http//www.wcoomd.org/~/media/wco/public/global/pdf/topics/research/research-paper-series/39_okazaki_big-data.pdf.

[23] WTO. (2015). World Trade Report 2015. Retrieved from https//doi.org/10.30875/1cee73f9-en.

[24] Fontagné, L., G. Orefice, and R. Piermartini. (2020). Making small firms happy? The heterogeneous effect of trade facilitation measures. Review of International Economics, 1–34. https//doi.org/10.1111/roie.12463.

[25] Li, Y. and J. Wilson. (2009). Trade Facilitation and Expanding the Benefits of Trade Evidence From Firm Level Data. In: ARTNeT Working Papers.

[26] Moïsé, E. and S. Sorescu. (2013). Trade Facilitation Indicators the Potential Impact of Trade Facilitation on Developing Countries' Trade. OECD Trade Policy Papers, 144. https//doi.org/https//doi.org/10.1787/5k4bw6kg6ws2-en.

[27] Beverelli, C., S. Neumueller, and R. Teh. (2015). Export diversification effects of the WTO trade facilitation agreement. World Development, 76, 293–310. https//doi.org/10.1016/j.worlddev.2015.07.009.

[28] Christ, N. and M.J. Ferrantino. (2011). Land transport for export the effects of cost, time, and uncertainty in sub-Saharan Africa. World Development, 39(10), 1749–1759. https//doi.org/10.1016/j.worlddev.2011.04.028.

[29] Arvis, J.F., J.-F. Marteau, and G. Raballand. (2010). The Cost of Being Landlocked. Washington, DC: The World Bank. https//doi.org/10.1596/978-0-8213-8408-4.

[30] Halaszovich, T.F. and A. Kinra. (2018). The impact of distance, national transportation systems and logistics performance on FDI and international trade patterns results from Asian global value chains. Transport Policy. https//doi.org/10.1016/j.tranpol.2018.09.003.

[31] Djankov, S., C. Freund, and C.S Pham. (2010). Trading on time. Review of Economics and Statistics, 92(1), 166–173. https//doi.org/10.1162/rest.2009.11498.

[32] WTO. (2018). The Relationship Between the Trade Facilitation Agreement and the Agreement on the Application of Sanitary and Phytosanitary Measures. Science. https//doi.org/10.1126/science.202.4366.409.

[33] White Paper: The Role of Port Community Systems in the development of the Single Window (June 2011). European Port Community Systems Association (EPCSA).

[34] Masahiko, F. (2021). Cross-border logistics practices, policies and its impact. Available from: https://doi.org/10.1016/B978-0-12-814060-4.00003-4.

Chapter 2
Maritime ports and cybersecurity

Kristen Kuhn[1], Jeptoo Kipkech[1] and Siraj Ahmed Shaikh[1,2]

Maritime ports play a vital role in global trade. Goods are loaded at a port, transported and unloaded at another port. Each port serves as a node in a global supply chain, which shapes trade and links states, economies, cultures and the unknown.

In many ways, maritime ports resemble cyberspace which also facilitates trade and serves as a connector. Like the Internet, the ocean – vast and deep – is a source of prosperity and peril. It can, then, be understood why, like the sea, cyberspace has recently been declared by NATO as a new domain of warfare [1]. The modern port exists in both sea and cyberspace, where these two domains of warfare overlap. The port is a physical infrastructure that faces traditional risks such as occupation. It is also a cyber-entity that faces new risks such as information leakage. While the traditional risks in ports have changed little, cyber risks are a new and growing threat. In 2013, cyber-incidents were the 15th leading global risk for organizations and, in 2019 – six years later – they were ranked second [2].

While cybersecurity threats and vulnerabilities are being addressed in the port industry, this has not yet been achieved to the required degree. This is in part because ports vary in structure and ownership, making standardization and regulation a challenge. Further, cyber risk is multiplied by the connectivity of information, communication and cyber–physical systems (CPSs), by which an isolated cyber-incident in a port may have cascading effects across the global port system. To be effective, cybersecurity must evolve rapidly and constantly alongside technology implemented in ports.

This chapter introduces the modern port in Section 2.1 and explores its importance in Section 2.2. It then presents cybersecurity in Section 2.3 and draws a connection between ports and cybersecurity in Section 2.4, highlighting ports as a cyber–physical environment. To illustrate the cyberthreat landscape, five known attacks in ports are explored in Section 2.5. This chapter then examines control mechanisms in place for cyber risk management for ports in Section 2.6 and reviews current cybersecurity guidelines and standards in Section 2.7. These concepts are encapsulated in the summary presented in Section 2.8, which mentions digital trends and the future of maritime ports and cybersecurity.

[1]Systems Security Group, Institute for Future Transport and Cities (IFTC), Coventry University, Coventry, United Kingdom
[2]Security, Risks Management and Conflict (SEGERICO) Research Group, Universidad Nebrija, Madrid, Spain

2.1 The modern port

A *port* is a shore-based installation for the dispatch of cargo between land and sea [3]. There are many different types of ports, all of which can be classified as either sea, lake, river, or canal depending on the type of business transacted [4]. For the purpose of this chapter which deals with maritime ports and cybersecurity, the term 'ports' refers to seaports. This section explores the port system in Section 2.1.1, port ownership models in Section 2.1.2 and port structure in Section 2.1.3.

2.1.1 Port system

In medieval times, a 'port' referred to a shore-based town whose main activity was trade [5]. A port had a harbour and could be recognized by its characteristic infrastructures, including dockyards, warehouses and customs houses. With the passage of time, the port attracted other forms of commercial and industrial infrastructures, such as banks, agencies and markets. These, in turn, attracted nearby goods and more extensive services; thus, the town expanded. This can be viewed in light of the *central place theory* which identifies such towns as geographical centres for markets and services [6]. These large towns then became cities, many of which today are major cities with a high concentration of civilians [7].

As port cities expand, so does the scope of their trade. Many necessities, along with secondary needs that sustain economies, are imported and exported by states through commercial ports. Ports which were previously isolated or limited to trade became part of a complex and interconnected transport network. At this point, it makes sense to combine the central place theory with the *network theory* [8]. The network theory views ports not only as geographically relevant centers, but as functionally relevant centers whose value lay in their links to other ports and transportation systems. The growth of ports relate to the degree they are interconnected and interdependent on other ports from multiple states [5]. This is the modern port system.

The modern port is a multimodal distribution hub that links transport by sea, river, road, rail and air routes. For many nations, ports are the main transport link with their trading partners and thus a focal point for motorways and railway systems [9]. This transport network is often referred to as the global supply chain. It is worth noting that in this network, ports themselves became a service in the mentioned chain. Take the case of Singapore, a major city that is also home to the world's second busiest port container terminal (PCT) [10]. For a small nation like Singapore with few natural resources or agricultural industry, the port itself is a fundamental economic asset and its security is essential to the nation's economic well-being [11].

2.1.2 Port ownership models

There are various models of port ownership, the basic forms of which can be viewed on a scale of public to private ownership, as seen in Table 2.1. On this scale, ports are classified into four models: service port, tool port, landlord port, and private port [13]. However, the modern port is often a combination of these categories [9].

Table 2.1 Examples are given for four port ownership models, from most public to most private, orientation of which relates to the ownership of infrastructure, superstructure, dockworkers, and other functions

Type	Infrastructure	Superstructure	Dockworkers	Other functions
Service port	Public	Public	Public	Mainly public
The Port of Colombo, in Sri Lanka, is a service port. It is publicly managed: the port owns, maintains and operates every asset. Service ports are often run by the ministry of transport, and cargo handling is managed by the port authority. Service ports are in decline, as many ports shift to landlord ports.				
Tool port	Public	Public	Private	Mainly public
The Port of Chittagong, in Bangladesh, is a tool port. The port authority owns and operates the port infrastructure and superstructure and makes them available to port operators. This can lead to conflicts regarding split of responsibility, as the port authority owns and operates the cargo handling equipment, but the private cargo handling firm usually signs the cargo handling contract with the cargo owner or shipowner.				
Landlord port	Public	Private	Private	Mainly private
The Port of Singapore, in Singapore, is a landlord port. It combines state and private ownership.[a] The port authority, or landlord, provides infrastructure, and the port operator, or tenant, provides superstructure.[a] The port is split into terminals which are leased for a set period of time. This is the dominant model in large and medium ports and is the most common growing model.				
Private port	Private	Private	Private	Mainly private
The Port of Southampton, in the United Kingdom, is a private port.[b] This is the only model in which the port land is owned privately and is rare as full privatization may be considered extreme. It implies a transfer of land, and often regulatory functions, from public to private ownership. This poses a risk that the port can be sold for non-port activities or raise security concerns.				

[a]Alderton [9].
[b]Monios [12].
Source: The World Bank [13].

Public ownership can be associated with a *port authority*, which is a government or private body that is responsible for port administration, construction and management, and sometimes responsible for port operations and security [13]. Port authorities may be established at all levels of government: national, regional, provincial, or local. Likewise, private ownership can be associated with a *port operator* which 'typically pursues conventional micro-economic objectives, such as profit maximization, growth, and additional market share' [13]. Port operators include terminal operators, cargo handling companies, and dockworker firms.

2.1.3 Port structure

In ports, cargo can be transferred between land and sea by two methods: ship gear or quay cranes [3]. Ship gear is managed by seafarers and accompanies the ship from port to port. Quay cranes are managed by *dockworkers* or labourers who load and unload vessels and are a part of either the port infrastructure (e.g. a paved terminal with deep-water access) or *superstructure* which is built on top of the infrastructure (e.g. cranes or cargo-handling equipment) [9]. There is contention between using ship gear and quay cranes regarding the split of responsibility. That is to say, quay cranes are generally accepted to be more efficient of the two [3]. Beyond the process of loading and unloading cargo, which happens on the *quayside*, where the land meets the sea, port structure varies. The goods arriving at or departing for the port are moved by other cranes and vehicles, where they are either stored or moved onto other modes of transport, such as trucks or trains to *hinterlands*, or surrounding area.

Ports are constantly changing in response to diverse external factors – from weather to technology to political agenda. This renders every port unique in structure. This subsection explores how the physical structure of the port is influenced by four factors: automation, cargo, ships, and transport network.

2.1.3.1 Automation

Most physical processes within ports, such as loading and unloading cargo, are performed with automated or semi-automated mechanical systems and machinery (e.g. ships, trucks, cranes and electronic gates) under the control of sophisticated software systems (e.g. industrial CPSs, supervisory control and data acquisition (SCADA) systems, and surveillance systems) [7]. Being a controversial issue, automation draws tension among many actors. All can agree, however, that port automation – to be followed by automated vessels – marks a new industry standard.

The level of port automation has greatly increased in the past couple of decades, with the ultimate realization being the smart port. A *smart port* can be defined as a fully automated port where all devices are connected via the Internet of Things (IoT) [14]. Automation carves out early adapters as more efficient and safer than ever before. With that said, as ports become increasingly interconnected and reliant on automation, they become more vulnerable to a new threat: cyber risk.

2.1.3.2 Cargo

The decision to use ship gear or quay cranes also depends on the nature of the cargo. A ship may be carrying textiles or foodstuffs, petroleum, lumber, or hazardous goods. Special cargo arrives at terminals built to accommodate their storage requirements, e.g. grain is stored in silos; liquefied natural gas is kept in pressurized tanks. General cargo, however, is stored in containers that arrive to a PCT. Containers are standard units adopted in the 1970s for transport of cargo on a global scale. This era is called containerization and revolutionized transport. It also illustrates that ports must change in response to changes in cargo-handling technology [9]. Two external forces that had a prominent effect on cargo are standardization (containers) and globalization (goods passing through ports vary in type).

The standardization of shipping containers also gave birth to the twenty-foot equivalent unit (TEU) which is a key unit of measurement throughout the maritime industry [15]. The TEU is the standard unit of measurement for shipping containers, vessel size, and capacity, including how much cargo passes through a port, thus calculating port activity. For instance, top US ports Los Angeles and Long Beach handle more than 16 million TEUs a year, while top ports in China Shanghai and Shenzhen handle more than 65 million TEUs a year [15].

2.1.3.3 Ships

While the container is an industry standard for cargo, vessels that carry cargo are far from being standardized. This presents a continuous challenge for ports, whose infrastructure must accommodate the ships they host. A quay crane, for instance, cannot unload a ship that is taller than its own height. This challenge is greater still, as ports last longer than ships. Most of the UK ports, for instance, were built between one and two centuries ago when ships were small [9].

In search of greater fuel efficiency and economies of scale, there is a tendency across the sector for larger vessels which move more goods through ports faster, increasing productivity and profit [3,16]. As container ships continue to grow, ports and port cities have adjusted to host them [17]. This concept was outlined in 1965: 'The ports of the world are obsessed with the problem of handling more and larger ships; despite recessions in trade and financial ups and downs the management of our ports, whatever form their control may take, can never evade this problem' [3,9].

Today, some container ships are nearly 400 meters long – the distance around an Olympic running track [17]. The rise of *mega-ship*, which can be defined as a ship with a capacity greater than 10,000 TEUs, took place in earnest as a response to the 2008 financial crisis [16]. Today, the mega-ship is challenging the port: the inability to expand because of a dearth of land suggests the end of the link between cities and ports [18]. Growing cities simply cannot accommodate growing ports so there is a demand for special port complexes.

2.1.3.4 Transport network

Since the design of ports and docks centuries ago, transportation has changed continuously, from horse and cart to autonomous vehicles [9]. The design and infrastructure of ports shift according to vehicles and cargo-handling technology. The transport connections to a port also determine the geographic layout. For example, the warehouse storing goods may be situated near tracks, with additional cranes that lift containers onto trains. The design of the port, and operational management, needs to be done so that congestion is minimized. The central place theory, which sees the port in a strictly spatial context, takes into account that growing ports infringe on cities – also growing, to occupy more space. However, this theory fails to account for other issues – such traffic congestion created from the port, which often also affects human mobility in surrounding areas [6].

2.2 The importance of ports

Cyber systems create benefits, but they also introduce risks. In the context of ports, *cyber risk* is the 'probability of a threat agent exploiting a vulnerability to cause harm to a computer, network, system, or utility, resulting in financial losses, disruption or damage to the reputation of an organization' [19]. Risks to ports affect not only the ports themselves, but also their customers, stakeholders and the global supply chain [19]. The risks can include, for example, financial loss, theft of cargo or information, strikes, and security malfunctions, which can lead to the shutdown of a port. This section explores ports as critical infrastructure (CI) in Section 2.2.1, ports as critical information infrastructure in Section 2.2.2, and the impact of port disruption in Section 2.2.3.

2.2.1 Ports as critical infrastructure

Ports are a critical part of a national transport infrastructure [9]. CI, including ports, constitutes a probable target for cyberattacks, given its importance in the functioning of society [20]. Ports are especially vulnerable to cyberthreats, as they are dependent on data systems, handle huge volumes of cargo and/or passengers, process an immense number of transactions with high monetary value, and involve a wide range of stakeholders.

Through history, commercial ports have been vital to a state and disruption of their services can cause damage. For this reason, commercial ports are considered CIs. As ports are the veins of global economy and cross-border trade, there exists a network of interconnected CIs that have 'physical and cyber multi-inter-dependencies, interacting with all sectors of the economy; therefore, their malfunctioning or disruption will have cascading effects on several other infrastructures or cross-border services that depend on them' [7].

2.2.2 Ports as critical information infrastructure

The advanced development of information systems, the evolution, and the diffusion of broadband communication have led to a wide adoption of information and communication technology (ICT) by CIs [7]. These ICT systems are critical for the normal functioning of the CI and are themselves a form of CI. Critical Information Infrastructures (CIIs) are ICT systems that are CIs for themselves or that are essential for the operation of one [21]. Examples of ICT systems, which may also serve CIIs, include communications systems as shown in Table 2.2.

Ports host CIIs and their disruption would have significant consequences [23]. 'The large amount of critical and sensitive data, the information and services that are managed daily, the number of entities called to be served, and the interdependencies with other infrastructures require effective security management' [7]. The malfunctioning or disruption of a single CII that is connected to a network could affect an entire network of users (CIs) and the services that depend on them. The drawback of making ICTs the backbone of CIs is that it leaves systems vulnerable to exploitation or manipulation from threats [24].

Table 2.2 Many ICT systems can be found in a port; some examples are included in the following table

ICT systems	Examples
Security	Vehicle access, building access and control gates
Communications	Mobile radio, email and websites for cargo and customs
Business	Terminal operation system and container terminal Management system and office systems, e.g. payroll
Terminal automation	Vessel scheduling software, yard equipment and maintenance, control systems for cranes and remote monitoring of equipment

Source: Boyes [22].

2.2.3 Impact of port disruption

Disruption refers to a disturbance in which material flows are stopped entirely, while *delay* refers to a disturbance in which the rate of material flows is slowed [25]. Both delay and disruption will postpone the time of arrival of goods and are risks for ports [26]. For the purpose of this chapter, disruption and delay will be referred to simply as 'disruption'. Some scenarios that cause port disruption include natural disasters, labour strikes, climate change and cyberattacks.

Port disruption has a domino effect, posing both costs and benefits to other ports. It is worth noting that ports compete for cargo, and there could be much to gain from a well-timed disruption at a neighbouring port. For instance, a five day shutdown of a US West Coast port would result in ships being rerouted to East Coast ports. To reach the US East Coast, these ships – most too large to cross the Panama Canal – would transverse the Suez Canal and this would delay their shipment by almost a week [27]. In this scenario, US East Coast ports benefit by gaining shipments, and the Suez Canal suffers due to unexpected congestion. The degree of coordination that goes into the flow of cargo highlights the interdependency of ports and means the impact of disruption is widely felt. This subsection explores five potential hazards of port disruption: congestion, economy, environment, geopolitics and safety.

2.2.3.1 Congestion

Port activity is increasingly required to fit perfectly into wider logistics chains, but congestion can prevent a match between ports and their network. When a port suffers *congestion*, it is understood that ships are queuing, or waiting to obtain a berth [28]. The busier the traffic, the greater the imposed cost [29]. Congestion costs include time loss, additional fuel consumption, inconvenience and possibly even accidents [29]. However, time loss is the largest contributing factor and this is transferred to others. According to Hapag-Lloyd, a shipping company, their liners often call at over a dozen ports per voyage [30]. Delays can impact operations and congestion may be felt elsewhere in the logistics chain.

2.2.3.2 Economy

Ports are a major economic multiplier for a state's prosperity [9]. In 2018, nearly 80 per cent of the world's trade volume and 70 per cent of its trade value passed through ports [31]. Ports are vital to trade and their disruption, in addition to the effect on port operations, would cause damage to trade flows, economies and various parties concerned [7,25]. The economy is a concern for many stakeholders ranging from states to the firms and people affected by maritime business around the globe [25]. Ironically, ports and shipping are among the most important causes for uncertainty in supply chains [32].

2.2.3.3 Environment

Environmental pollution is another potential hazard of port disruption. Oil and gas tankers compose one of the highest risks in a port environment, with the majority of such accidents involving tanker vessels, barges, platforms and petroleum shore stations [33]. Further, some polluted areas are caused by illegal human activities such as tanker cleaning. Every year, diesel, oil, petrol and other toxic chemicals are released into the sea, causing damage to wildlife, habitats and ecosystems. While illegal dumping is not always accidental, such accidents could have catastrophic effects on the environment. Minimizing the opportunity for accidents means actively working to ensure security of ports and ships, including the cybersecurity of their physical systems which may, if vulnerable, be leveraged by adversaries. To minimize the possible damage to the environment, it is essential to develop and implement robust cybersecurity measures.

2.2.3.4 Geopolitics

Protection of CI has always been at the centre of homeland security. After all, if you do not have food, water, energy, power and communication, you do not have a country [34]. As CI, ports are important to a state both strategically and politically. The value of a port to the economy of a country has been demonstrated during wartime: 'Ports are a primary object in each campaign, for the enemy knows well the crippling effect on his opponent of capturing, or even putting out of action, the ports through which supplies, arms and men reach the fighting zone' [3]. Port disruption represents geopolitical hazard for states by compromising their capabilities.

2.2.3.5 Safety

Ports are dangerous places to work for both dockworkers and seafarers, who undergo extensive safety training to minimize workplace accidents. Maritime safety has many influencing human factors, including fatigue, stress, teamwork, communication and safety culture [35]. Port disruption can affect these factors, sometimes interfering with basic services like communication, to increase the safety risks to people [7]. Further, the disruption of CIIs could impact safety [23]. Also, serious accidents themselves can disrupt ports and are detrimental to the growth and efficiency of the port, not to mention its reputation [25].

2.3 Cybersecurity

Cybersecurity is 'the protection of information systems (hardware, software, and associated infrastructure), the data on them, and the services they provide, from unauthorised access, harm or misuse' [36]. This includes intentional harm by a system operator or accidental harm as a result of failing to follow security procedures.

Mission-critical information assets, the 'crown jewels' of an organization, are of great value and would cause a major business impact if compromised. Cybersecurity measures need to protect these crown jewels against a range of low-level crimes to national security concerns. This is accomplished through an understanding of the vulnerabilities and threats to business. This section analyses fundamental cybersecurity properties of information systems in Section 2.3.1, which are essential in preventing exploitation of vulnerabilities and threats in Section 2.3.2, which lead to cyberattacks in the maritime sector, which are defined in Section 2.3.3.

2.3.1 Cybersecurity attributes

Desirable characteristics of a secure system have come to be defined by the *CIA triad* [37] which includes three core cybersecurity attributes: confidentiality, integrity and availability. *Confidentiality* focuses on the control and authorization of information access and disclosure with respect to privacy. A critical cybersecurity concept is the *principle of least privilege*, whereby access to information and assets should be granted only on a need-to-know basis. *Integrity* protects against improper information modification or destruction and maintains consistency, accuracy and trustworthiness of data or system. *Availability* ensures timely and reliable accessibility and use of data, asset information, systems and associated processes.

The CIA triad is the heart of information assurance standards and guides policy for information security in an organization. Organizations should have management, operational and technical security controls in place to protect the confidentiality, integrity, and availability of their systems and information.

Furthermore, effective and holistic cybersecurity requires information systems to be built on four pillars: people, processes, technologies, and physical aspects [38]. Ultimately, when combined into a single and integrated framework, this holistic strategy will yield the most effective cyber defences in ports.

2.3.2 Vulnerabilities and threats

The exponential growth of the Internet and interconnected devices has created manifold benefits for society and the economy, but with these benefits come new cyber vulnerabilities and threats.

Vulnerability is a weakness in an information system, its security procedures, internal controls or implementation that could be exploited by a threat [39]. Many common vulnerabilities relate to hardware, software, networks, personnel, and

Table 2.3 Categorization of common cyberattacks in the maritime industry based on two factors: (1) threat sources intent and (2) target

	Intentional	**Unintentional**
Targeted	Brute force Denial of service Spear phishing	Victim to social engineering Escaped proof of concept Runaway penetration test
Untargeted	Malware Phishing Water holing Scanning	User error

Source: BIMCO, CERT-UK [42,43].

organizations [40]. Most information system vulnerabilities are associated with security controls that either have not been applied (intentionally or unintentionally) or have been applied, but retain some weakness [41]. Organizations may need to reassess existing security controls over time to determine the effectiveness of their controls.

A *threat*, on the other hand, is a circumstance or event with the potential to adversely impact organizational operations and assets, individuals, other organizations, or the nation through unauthorized access, destruction, disclosure, modification, or denial of service to an information system [39]. Threats can be intentional, accidental, untargeted, or targeted in nature as seen in Table 2.3.

A situation or event caused by a threat that has the potential for adverse impact is referred to as a *threat event* [39]. When a threat event occurs, it becomes an incident that can jeopardize the confidentiality, integrity, or availability of an information system or constitute a violation of security policies, procedures, or their use [44].

The level of *impact* from a threat event is the magnitude of harm that can be expected to result from the consequences of unauthorized disclosure, modification, destruction, or loss of information or information system availability [45]. Organizations can experience the impact of adverse events at the information system level (e.g. failing to perform as required), at the business level (e.g. failing to meet business objectives), and at the organizational level (e.g. failing to comply with legal or regulatory requirements, damaging reputation, or relationships) [41].

A cybersecurity threat is actualized by a *threat actor*, who is a person or an entity responsible for an incident and can be internal or external to an organization [46]. Threat actors have a variety of motivations and capabilities that require the use of different risk mitigation and control techniques [39]. Some actors are motivated by financial gain, while others are motivated by political, ideological, or religious reasons. Threat actors include hacktivists, criminals, disgruntled insiders and states.

Organizations often categorize the tactics, techniques and procedures of adversaries to create *threat scenarios* that describe how a threat can orchestrate and impact their business. For adversarial threats, an assessment of likelihood of occurrence and the success of an attack are typically based on the adversary's

Table 2.4 The common threat actor groups in the maritime industry, their motivations, the groups they exist within, and their targets

Threat actor groups	Motivations	Threat actors	Targets
Criminals	Economic/financial Information advantage Reputation	Individuals Organizations	Assets Individuals Organizations
Espionage	Commercial/industrial Intellectual property Competition	Organizations Nations	Governments Individuals Organizations
Hacktivists	Challenge/egoism Ideological/political Social change	Activists Individuals	Governments Individuals Organizations
Insider threats	Financial gain Revenge	Contractors Employees Partners	Physical/process Technical failures Poor operational design
Terrorists/war[a]	Political/social change Fear Religion/ideology	Hackers Individuals Terrorists States	Individuals Infrastructures Organizations Public/critical targets Governments/military

[a]The authors have combined the categories cyberterrorist and cyberwar.
Source: Boyes [61].

intent, capability and targeting [39,41]. Table 2.4 analyses the main threat actor groups, their motivations, the groups they exist within, and their targets.

2.3.3 Cyberattacks

Of the many challenges faced by the modern port, one of the most complex and potentially damaging is the threat of a large-scale cyberattack against information communication technology and cyber infrastructure.

Cyberattacks include nefarious activities that target both information technology (IT) systems or cyber assets, operational technology (OT) systems or physical assets, and computer networks or personal computer devices in an attempt to compromise, destroy, or access systems and data [45]. Breaches result in property damage or theft, data damage, loss of income due to outages and operational failure, website defacement, and cyber extortion. The consequences of a cyberattack depend on the nature of the attack, incident complexity, and established industry procedures.

An *attacker* is a party who attacks a host, network, or other IT resources [44]. Not all attacks are intentional, as some result from users who accidentally or unintentionally violate security policies or requirements, to the point of compromising security, as seen in Table 2.3. Attackers who intend to exploit vulnerabilities are motivated by various reasons, ranging from the desire to make political or social statements to financial gain and cyberwarfare. The skill sets of attackers vary widely as do their motivations. Attackers commonly violate cyber-security attributes in the CIA triad: confidentiality, integrity, and availability.

Attackers utilize paths to gain access to a target known as *attack vectors* [47]. Each attack vector can be thought of as comprising a source of malicious content, a potentially vulnerable processor of that malicious content, and the nature of the malicious content itself. An example of a threat vector is a malicious email attachment (content) in an email client (source) rendered by a vulnerable application (processor). The next section explores common attack vectors found in ports, which have led to significant breaches in recent years, as discussed further in Section 2.5.

2.3.3.1 Advanced persistent threat

This attack is widely acknowledged to be the most sophisticated and potent class of security threat. Advanced persistent threats (APTs) are specifically designed to quietly and slowly spread to other hosts, gathering information over prolonged periods of time and eventually leading to exfiltration of sensitive data and cause other negative impacts [48]. The threat actors involved are technological experts who are well trained, well funded, organized, and capable of utilizing a range of technologies to achieve their objectives.

2.3.3.2 Backdoor

A *backdoor* is an undocumented way of bypassing normal authentication procedures that enable remote access to a system [45]. Although some backdoors are secretly installed, others are deliberate and widely known. These backdoors were originally designed to provide the manufacturer with a way to restore user passwords.

2.3.3.3 Malware

Malware, or malicious software, is designed to gain access or cause damage to a computer, server, or network without the knowledge of the victim [45]. Malware steals resources from a computer and exploits known deficiencies and problems of the network such as outdated or unpatched software. Examples of malware include viruses, Trojan horses, ransomware, spyware, and worms. According to the *2018 Maritime Cyber Security Survey* by IHS Markit and BIMCO, 77 per cent of cyberattacks in the maritime industry are malware attacks [49].

2.3.3.4 Phishing

Phishing, the practice of sending emails to a number of potential targets asking for particular pieces of sensitive information, is a social engineering technique designed to deceive the user into disclosing sensitive or confidential information [45]. *Spear phishing*, on the other hand, targets specific individuals with personalized emails containing malicious software or links.

2.3.3.5 Ransomware

Ransomware is a type of malware that locks a computer, holding it 'hostage' by means of file encryption and forcing victims to pay money in order to get their files back [50]. Ransomware has become a lucrative business with increasing popularity.

2.3.3.6 Social engineering

Social engineering is a non-technical practice used to manipulate individuals within an organization into breaking security procedures [45]. There are many social engineering tactics and mediums of implementation, such as email, web, phone or USB drives. Phishing is an example of social engineering.

2.3.3.7 Virus

A *virus* is a computer programme that can self-replicate and infect a computer without permission or knowledge of the user [45]. A virus might corrupt or delete data on a computer, spread to other computers by attaching itself to an active host programme or an already-infected programme, and execute code when a user launches these programmes.

2.3.3.8 Worms

A computer *worm* makes as many possible copies of itself from computer to computer [45]. It can self-replicate without any human interaction and does not need to attach itself to a programme in order to cause damage. Worms can modify and delete files and even inject additional malware into the computer. In contrast to viruses that require an already-infected host file to propagate, worms are stand-alone malware and do not require a host programme or a human to spread.

To protect users from being victimized by the previously mentioned attacks, protective approaches are needed not only to detect these malicious programmes but also to prevent them from inflicting damage and compromising the CIA triad in the first place. To do so, a profound understanding of the nature of attacks is required for in-depth examination of vulnerabilities to develop effective defensive solutions.

2.4 Ports and cybersecurity

Unlike the sea and the port, cyberspace is not a geographical domain [1]. However, physical and cyber domains not only affect each other but also interact and intersect to create a unique cyber–physical domain. Adopting cybersecurity attributes provides a modern framework to facilitate collaboration between security and safety specialists. When applying these attributes to CPSs, we gain insight into cyber vulnerabilities. This section explores ports as a cyber–physical environment in Section 2.4.1 and the cybersecurity attributes specific to ports in Section 2.4.2.

2.4.1 Port as a cyber–physical environment

A CPS is an 'integration of computation with physical processes. Embedded computers and networks monitor and control the physical processes, usually with feedback loops where physical processes affect computations and vice versa' [51]. CPSs are systems of collaborating computational entities which link physical and virtual surroundings to achieve a global behaviour [52]. CPSs are composed of a set of networked agents interacting with the physical world; these agents include sensors, actuators, control processing units, and communication devices that enable automation [53,54]. These sectors control industrial control systems through the use of SCADA systems, distributed control system, and IoT, which are based on the nature of individual systems [55].

The global supply chain is a complex CPS composed of distinct and interconnected IT and OT systems [7]. The difference is that IT manages the flow of digital information, while OT deals with machines and physical processes [56]. A traditional juxtaposition would be that of software (IT) and hardware (OT) [56]. However, this does not stress the interaction between cyber and physical elements. It is not sufficient to understand IT and OT components separately. Rather, CPS is about the intersection of physical and cyber [51].

The fundamental ICT systems of ports simplify and accelerate processes and form a foundation for an intelligent transport system [57]. Ports use ICT systems to manage, store, and share information, ensuring swift and seamless product/data exchange from the producer to the end consumer during the provision of services [7]. Ports use a wide array of ICT systems to perform vital functions, including communications, security, business, and terminal automation, as seen in Table 2.2.

As the cyber–physical environment grows in ports, so does vulnerability. The smartening of ports does not necessarily increase threats (the number, capability, or intent of threat actors), but it does increase vulnerabilities (in the form of a larger and more complex attack surface) and the potential consequences (by allowing attacks on networks and data to cause physical damage in the real world) [58].

Likewise, it is key to acknowledge ‘the three sides of the coin’: ship, shore and their connections [59]. As ports and ships become smarter, their communication and information-sharing efforts grow in sophistication. The harmonized collection, integration, exchange, presentation and analysis of marine information on board and ashore by electronic means can be referred to as *e-Navigation* [59]. Ship-borne users of e-Navigation include offshore energy vessels, fishing vessels and commercial tourism crafts; shore-borne users encompass ship owners and operators, port authorities, insurance and financial organizations. For instance, congestion may lead to port disruption, but e-Navigation may help to reduce delays by means of modulating a ship’s journey. While e-Navigation increases efficiency in ports and works to protect the marine environment, it also raises cyber risks from the susceptibility of port systems to cyberattacks.

2.4.2 Cybersecurity attributes of ports

The CIA triad alone is not adequate to manage infrastructure susceptibilities. The *Parkerian hexad* provides a fortified approach to encompass both information insurance and engineering good practice [60]. This model includes the three CIA attributes and three additional attributes: authenticity, utility and possession/control. Furthermore, an extended version of the Parkerian hexad, by Boyes, includes two additional attributes: safety and resilience. These eight attributes demonstrate a comprehensive cybersecurity model for ports [38] as shown in Table 2.5.

In this model, the attributes from Table 2.5 address four distinct operations, whereby controlled access to port systems falls under the first two attributes: (1) confidentiality and (2) possession/control. Port system configuration, information

Table 2.5 This table shows the eight maritime cybersecurity attributes by Boyes: an extended version of the CIA Attributes and the Parkerian hexad, including two additional attributes: safety and resilience

Attributes	Definition
Confidentiality	The control and authorization of information access and disclosure with respect to privacy.[a] Systems should prevent unauthorized access to sensitive data, and personal data should be handled in accordance with the General Data Protection Regulation (GDPR)[b]
Possession/ control	Systems shall be designed, implemented, operated, and maintained so as to prevent unauthorized control, manipulation, or interference
Integrity	Protect against improper information modification or destruction and maintain consistency, accuracy, and trustworthiness of data or system[a,b]
Utility	Ensure asset information and systems remain usable and useful across the life cycle of the asset
Authenticity	Ensure system inputs and outputs, and that the state of any associated processes and data are genuine and have not been tampered with or modified
Safety	The design, implementation, operation and maintenance of systems and related processes to prevent harmful states that may lead to injury, loss of life, or unintentional physical or environmental damage[c]
Resilience	The ability of the asset information and systems to transform, renew and recover in a timely way in response to adverse events[c]
Availability	Ensure timely and reliable accessibility and use of data, asset information, systems, and associated processes[a,b]

[a]Dukes [37].
[b]Parker [60].
[c]Boyes *et al.* [38].
Source: Boyes [63].

quality, and validity are included in the next three attributes: (3) integrity, (4) utility and (5) authenticity. Finally, the continuity of port operations, safety of people, and assets are encompassed by the last three attributes: (6) safety, (7) resilience and (8) availability. These four operations are shown in Figure 2.1.

2.5 Attack scenarios

There are many known cyberattacks in maritime ports; however, only a fraction of the attacks that occur are publicly known. 'Criminal organizations have hacked the computers and networks of shipping lines and their agents, port authorities, and port operators to obtain information on ship-sailing schedules, ports of call, the specific cargoes and containers on-board, load/manifest data, trucker information, cargo release information in destination ports, and which security measures are in place' [62]. The 2018 Maritime Cybersecurity Survey [49] reveals that phishing

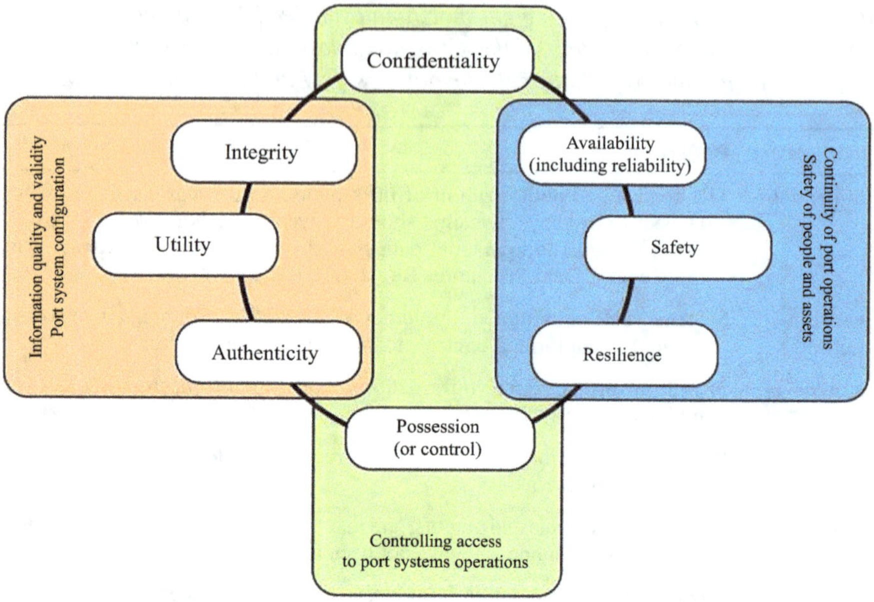

Figure 2.1 The diagram shows the four operations addressed by the eight maritime cybersecurity attributes by Boyes

(49 per cent) and malware (44 per cent) were the most common forms of incidents faced in the maritime sector, mostly leading to service disruption (49 per cent) and system downtime (44 per cent). This section investigates common attack vectors that have been observed in the ports and provides five real cyberattack scenarios, summarized in Table 2.6, to illustrate the cyberthreat landscape.

2.5.1 APT40

APT40 is a cyberespionage threat actor group believed to be sponsored by the Chinese state that has been targeting engineering, transportation and defence industries that overlap with maritime technologies. The earliest public reports from Proofpoint show a pattern of sending targeted emails or spear phishing, to a number of US shipbuilding companies and organizations with maritime links which, if successful, would have resulted in backdoor software being installed on the target machine. The actor then used this access to move laterally within the organization and used information gleaned (e.g. account credentials) to help them target other organizations. FireEye reports that they expect the activities of this group to be continued in, at least the near and medium term despite the recent public attention [64].

2.5.2 The Port of San Diego

In 2018, five days after an attack on the Port of Barcelona [65], the Port of San Diego's IT systems were disrupted by ransomware (malware) that prompted

Table 2.6 Cyberattacks

Year[a]	Attack scenario	Threat actor	Attack type
2019	APT40	Cyber espionage	Phishing Backdoors Web server exploitation Web compromise
2018	The Port of San Diego	Criminals	Ransomware (malware)
2017	The Maersk NotPetya Attack	Terrorists/war	Worm (malware)
2014	The Danish Port Authority	Terrorists/war	Phishing Backdoors Virus
2013	The Port of Antwerp	Criminals	Phishing

[a]Year attack was discovered.

investigations by the Federal Bureau of Investigation and the Department of Homeland Security [66]. Once inside the victim's network, the attacker used the programme to encrypt valuable data and then demanded a ransom payment in Bitcoin in exchange for the decryption key [67]. The threat actor group allegedly attacked more than 200 targets, including hospitals, health-care companies and public institutions, with the Port of San Diego being their most recent victim [67]. The indictment alleges that two cybercriminals have collected more than six million dollars in ransom payments to date; however, the port did not pay the ransom demand. The attack took down non-critical administrative systems for a brief period and did not affect commercial port operations. According to the Port of San Diego, no data loss occurred as a result of the attack because the port's IT team had backups in place [67].

2.5.3 The Maersk NotPetya Attack

The Maersk NotPetya Attack took its name from its resemblance to *Petya*, a ransomware that surfaced in 2016, used to extort victims to pay for a key to unlock their files [68]. However, the ransom messages of NotPetya were only a ruse: the goal was purely destructive. The attack included a worm (malware) that would irreversibly encrypt master boot records of computers, the part of a machine that tells it where to find its own operating system. Any ransom payment that victims tried to make was futile: no key existed to reorder the contents of their computer.

The attack, which originated in Ukraine and has been attributed to a threat actor group tied to the Russian state [69], was likely more explosive than its creators had intended [68]. It cost AP Moller–Maersk, which was not even the intended target of the attack, at least USD 300 million [69]. It even spread back to Russia, striking the state oil company Rosneft [68]. 'But the story of NotPetya isn't truly about Maersk, or even about Ukraine. It's the story of a nation-state's weapon of war released in a medium where national borders have no meaning, and where collateral damage travels via a cruel and unexpected logic: where an attack aimed at Ukraine strikes Maersk, and an attack on Maersk strikes everywhere at once' [68]. The release of NotPetya was an act of cyberwar and has been included in the 'Terrorist/war' threat actor group in Table 2.6.

2.5.4 The Danish Maritime Authority

In April 2012, the Danish Maritime Authority was subjected to a cyberattack. The phishing attack was carried out when a computer of an employee at the Danish Maritime Authority was infiltrated by an email with an infected PDF attachment containing a virus [70]. When the employee opened the infected PDF file, hackers were given back-door access to the contents of his computer and to the rest of the Maritime Authority's network. The virus spread through the port network and successfully reached other Danish government institutions [71]. The attackers stole sensitive information from Danish shipping companies and the merchant navy [70].

The Danish authorities eventually discovered the attack two years later and stopped the infiltration by shutting down the entire system for several days before reopening it with new antivirus programmes to further prevent such attacks [70]. It was difficult to assign blame without the necessary evidence, but according to the Danish Defence Intelligent Service report, the attack was sponsored by another state [72] and thus is classified under the 'Terrorist/war' threat actor group in Table 2.6.

2.5.5 The Port of Antwerp

Over a two-year period, from 2011 to 2013, the Port of Antwerp suffered an attack, whereby the movement and location of containers was intercepted and controlled by drug traffickers, who were able to hide illegal drugs amidst legitimate cargo [73]. These attackers are included in the criminal threat actor group in Table 2.6.

The group sent malicious software via email, by means of a *phishing attack*, to the staff, which enabled them to get remote access to the port's data [74]. Although this first attempt was identified and a firewall was installed to prevent further attack, the hackers managed to break into port facilities and fit keylogging devices on the legitimate computers [74]. In this way, they gained wireless access to keystrokes typed by staff and captured screenshots from their computers [74]. When containers began to vanish entirely, the attack became apparent to the Port of Antwerp; and they went to official authorities to 'uncover the mystery of the missing cargo' [73]. The drug smuggling operation was discovered and more than 1,000 kg of cocaine and heroin was seized [64]. To prevent such further attacks, the Port of Antwerp installed a firewall, but it is worth mentioning that the hackers were able to breach that remotely [73].

This was one of the earliest cyberattacks in the maritime industry to be made public, being disclosed by Europol in 2013 [64]. It illustrates how cyberspace is another realm in which traditional crimes can be committed. It also highlights the intertwinement of cyber and physical systems [73].

2.6 Cyber risk management for ports

A 2018 survey by the Global Maritime Forum on issues in the maritime industry ranked cyberattacks and data theft as the highest in likelihood, the third highest in impact, and the lowest in industry preparedness [75]. *Cyber risk*, or information security risk, arises from loss of confidentiality, integrity, or availability of information or information systems and reflects potential adverse effects to organizational operations (i.e. mission, functions, image, or reputation), assets, individuals, other organizations and the state [39]. This section explores risk assessment in Section 2.6.1, risk management in Section 2.6.2, and risk strategy for ports in Section 2.6.3.

2.6.1 Risk assessment

Allianz's 2019 Risk Barometer [2] classified cybersecurity as the most feared trigger of business interruption. A *risk* identifies the conditions under which external or internal threats can exploit existing vulnerabilities to cause an incident and damage assets [76]. Also called cybersecurity assessment, *cyber risk assessment* is the process of identifying, estimating, and prioritizing information security risks [39]. The ISO/IEC 27001:2013 provides a consolidated standard framework for information security management and says that risk assessment is characterized by three risk activities, outlined in Table 2.7: identification, analysis and evaluation [77].

The complexity of infrastructure, paired with rapidly evolving security and safety requirements, makes it essential to have a proactive, comprehensive and integral approach to identifying risk in the ICT and physical system of ports [78]. *Risk identification* is the process of recognizing and describing risks [77]. Once the risks are identified, they are classified according to likelihood and impact. *Risk analysis* includes a review of identified risks to provide a quantitative estimate for the likelihood of a specific risk and the related impact on assets [77]. During *risk evaluation*, each risk is compared against an evaluation criteria, where risks are measured against security requirements indicating the required security measures [77,79]. This is where it is decided how risks will be treated. Steps will then be taken to exploit opportunities, counter threats and protect the global supply chain [80].

Table 2.7 ISO 27001:2013 provides a standard framework for information security management, where risk assessment is characterized by three activities: identification, analysis, and evaluation

Risk assessment	Definition
Identification	Recognize and describe risks
Analysis	Review identified risks and estimate the likelihood and impact
Evaluation	Compare each risk against the evaluation criteria

Source: ISO [77].

2.6.2 Risk management

Risk management aims to protect business assets (physical and cyber) and minimize costs in the case of failures. Risk management is based on the experience and the knowledge of best practice methods that consist of an estimation of the risk situation based on the business process models and organizational infrastructure. 'These models support the identification of potential risks and the development of appropriate protective measures' [7].

Cyber risk management consists of the process of identifying, analysing, assessing, and communicating a cyber risk. It also includes accepting, avoiding, transferring, or mitigating cyber risk to a desired level. It takes into consideration the costs and advantages of actions taken by stakeholders. It aims to support safe and secure shipping that is operationally resilient to cyber risks, and it extends from senior management level to all operators in the port facilities [7]. In light of increasing risk, the International Maritime Organization (IMO) issued high-level guidelines on maritime cyber risk management in 2017 to support safe and secure shipping from current and emerging cyberthreats. Other guidelines include the 2018 NIST Framework for Improving Critical Infrastructure Cybersecurity [81], which present five functional elements of effective cyber risk management: identify, protect, detect, respond, and recover.

The maritime industry significantly focuses on the security and risk management of physical security [7]. However, the emerging landscape of IT-empowered CII requires a shift in how the industry thinks about risk assessment [82]. When cyber risk management is effective, it considers safety and security effects consequent to the revelation or exploitation of vulnerabilities in IT systems. Cyber risk management should be durable and evolve as a natural extension of already existing practices and strategies of safety and security management [59].

2.6.2.1 Coordinated risk management

Collaboration is prudent to 'achieve better validation of the source of critical elements of software and hardware, particularly for systems that contain high value, sensitive information' [80]. Most of the existing risk management methodologies tend to focus too much on aspects of physical security and pay limited attention to CIIs, ignoring the complex nature of IT systems and assets used in the maritime sector, along with their interrelationships [82]. In order to reduce potential attack avenues, ports must have robust monitoring of and proper coordination among agencies [80]. A holistic method for managing security risks can help ports to check their compliance with existing legal, regulatory and standardized regime to detect possible violations, gaps, and a need for new regulations and directives [78].

2.6.3 Risk strategy for ports

As ports and port operators are scattered around the globe, it is difficult to develop an overall strategy for all members of the port network. For example, the offices of a big container shipping line can be spread across 150 countries and the shipping line may operate 300 vessels [20]. This subsection examines cyber insurance, investment decisions, and reporting.

2.6.3.1 Cyber insurance

The cyber insurance market is emerging but may be considered relatively immature [83]. How to set premiums is a key question to the development of a more mature cyber insurance market. Setting premiums is particularly challenging for cyber risk, as there is limited information sharing regarding cyber-incidents, leading to a lack of actuarial data from past events and lack of normative standards [84]. The ability to model cyber risk is currently limited but will improve substantially as more data is accumulated and shared [83]. Other cyber insurance challenges include a lack of legal framework with uncertainty in liability and lack of cyber standards.

As new technologies are implemented and reliance on them grows, cyber risks are increasing and so is cyber-related loss [83]. Thus, it is likely that full cybersecurity for transportation infrastructure is not achievable solely through technological improvements [83]. Therefore, in addition to attempting to prevent attacks and reduce cyber risks, transportation managers should also prepare themselves financially for inevitable losses through self-insurance and insurance [84]. Cyber insurance is currently available, but limited, and expansion of cyber insurance coverage is needed to manage risks [83].

2.6.3.2 Investment decisions

Port terminals are both privately and publicly held, most often through leases that last anywhere from 25 to 50 years. To invest in automating a terminal, a process that takes about 10 years, a shipping company must be sure they will get a return on their investment before their lease expires. If there exists a large enough threat to this profit margin by means of cyberattacks which could shut down business communications, disable physical security systems, and more, then port operators may reconsider leasing ports.

2.6.3.3 Reporting

An issue regarding cybersecurity awareness is the tendency of exploited organizations to refrain from reporting cyberattacks due to fear of damage to their reputation [85]. This has made the government responsible for making sure that cyberattacks are reported. Today, it is often the case that national strategies are installed to legally require data breaches to be reported, and when they are not there are penalties – often in the form of fines. Some examples dictating these terms are the UK GDPR [86] and EU Directive 2016/1148 [87]. These are further explored in Section 2.7.2.

2.7 Cybersecurity guidelines and standards

Over the past decade, governments, organizations and individuals have developed many strategies to strengthen cybersecurity in ports. Cybersecurity guidelines provide a framework for the risk assessment process whereby threats and vulnerabilities are identified to determine the risks they pose and to establish an effective treatment plan [7]. The International Chamber of Commerce [88] categorizes these

Table 2.8 Definitions and examples of high-level cybersecurity documentation in the maritime sector

Documentation[a]	Definition	Example
Guidelines	High-level vision statements	IMO Shipping Regulations
National strategies	Articulate cybersecurity approach in national or legal context	EU Directives and GDPR
Frameworks	Prioritize or evaluate resources that help benchmark progress	NIST Cybersecurity Framework
Standards of practice	Defines measures to enhance the security of port facilities and ships	ISPS Code, ISO/IEC 27001, and Common Criteria

[a]Technical standards not included.

instruments into five groups: guidelines, national strategies, frameworks, standards of practice, and technical standards. These documents are summarized in Table 2.8. This section examines each group and gives authoritative examples of recommended practice in the maritime sector.

2.7.1 Guidelines

Guidelines are high-level recommendations that scope concern for cybersecurity and provide a charter for individuals, organizations and states [88]. There exist multiple maritime industry cybersecurity best practice guidelines from organizations such as IET [38], IMO [89], and IALA [90]. This section will analyse the IMO regulations for shipping.

2.7.1.1 IMO Shipping Regulations

The IMO is a specialized UN agency and is the primary regulator for shipping bodies [89]. In 2016, the IMO published guidelines for maritime cybersecurity management to provide high-level guidance related to the development and implementation of cyber risk management to safeguard the maritime sector from cyberthreats [7].

In 2017, the IMO Facilitation Committee (FAL) and the Maritime Security Committee (MSC) issued guidelines for maritime cyber risk management in *MSC-FAL.1/Circ.3* [89]. With regard to ports, these guidelines place an obligation on shipowners, operators, and stakeholders to adopt a risk management approach to the financial consequences of a full or partial loss of availability, integrity and confidentiality of sensitive data.

Additionally, the 2017 *Resolution MSC.428(98) Maritime Cyber Risk Management in Safety Management Systems* [91] encouraged administrations to formalize cybersecurity requirements in existing safety management systems (as defined in the ISM Code) no later than January 2021. These guidelines form the foundation of high-level statements across the maritime sector. It is worth noting that the IMO has yet to develop formal cybersecurity regulations specific to the maritime sector.

2.7.2 National strategies

National strategies articulate an approach to cybersecurity tailored to a specific national or legal context [88]. These high-level documents are often based on guideline, which means that since the IMO has yet to develop cybersecurity regulations specific to the maritime sector, national authorities by and large have yet to do this as well [92]. This section describes national strategies in ports, namely, the EU Directives concerning cybersecurity and General Data Protection Regulation (GDPR).

2.7.2.1 EU Directives and GDPR

In July 2016, the *EU Directive 2016/1148 (NIS Directive)* proposed a wide-ranging set of measures to boost the level of cybersecurity of network and information systems to secure services vital to the EU economy and society [87]. With respect to port facilities and ports, this Directive states that EU countries must take actions to improve cybersecurity capabilities that cover all operations, including radio and telecommunication systems, computer systems and networks. It also ensures maritime operators to take into account international codes and cybersecurity recommendations developed, especially IMO guidelines. Similarly, there are punishments applied by the state when these laws are not followed.

The 2018 GDPR, also known as *Regulation (EU) 2016/679*, requires all industries, including the maritime industry to process personal data relating to EU data subjects, residents or citizens securely using appropriate technical and organizational measures [86]. With the adoption of the GDPR, organizations across the EU are required to report data breaches to the Information Commissioner's Office which promotes security and accountability. A violation of GDPR provisions attracts penalties of up to 20 million euros, or in the case of an undertaking, 4 per cent of the organization's annual turnover, whichever is higher.

Under the *EU Cybersecurity Act*, which was developed in 2019, the position of ENISA is strengthened with regard to cybersecurity matters for EU member states, as the act defines an EU-wide cybersecurity certification framework for ICT products, services and processes [93]. This framework will provide a comprehensive set of rules, technical requirements, standards and procedures in order to attest that ICT products and services can be trusted on the basis of EU requirements.

The *European Maritime Single Window (Regulation 2019/1239)* [94] ensures competitiveness, efficiency, and environmental sustainability of the European maritime transport sector. It is necessary to reduce the administrative burden on ships, to facilitate the use of digital information and to contribute to the integration of the sector to the digital multimodal logistic chain. This is achieved through the submission of data elements required by both maritime and customs authorities using a harmonized cargo data set.

2.7.3 Frameworks

Developing national strategies further, *frameworks* gather a catalogue of prioritized or evaluated resources that help organizations to benchmark their maturity and progress in addressing cybersecurity risks [88]. An example of a framework that guides ports and port facilities is the National Institute of Standards and Technology (NIST) Cybersecurity Framework.

2.7.3.1 NIST Cybersecurity Framework

In 2018, the US National Institute of Standards and Technology (NIST) published a revised version of the 2014 framework for improving CI cybersecurity, often referred to as the *NIST Cybersecurity Framework* [81]. Created through the collaboration between industry and government, the voluntary framework consists of standards, guidelines, and practices to reduce cyber risks to CI. As ports are CIs, the NIST Framework can be applied as a prioritized, flexible, repeatable, and cost-effective approach to port cyber risk management. The NIST core functions, as seen in Table 2.9, allow organizations to view the life cycle of their management of cybersecurity risks [81].

2.7.4 Standards of practice

Standards of practice are documents that guide or govern organizational processes to ensure the robust and consistent operation of cybersecurity best practices [88]. This section describes well-known examples such as the International Ship and Port Facility Security (ISPS) Code, ISO/IEC 27001, and Common Criteria.

2.7.4.1 The ISPS Code, ISO/IEC 27001, and Common Criteria

The ISPS Code was an amendment to the Safety of Life at Sea Convention in 2002 and defines measures to enhance the security of port facilities and ships [95]. The aim of the ISPS Code is to enhance the maritime security both on board ships and in ports. The traditional approach to maritime security was container security; however, the ISPS Code has recently included port cybersecurity with regard to access control and authentication requirements that will help the port authorities to secure

Table 2.9 NIST core functions

NIST core functions	Definitions
Identify	Covers all personnel roles and responsibilities that need to be defined for cyber risk management and all systems, assets, data, and capabilities that when endangered can pose risks to the port and its ships
Protect	Implementation of risk control processes and measures and the contingency planning to protect cyber activities and confirm the continuity of operations
Detect	Development and implementation of all necessary activities that a port and its facilities need to detect a cyberattack in time
Respond	Activities that are needed to provide resilience and to restore systems necessary for operations and services
Recover	Identification of measures that is necessary for backup or restore of cyber systems for operations

Source: NIST 2018 Framework [81].

their ICT systems and to mitigate existing and upcoming cyber risks in a better way.

The ISPS Code has three security levels, ranging from low to high in proportion to the nature and scope of the incident or perceived security threat. The code requires that ports and port authorities develop and implement improved Port Facility Security Plans (PFSP) for each operational level that outlines the measures to be put in place to address threats and the countermeasures [95]. PFSP is based on the Port Facility Security Assessment and on a risk-analysis scheme that is implemented by governments and authorized security organizations to identify major assets, possible threats, and countermeasures.

The ISO/IEC 27000 family of standards [77] helps organizations keep information assets secure. The ISO/IEC 27001:2013 standard provides requirements for an information security management system. The maritime industry implements this standard to:

- examine the port's information security risks, taking account of the threats, vulnerabilities, and impacts;
- design and implement information security controls and forms of risk treatment; and
- adopt an overarching management process to ensure that information security controls continue to meet the port's information security needs on an ongoing basis.

Common Criteria, also known as ISO/IEC 15408 [96], is an international standard that provides independent and objective validation of the reliability, quality and trustworthiness of IT products. This demonstrates that manufacturers have met the necessary security requirements to sell their products in valuable markets, providing increased confidence. Common Criteria sets specific information assurance goals, including strict levels of integrity, confidentiality and availability for systems and data, accountability at the individual level, and assurance that all goals are met.

2.8 Summary

The global port system is a mixed bag, where ports vary in structure and ownership. Each port, acting as a value multiplier in the system, is a CI and includes critical ICT systems. The disruption of a single port could have a severe impact not only on port operations but also act as a domino effect across the system.

The interconnected networks of both information and CPSs introduce new vulnerabilities and threats to ports, namely, the threat of a large-scale cyberattack on ICT systems and key infrastructure. It is imperative to act with urgency and purpose to protect the cyber domain from crippling attacks and disruption. Understanding ports as a cyber–physical environment allows us to classify cybersecurity attributes in ports and manage cyberthreats.

There are instances in which vulnerabilities are not well managed, as presented in the five cyberattack scenarios. These highlight the importance of cyber risk management in ports and the need for a coordinated strategy. Existing guidelines and standards have strengthened cybersecurity in ports, but there is much work to be done. It is an uphill battle, as the proportion, complexity and evolving nature of cyber mean that there is no across-the-board response to cyber-incidents.

This aim is further challenged by the continuous evolution of the modern port and the increasingly faster pace at which mainstream technology is adopted. Amidst a technological revolution, the entire maritime industry is exploring, developing, and implementing new systems and innovative technologies to achieve greater efficiency. Some technological trends shaping the port industry include autonomous ships, blockchain applications, cargo and vessel tracking [97]. But cyber and ports are not the only entities embracing the future. Just as ports evolve and become smarter, so do cyberattackers. It remains a constant spar for the cutting edge which, in turn, propels the quest for the next wave of new technology and improved efficiency. While the future can only be predicted, it is certain that ports and cyberspace are two realms that cannot again be separated.

References

[1] Ablon L, Binnendijk A, Hodgson Q, *et al.* Operationalizing Cyberspace as a Military Domain: Lessons for NATO. RAND Corporation; 2019. Available from: https://www.rand.org/pubs/perspectives/PE329.html.

[2] Allianz Risk Barometer. Allianz Global Corporate and Specialty SE; 2019. Available from: https://www.agcs.allianz.com/content/dam/onemarketing/agcs/agcs/reports/AGCS-Safety-Shipping-Review-2019.pdf.

[3] Oram RB. Cargo Handling and the Modern Port: The Commonwealth and International Library of Science Technology Engineering and Liberal Studies. vol. 309. Oxford, England: Elsevier; 2014.

[4] Fair M. Port Administration in the United States. Cambridge, MD: Cornell Maritime Press; 1954.

[5] Antunes C. Early Modern Ports: 1500–1750. Institute of European History; 2010. Available from: http://www.ieg-ego.eu/antunesc-2010-en.

[6] Christaller W and Baskin C. Die Zentralen Orte in Süddeutschland. Central Places in Southern Germany; Translated by Carlisle W. Baskin. Englewood Cliffs, NJ: Prentice-Hall; 1966.

[7] Polemi N. Port Cybersecurity: Securing Critical Information Infrastructures and Supply Chains. Amsterdam, Netherlands: Elsevier; 2017.

[8] Hohenberg P and Lees L. The Making of Urban Europe, 1000-1994. Cambridge, MA: Harvard University Press; 1995.

[9] Alderton P. Port Management and Operations. London, England: Informa; 2008.

[10] Lloyd's List's One Hundred Container Ports 2018 Edition. Lloyds List Intelligence; 2018. Available from: https://lloydslist.maritimeintelligence.informa.com/one-hundred-container-ports-2018.

[11] Magnuson S. Security Around Singapore Critical to World Economy. National Defense Magazine; 2007.

[12] Monios J. Port governance in the UK: Planning without policy. Research in Transportation Business & Management. 2017;22:78–88.

[13] The International Bank for Reconstruction and Development/The World Bank. Module 3: Alternative port management structure and ownership models. In: Port Reform Toolkit: Second Edition; 2007.

[14] Yang Y, Zhong M, Yao H, *et al.* Internet of things for smart ports: Technologies and challenges. IEEE Instrumentation & Measurement Magazine. 2018;02(21):34–43.

[15] The History of the TEU (Twenty-foot Equivalent Unit); 2019. Available from: https://www.icontainers.com/us/2019/08/06/history-of-teu-twenty-foot-equivalent-unit/.

[16] Mega-Ships; 2019. Available from: https://www.joc.com/special-topics/mega-ships.

[17] Container Ship Design. World Shipping Council; 2019. Available from: http://www.worldshipping.org/about-the-industry/liner-ships/container-ship-design.

[18] Ship Technology. Mega Container Ships and How They Are Changing Ports; 2016. Available from: https://www.ship-technology.com/features/featuremega-container-ships-and-how-they-are-changing-ports-4974826/.

[19] Habash R, Groza V, and Burr K. Risk management framework for the power grid cyber-physical security. Current Journal of Applied Science and Technology. 2013:1070–1085.

[20] Jensen L. Challenges in maritime cyber-resilience. Technology Innovation Management Review. 2015;5(4):35.

[21] Council Directive 2008/114/EC of 8 December 2008 on the identification and designation of European critical infrastructures and the assessment of the need to improve their protection. Official Journal of the European Union; 2008.

[22] Boyes H. Maritime Cyber Security–Securing the Digital Seaways. The Institution of Engineering and Technology; 2013. p. 56–63.

[23] Alberts C and Dorofee A. OCTAVE Method Implementation Guide Version 2.0. Volume 2: Preliminary Activities. Pittsburgh, PA: Carnegie Mellon University; 2001.

[24] Carrapico H and Barrinha A. The EU as a coherent (cyber) security actor? JCMS: Journal of Common Market Studies. 2017;55(6):1254–1272.

[25] Lam J and Su S. Disruption risks and mitigation strategies: An analysis of Asian ports. Maritime Policy & Management. 2015;42(5):415–435.

[26] Omer M, Mostashari A, Nilchiani R, *et al.* A framework for assessing resiliency of maritime transportation systems. Maritime Policy & Management. 2012;39(7):685–703.

[27] Cohen S. Economic Impact of a West Coast Dock Shutdown. vol. 10; 2019.

[28] Jansson J and Shneerson D. *Port Economics*. vol. 8. Cambridge, MA: MIT Press; 1982.

[29] Meersman H, Van de Voorde E, and Vanelslander T. Port congestion and implications to maritime logistics. In: Maritime Logistics: Contemporary Issues. Bingley, England: Emerald Group Publishing Limited; 2012. p. 49–68.
[30] Hapag-Lloyd. Container Ships Soon to Sail Unmanned? 2015. Available from: https://www.hapag-lloyd.com/en/news-insights/insights/2015/08/container-ships-soon-to-sail-unmanned_42532.html.
[31] Review of Maritime Transport 2018. United Nations Conference on Trade and Development (UNCTAD); 2018.
[32] Sanchez-Rodrigues V, Potter A, and Naim M. Evaluating the causes of uncertainty in logistics operations. The International Journal of Logistics Management. 2010;21(1):45–64.
[33] Silgado D. Cyberattacks: A Digital Threat Reality Affecting the Maritime Industry. Malmö, Sweden: World Maritime University Dissertations; 2018.
[34] Lewis T. Critical Infrastructure Protection in Homeland Security: Defending a Networked Nation. Hoboken, NJ: John Wiley & Sons; 2014.
[35] Hetherington C, Flin R, and Mearns K. Safety in shipping: The human element. Journal of Safety Research. 2006;37:401–11.
[36] National Cyber Security Strategy 2016–2021. UK Government; 2016. Available from: https://www.enisa.europa.eu/topics/national-cyber-security-strategies/ncss-map/national_cyber_security_strategy_2016.pdf.
[37] Dukes C. Committee on National Security Systems (CNSS) glossary. In: Tech Rep. Fort Meade, MD, USA: CNSSI; 2015. p. 4009.
[38] Boyes H, Isbell R, and Luck A. Good Practice Guide Cyber Security for Ports and Port Systems. London: Institution of Engineering and Technology; 2020.
[39] National Institute of Standards and Technology. Special Publication 800-30 Guide for Conducting Risk Assessments; 2012. Available from: https://nvlpubs.nist.gov/nistpubs/Legacy/SP/nistspecialpublication800-30r1.pdf.
[40] International Organization for Standardization. ISO/IEC 27005:2018—Information Security Risk Management; 2018.
[41] National Institute of Standards and Technology. Special Publication 800-39 Managing Information Security Risk: Organization, Mission, and Information System View; 2011. Available from: https://nvlpubs.nist.gov/nistpubs/Legacy/SP/nistspecialpublication800-39.pdf.
[42] The Guidelines on Cyber Security Onboard Ships. BIMCO; 2017. Available from: https://www.ics-shipping.org/docs/default-source/resources/safety-security-and-operations/guidelines-on-cyber-security-onboard-ships.pdf?sfvrsn=20.
[43] Common Cyber Attacks: Reducing The Impact. GHCQ & CERT-UK; 2015. Available from: https://assets.publishing.service.gov.uk/government/uploads/system/uploads/attachment_data/file/400106/Common_Cyber_Attacks-Reducing_The_Impact.pdf.
[44] Stouffer K, Falco J, and Scarfone K. Draft NIST Special Publication 800-82: Guide to Industrial Control Systems (ICS) Security. National Institute of

Standards and Technology; 2011. Available from: https://nvlpubs.nist.gov/nistpubs/SpecialPublications/NIST.SP.800-82r2.pdf.

[45] CNSS. CNSSI 4009 National Information Assurance (IA) Glossary. Committee on National Security Systems (CNSS); 2015.

[46] Bodeau D, McCollum C, and Fox D. Cyber Threat Modeling: Survey, Assessment, and Representative Framework. HSSEDI, The Mitre Corporation; 2018.

[47] Souppaya M and Scarfone K. Special Publication 800-154 Guide to Data-centric System Threat Modeling. National Institute of Standards and Technology; 2016. Available from: https://csrc.nist.gov/csrc/media/publications/sp/800-154/draft/documents/sp800_154_draft.pdf.

[48] ISACA. Cybersecurity Fundamentals Glossary; 2016.

[49] Fairplay and BIMCO. Fairplay and BIMCO Maritime Cyber Survey 2018; 2018.

[50] NCSC glossary; 2016. Available from: https://www.ncsc.gov.uk/information/ncsc-glossary.

[51] Lee E and Seshia S. Introduction to Embedded Systems: A Cyber-Physical Systems Approach. Cambridge, MA: MIT Press; 2016.

[52] Monostori L. In: Chatti S and Tolio T, editors. Cyber-Physical Systems. Berlin, Heidelberg: Springer Berlin Heidelberg; 2018. p. 1–8.

[53] Cardenas A. Cyber-physical systems security knowledge area. The Cyber Security Body of Knowledge. 2018;1:1–4.

[54] Brinkmann M and Hahn A. Testbed architecture for maritime cyber physical systems. In: 2017 IEEE 15th International Conference on Industrial Informatics (INDIN). IEEE; 2017. p. 923–928.

[55] Colbert E, Sullivan D, and Kott A. Cyber-physical war gaming. Journal of Information Warfare. 2017;16(3):119–133.

[56] Coolfire. What Is The Difference Between IT And OT? 2019.

[57] Fok E. An introduction to cybersecurity issues in modern transportation systems. ITE Journal. 2013;83(7):18–21.

[58] Habibzadeh H, Nussbaum BH, Anjomshoa F, *et al.* A survey on cybersecurity, data privacy, and policy issues in cyber-physical system deployments in smart cities. Sustainable Cities and Society. 2019;50:101660. Available from: http://www.sciencedirect.com/science/article/pii/S2210670718316883.

[59] International Maritime Organization (IMO). Strategy for the Development and Implementation of E-Navigation; 2019. Available from: http://www.imo.org/en/OurWork/Safety/Navigation/Pages/eNavigation.aspx.

[60] Parker DB. Fighting Computer Crime: A New Framework for Protecting Information. New York, NY, USA: John Wiley & Sons, Inc.; 1998.

[61] Boyes H, Isbell R, and Luck A. Code of Practice: Cyber Security for Ports and Port Systems. London: Institution of Engineering and Technology; 2016. p. 28.

[62] Sen R. Cyber and Information Threats to Seaports and Ships; 2016. p. 281–302.

[63] Boyes H. Cybersecurity and cyber-resilient supply chains. Technology Innovation Management Review. 2015;5(4):28.

[64] Snape J. Marine and Offshore Cyber Briefing: Threat Case Studies. Nettitude; 2019. Available from: https://cdn2.hubspot.net/hubfs/3021880/Ebook%20PDFs/NETT 2019 MO TCS%20EC.pdf.
[65] Hacking attack in Port of Barcelona. Accessed: 2020-03-20. Available from: https://www.securitynewspaper.com/2018/09/26/hacking-attack-in-port-of-barcelona/.
[66] BBC. San Diego Port Hit by Ransomware Attack. BBC; 2018.
[67] Coolfire. Iranian Hackers Indicted for Port of San Diego Cyberattack. The Maritime Executive; 2018.
[68] Greenberg A. The Untold Story of NotPetya, the Most Devastating Cyberattack in History. Wired; 2018. Available from: https://www.wired.com/story/notpetya-cyberattack-ukraine-russia-code-crashed-the-world/.
[69] Juliano M. Maritime Braces for the Next Cyber Breach. TradeWinds; 2018. Available from: https://www.tradewindsnews.com/safety/maritime-braces-for-the-next-cyber-breach/2-1-437319.
[70] State-sponsored Hackers Spied on Denmark. Accessed: 2019-11-25. Available from: https://www.thelocal.dk/20140922/denmark-was-hacked-by-state-sponsored-spies.
[71] Miller R. Telematics Solutions in Maritime and Inland Waterway Transport. Cambridge University Scholar; 2019.
[72] CyberKeel. Maritime Cyber-Risks. Copenhagen, Denmark: CyberKeel; 2014.
[73] Nguyen L. Collaboration in the Shipping Industry: Innovation and Technology. KNect365; 2018. Available from: https://knect365.com/article/pdfs/91705d00-6d9d-4ba3-98a4-9b10c92ad520_Shipping2030_report_Feb16-2018_.pdf.
[74] Police Warning After Drug Traffickers' Cyberattack. Accessed: 2019-11-25. Available from: https://www.bbc.co.uk/news/world-europe-24539417.
[75] Global Maritime Issues Monitor 2018; 2018. Available from: https://www.globalmaritimeforum.org/global-maritime-issues-monitor-2018.
[76] Refsdal A, Solhaug B, and Stølen K. Cyber-risk management. In: Cyber-Risk Management. Cham, Switzerland: Springer; 2015. p. 33–47.
[77] International Organization for Standardization. ISO/IEC 27001:2013—Information Security Management Systems; 2013.
[78] Papastergiou S, Polemi N, and Karantjias A. CYSM: An innovative physical/cyber security management system for ports. In: International Conference on Human Aspects of Information Security, Privacy, and Trust. Cham, Switzerland: Springer; 2015. p. 219–230.
[79] Rocchetto M, Ferrari A, and Senni V. Challenges and opportunities for model-based security risk assessment of cyber-physical systems. In: Resilience of Cyber-Physical Systems. Cham, Switzerland: Springer; 2019. p. 25–47.
[80] Chertoff M. The cybersecurity challenge. Regulation & Governance. 2008;2(4):480–484.

[81] National Institute of Standards and Technology. Framework for Improving Critical Infrastructure Cybersecurity; 2018. Available from: https://nvlpubs.nist.gov/nistpubs/CSWP/NIST.CSWP.04162018.pdf.
[82] Polatidis N, Pavlidis M, and Mouratidis H. Cyberattack path discovery in a dynamic supply chain maritime risk management system. Computer Standards & Interfaces. 2018;56:74–82.
[83] Tonn G, Kesan JP, Zhang L, and Czajkowski J. Cyber risk and insurance for transportation infrastructure. Transport Policy. 2019;79:103–114.
[84] Toregas C and Zahn N. Insurance for Cyber Attacks: The Issue of Setting Premiums in Context. Washington, DC: George Washington University. 2014.
[85] Meyer-Larsen N and Müller R. Enhancing the cybersecurity of port community systems. In: International Conference on Dynamics in Logistics. Cham, Switzerland: Springer; 2018. p. 318–323.
[86] Guide to the General Data Protection Regulation. United Kingdom; 2018. Available from: https://ico.org.uk/for-organisations/guide-to-data-protection/guide-to-the-general-data-protection-regulation-gdpr/.
[87] Directive 2016/1148 of the European Parliament and of the Council of 6 July 2016 concerning measures for a high common level of security of network and information systems across the union. Official Journal of the European Union. 2016; L 194:1–30.
[88] International Chamber of Commerce. ICC Cyber Security Guide for Business; 2015.
[89] International Maritime Organization (IMO). MSC-FAL.1/Circ.3 Guidelines on Maritime Cyber Risk Management; 2017.
[90] IALA. Identity Management and Cyber Security; 2020. Available from: https://maritimeconnectivity.net/docs/Identity%20Management% 20and%20Cyber%20Security.pdf.
[91] International Maritime Organization (IMO). Resolution MSC.428(98) Maritime Cyber Risk Management In Safety Management Systems; 2017.
[92] Karlsson J. The Future of Maritime Cybersecurity. Secure State Cyber; 2019. Available from: https://securestatecyber.com/cyberbloggen-en/the-future-of-maritime-cybersecurity/.
[93] The EU cybersecurity certification framework; 2019. Available from: https://ec.europa.eu/digital-single-market/en/eu-cybersecurity-certification-framework.
[94] Regulation (EU) 2019/1239 of the European Parliament and of the Council of 20 June 2019 establishing a European Maritime Single Window environment. Official Journal of the European Union. 2019; L 198.
[95] International Maritime Organization. International Ship and Port Facility Security Code; 2003. Available from: http://www.imo.org/en/about/conventions/listofconventions/pages/default.aspx.
[96] Common Criteria; 2017. Available from: https://www.commoncriteriaportal.org.
[97] UNCTAD: 7 Key Trends Impacting the Shipping Industry's Future. World Maritime News; 2018.

Chapter 3
e-Navigation and shore-based monitoring systems

Jens Schröder-Fürstenberg[1]

The advent of the next-generation electronic navigational systems will be confronting mariners on-board ships with new opportunities and new challenges.

Could the shift from traditional paper-chart-based navigation towards navigation supported by electronic charts be considered as a paradigm shift of navigation, the development of the next-generation electronic navigational systems offers much more possibilities?

This chapter describes the current composition of information needed to navigate a ship safely based either on paper documents or on electronic systems.

It provides an overview of the achievements and the ongoing work of the main involved parties, the International Maritime Organization (IMO) and the International Hydrographic Organization (IHO). The IMO as the main legal body has introduced the e-navigation concept and associated services. In parallel, the IHO has developed the S-100 Standard as the Universal Hydrographic Data Model which, in turn, has been accepted as the underlying basis of IMO's e-navigation. The International Electrotechnical Commission (IEC) as the third actor binds the two components by developing test standards to approve on-board systems able to provide S-100 compliant information supporting the e-navigation concept. The combination of e-navigation services with future S-100-based navigational systems offers wide more optimisation potential than currently used on-board and ashore. This includes the extensive data exchange between the involved parties.

Shore-based monitoring systems planned in an e-navigation environment are either under development or are at a conceptual status. This chapter collects and provides the ideas and concepts behind some of them.

The chapter further describes certain additional services and features required to make e-navigation and S-100 operational.

The chapter neither indents to assess the value of the e-navigation concept at all nor the value of particular e-navigation components. The IMO appreciated the idea behind e-navigation before they initiated the development of the e-navigation concept.

[1]Federal Maritime and Hydrographic Agency (BSH), Rostock, Germany

At the end, the chapter risks a view in the long-term future regardless of the current legal and technical limitations.

3.1 Navigation, the past and the present

3.1.1 What was first, nautical publications or navigational charts?

Substantial parts of nautical publications are Sailing Directions (SDs). They have been used to provide information, which either cannot or not sufficiently charted. A kind of SDs can be traced back to antiquity. One of the first written 'Sailing Directions' has been given by Circe to Homer to avoid a specific island: 'So being much troubled I said to the men, "My men, I know you are hard pressed, but listen while I tell you the prophecy that Teiresias made me, and how carefully Aeaean Circe warned me to shun the island of the blessed sun-god, for it was here, she said, that our worst danger would lie. Head the ship, therefore, away from the island [1]"' (Figure 3.1).

The Phoenician Scylax, a Persian emperor of the sixth century B.C., who was in the service of Darius the Great wrote the first ancient records. He describes

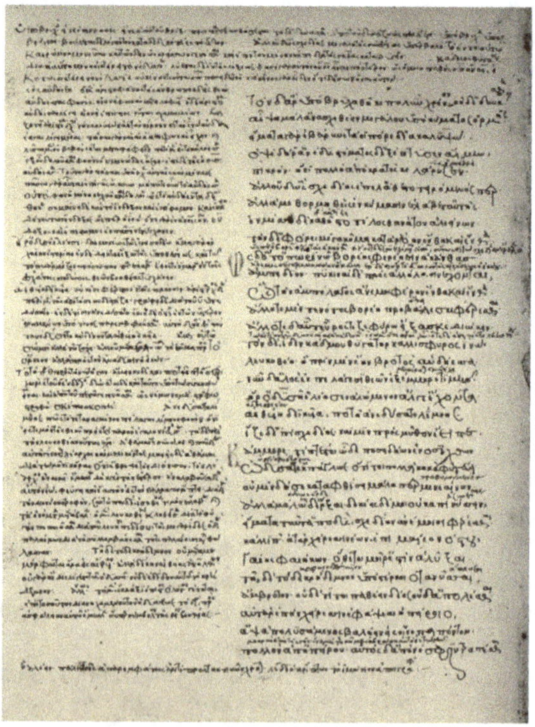

Figure 3.1 (a) Homer Odyssee [13] and (b) historic chart Piri Reis [14]

Figure 3.1 (*Continued*)

inter alia the day of voyage, the favourable weather and wind [2]. In the early sixteenth century, Portuguese seafarers travelled frequently to Asia on the basis of detailed SDs and charts [3]. A combination of both made the voyage around Africa safer.

Sailing routes for Northern European waters were described first around the eighth century. The first detailed sea charts were published around the thirteenth/fourteenth century to supplement SDs.

That system changed in the middle of the nineteenth century. Sea charts became more accurate and scientific principles were used to portrayal sea and land features. SDs started to complement sea charts.

3.1.2 Paper charts and electronic charts

Charts designated as official navigational ones have been prepared according to standards defined by the IHO. The IHO has assigned numbers and letters to their publications. The numbering schema is coincidental, whereas the letters identify the purpose of the publication:

- B – Bathymetric Publications;
- C – Capacity Building Publications;
- M – Miscellaneous Publications (including basic documents);
- P – Periodic Publications; and
- S – Standards and Specifications [4].

The standard for paper charts is S-4 (Regulations for International (INT) Charts and Chart Specifications of the IHO). The most relevant standards for the provision of charted information in Electronic Chart Display and Information Systems (ECDIS) are S-57 and S-52. The S-57 standard (Transfer Standard for Digital Hydrographic Data) defines the content of electronic charts, whereas the S-52 standard (Specifications for Chart Content and Display Aspects of ECDIS) specifies the data portrayal.

Paper charts could have different horizontal reference systems. This circumstance required a mariner's attention during the transfer of positions from one chart to another.

Electronic charts use a common horizontal reference system. This is necessary because two or more adjacent charts could be displayed on one single screen and that gaps or overlaps caused by different horizontal reference systems must be avoided (Figure 3.2).

3.1.3 Nautical publications

Hydrographic offices publish a whole set of nautical publications with the aim to provide the mariner the best available information on the planned route. The information provided is either based on own sources or on external ones. Have hydrographic offices taken a provision of information for the whole world for granted in the past, economic reasons and agreements among the coastal states led to the current situation where most hydrographic offices provide information for waters under their coastal state legislation. A data exchange mechanism between hydrographic offices avoids spatial gaps in information provision. The provision of nautical publications' content is according to the IHO's resolutions.

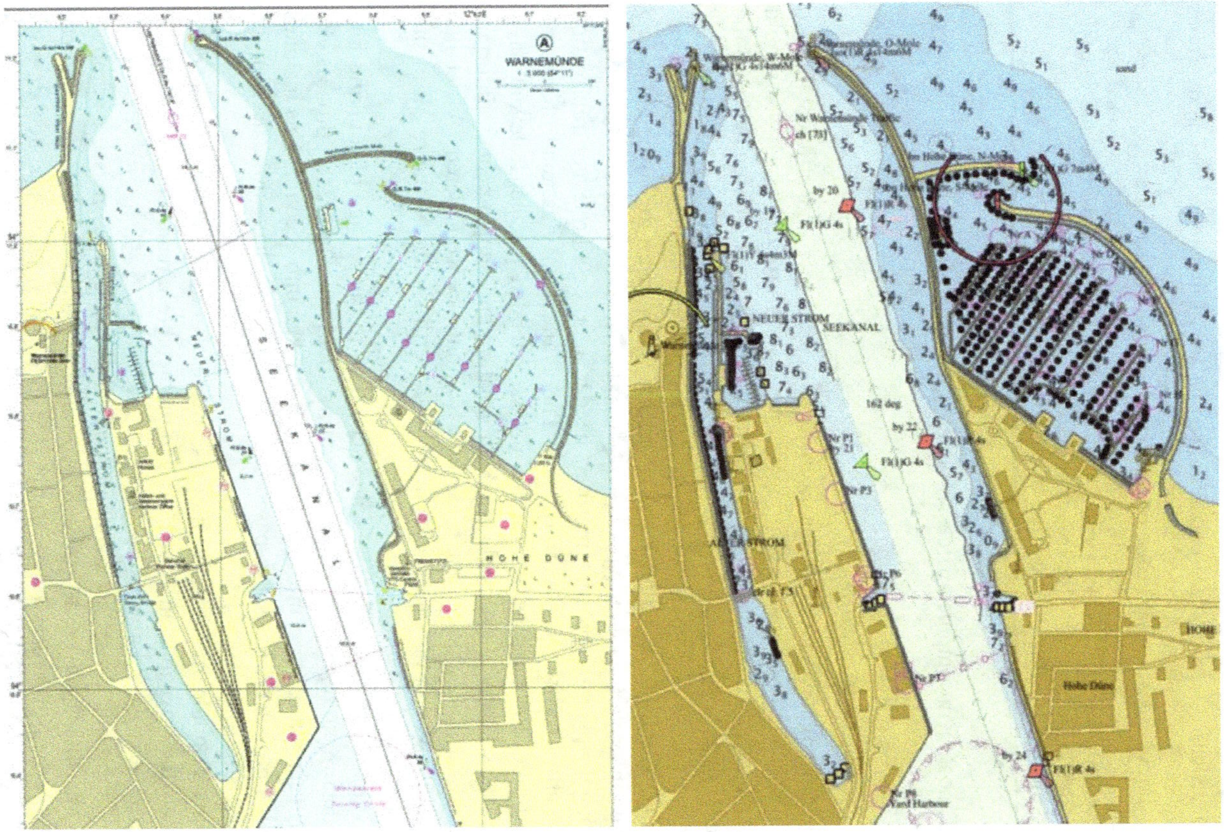

(a) (b)

Figure 3.2 (a) Paper chart snippet and (b) ENC snippet. © 2020 Bundesamt für Seeschifffahrt und Hydrographie, Hamburg/Rostock

Table 3.1 List of nautical publications

Publication	Publication content
Sailing Directions	Description of coastal area, harbours and sailing directions
List of Lights	Description of light characteristics
List of Radio Signals	Provision of information on radio services
Tide Tables	Provision of predicted tide information for selected ports
Tidal Stream Atlases	Provision of predicted tidal stream directions and values
Mariners' Handbook	Provision of general nautical information
Nautical Almanacs	Provision of celestial information

The most hydrographic offices produce the following nautical publications (Table 3.1).

3.1.4 Navigating a ship

The objective of navigation on sea is to bring a ship safe and efficient from a starting point to its destination. During the navigation process, a mariner should consider the safety of life at sea, the protection of the marine environment and the safety and efficiency.

Three principle steps describe the navigation process: planning, execution and monitoring. The IMO has collected all necessary information for voyage planning in Resolution A.893(21), see [5].

During the planning process, mariners need to appraise all relevant information on ships' characteristics and performance. The available navigational charts and nautical publication must be appropriate for the intended voyage. Relevant navigational charts and nautical publication information and supplementary information should be assessed and their usefulness should be considered. A detailed voyage plan covering the passage from berth to berth should be the result of this planning process.

The ship should execute the voyage according to the voyage plan taking into account the time of departure and estimated time of arrival. During this process, the appearing relevant nautical information should be taken into account. This includes information provided from shore-based stations which may have influences on the voyage execution.

Mariners should monitor the progress made according to the voyage plan. Changes to the voyage plan based on non-predictable circumstances during the planning process should always be recorded.

Navigating and operating a ship in modern times requires massive administrative work, intensive ship-to-ship, ship-to-shore and shore-to-shore communications, the record of events relevant for safe navigation, port approach or pilot arrangements or the exchange of cargo information. It also relates to work with incompatible systems and technologies.

Considering that the predicted increase of communications and the uncoordinated development of systems could lead to shipping accidents with negative effects on humans and on the environment. Mariners, companies, international organisations and national bodies approached IMO to develop a solution that reduces the risks significantly. The idea of e-navigation was born.

3.2 The e-navigation concept

3.2.1 What was the motivation to initiate the e-navigation concept?

The IMO started to discuss e-navigation in 2005. At this time, many IMO member states had concerns that new navigation technologies will evolve and developed in an uncoordinated way and that this produces rather confusion than simplification for the mariner.

IMO considered feedback from mariners who had new equipment on-board which was not integrated with other hardware. More worse was that the equipment provided information which was not useful for navigational purposes.

The IMO agreed to start developing an e-navigation strategy to address these circumstances and to install a mechanism which keeps control on new development and ensures that new electronic systems would be integrated on-board in a controlled/supervised way.

The term 'e-navigation' does not mean 'electronic navigation' or 'enhanced navigation' but it describes the support of navigation by electronic means to simplify mariner's workload and to harmonise the data collection process and data integration. It means that electronic systems on-board and ashore provide/exchange information in a systematic and coherent way supporting the mariners in the decision-making processes related to berth-to-berth navigation, to improve the safety of navigation and the protection of the marine environment.

e-Navigation was expected to be driven by user needs. However, IMO intended to avoid the risk of over-reliance on technology and that e-navigation solutions may result in an inappropriate loss of good seamanship, replacement of human element by technology, degradation of bridge resource management and system failures.

The prioritised e-navigation solutions are:

S1: improved, harmonised and user-friendly bridge design;
S2: means for standardised and automated reporting;
S3: improved reliability, resilience and integrity of bridge equipment and navigation information;
S4: integration and presentation of available information in graphical displays received via communication equipment; and
S9: improved Communication of Vessel Traffic Service (VTS) Portfolio (not limited to VTS stations) [6].

During the recent years and on the basis of the strategic implementation plan (SIP), see Section 3.2.2, IMO is working on regulations and guidance to support the implementation of the e-navigation concept worldwide.

3.2.2 Strategic implementation plan

IMO developed an SIP to make the e-navigation concept real. This SIP combines expectations from ship and shore. It also underlines the importance of communication elements for data and information exchange. The completion year of the assessment of required tasks based on the prioritised e-navigation solutions (see Section 3.2.1) was 2019. These tasks permit industry to start the work on e-navigation products and services in a coordinated and harmonised way.

> Solutions S2, S4 and S9 focus on efficient transfer of marine information and data between all appropriate users (ship-ship, ship-shore, shore-ship and shore-shore). Solutions S1 and S3 promote the workable and practical use of the information and data on-board [6].

S9 solutions may be based on S4 results, whereas relations between S2 and S9 are rather limited and not worth being discussed further.

IMO assigned tasks to each of the solutions. These tasks defined the expected deliverables and the implementation schedule (see Appendix A). Most of the tasks have been completed and several resolutions, standards, regulations, recommendations and guidelines have either experienced revisions or been completely newly drafted, taking into account current technology which is able to provide more sophisticated solutions. Other tasks are still under development and their completion is expected soon. The most significant outputs related to the ship–shore, shore–ship communications and consequently, for shore-based monitoring systems are as follows:

- single-window solution for reportable information;
- single entry of reportable information in single-window solution;
- automated collection of internal ship data for reporting;
- automated or semi-automated digital distribution/communication of required reportable information, including both 'static' information (e.g. fixed ship's particulars such as length overall and breadth) and 'dynamic' information (e.g. draught of the vessel);
- national reporting requirements to apply standardised digital reporting formats;
- integration and presentation of available information in graphical displays received via communication equipment;
- implementation of a common maritime data structure (CMDS);
- standardised interfaces for data exchange;
- provision of mapping of specific services to a specific user with status and access requirements;
- provision of system for automatic source and channel management on-board for the selection of most appropriate communication means (equipment);
- routing and filtering of information on-board;
- provision of a quality assurance process to ensure that all data are reliable and are based on a consistent common reference system or converted to such before integration and display;

- implementation of harmonised presentation concept of information exchanged via communication equipment, including standard symbology and text support, taking into account human element and ergonomics design principles to ensure a useful presentation and prevent overload. Harmonise displays;
- development of a holistic presentation library as required to support accurate presentation across displays and harmonise displays;
- provision of alert functionality of inertial navigation system (INS) concepts to information received by communication equipment and integrated into INS;
- harmonisation of conventions and regulations for navigation and communication equipment; and
- development of Maritime Service Portfolios to refine services and responsibilities.
  (adapted from [6])

As mentioned earlier, one significant factor supporting the implementation of solution S9 (to improve the communication of VTS Portfolio (not limited to VTS stations)) was the introduction of Maritime Service Portfolios.

It is important that services provided by electronic means should be described in harmonised ways separately. In the meantime, the term 'Maritime Service Portfolio' has been replaced by 'maritime services in the context of e-navigation' to avoid misinterpretations.

3.2.3 Maritime services in the context of e-navigation

Maritime services should be provided for the following areas:

- port areas and approaches;
- coastal waters and confined or restricted areas;
- open sea and open waters;
- areas with offshore and/or infrastructure developments;
- polar areas; and
- other remote areas [6].

The initial list of proposed Maritime Service Portfolios (see [6]) has been improved in both level of details and description of the services during the last view years. The current list of maritime services in the context of e-navigation (2019) is presented in Table 3.2(a).

The purpose of each maritime service in the context of e-navigation is shown in Table 3.2(b).

Each maritime service has an association to at least one technical service. All involved stakeholders agreed that the IHO S-100 Standard (Universal Hydrographic Data Model) is one of the basic components of technical services. Using S-100 ensures that all relevant stakeholders are using the same data language for data transmission, reception and processing. However, noting that the primary focus of S-100 is hydrographic information, and taking into account that some e-navigation information is not hydrographic, it has been recognised that S-100 will possibly not be able to provide data model descriptions for all kind of data

Table 3.2(a) List of maritime services in the context of e-navigation (2019)

Maritime service in the context of e-navigation	Responsible domain coordinating body
MS 1 VTS Information Systems	IALA
MS 2 VTS Navigational Assistance Service	IALA
MS 3 VTS Traffic Organisation Service	IALA
MS 4 Port Support Service	IMO FAL Committee
MS 5 Maritime Safety Information	IHO, WMO
MS 6 Pilotage Service	IMPA, IMO FAL Committee
MS 7 Tug Service	IMO FAL Committee
MS 8 Vessel Shore Reporting	IMO FAL Committee
MS 9 Telemedical Assistance Service	IMHA
MS 10 Maritime Assistance Service	IMO FAL Committee
MS 11 Nautical Chart Service	IHO
MS 12 Nautical Publication Service	IHO
MS 13 Ice Navigation Service	WMO
MS 14 Meteorological Service	WMO
MS 15 Real-Time Hydrographic and Environmental Information Service	IHO
MS 16 Search and Rescue Service	Open (potentially SAR experts)

exchanged in maritime business, such as medical or comprehensive data used by custom, sender and receiver for cargo tracking.

It is envisioned that users should combine information from several maritime services and, therefore, from different domain stakeholders to develop an information status which turns out satisfactory. Building this information status requires an interoperability of technical services based on a coordinated approach of the domain control bodies. They have to coordinate the data models and required communication methods.

It has been recognised further that technical services could be dependent on different S-100-based products. This circumstance underlines the importance of data model and communication methods' harmonisation.

New maritime services are under discussion (2020). However, given by the fact that the responsible IMO body expected an output on 'Consideration of the descriptions of maritime services in the context of e-navigation' by 2021, IMO recommended initiating further afterwards (Figure 3.3).

3.3 The S-100 idea

3.3.1 What was the motivation to replace the current standards?

IMO's SOLAS Chapter V (Safety of Navigation) defines that a specially complied database can be used to reproduce charts and publications. That statement is the legal basis for coastal states to start the production of electronic navigational charts

Table 3.2(b) Purpose of Maritime Services in the context of e-navigation (2019), partly adapted from [7]

Maritime Service in the context of e-navigation	Purpose
MS 1 VTS Information Systems MS 2 VTS Navigational Assistance Service	Provide data in a digital format to support VTS Information Service (INS), to Navigational Assistance Service (NAS) and to Traffic Organisation Service (TOS)
MS 3 VTS Traffic Organisation Service	Create the means to reduce administrative burden and information overload, reduce miscommunication due to external interference, simplify work procedures, promote sustainable shipping and increase navigational safety
MS 4 Port Support Service	The purpose of this maritime service has not yet been defined
MS 5 Maritime Safety Information	Provide the mariner with information related to navigational and meteorological warnings, meteorological forecasts and other urgent safety-related messages
MS 6 Pilotage Service	Provide information related to the pilotage service when planning an operation before the pilot boards the vessel
MS 7 Tug Service	Provide access to all necessary tug-related information required by ships heading to port, in order to optimise transit times and promote efficient movement of goods and persons by using modern technology and common standards
MS 8 Vessel Shore Reporting	Provide secure submission and distribution of reports required by shore-based authorities in the harmonised standard and in the required timeframe; reduce the administrative burden on-board the ship and ashore; reduce the number of human errors and missing information by automating the reporting processes as much as possible; and provide real-time access to information to relevant stakeholders in a secure manner
MS 9 Telemedical Assistance Service	Provide decision support and advice to the seafarer on-board responsible for medical care whenever the provision of treatment cannot wait
MS 10 Maritime Assistance Service	Provide services to manage communications between the coastal state, ship's officers requiring assistance and other responsible maritime organisations
MS 11 Nautical Chart Service	Provide information to be used for safe navigation. The information provided as part of a Nautical Chart Service must complement information provided as part of other services such as Nautical Publication Services and Real-Time Hydrographic and Environmental Information Services.
MS 12 Nautical Publication Service	Provide the mariner with information to complement ENCs/Nautical Charts for advance planning and to navigate a ship safely during the intended voyage
	Provide information which is continuously updated and which is required for voyage planning and execution

(Continues)

Table 3.2(b) (*Continued*)

Maritime Service in the context of e-navigation	Purpose
MS 13 Ice Navigation Service	Provide ice-related information in high seas areas, coastal, offshore and local waters for decision-making, whether or not proceed with ice navigation
MS 14 Meteorological Service	Provide forecasts and warnings for the high seas and forecasts and warnings for coastal, offshore and local areas (including ports and harbours)
MS 15 Real-Time Hydrographic and Environmental Information Service	Provide real-time water level and surface current data intended for situational awareness, hazard avoidance, works on renewable marine energy and route planning Provide information on current and tide in complement to ENCs/Nautical Charts
MS 16 Search and Rescue Service	Provide search and rescue functions

(ENCs). As further stated by IMO, the information provided by the database should fit the IHO's resolutions and requirements.

SOLAS Chapter V, Regulation 2

Nautical Chart or nautical publication is a special-purpose map or book, or a specially compiled database from which such a map or book is derived that is issued officially by or on the authority of a government, authorised hydrographic office or other relevant government institution and is designed to meet the requirements of marine navigation.*

First adopted in 1992, S-57 is the current IHO standard used to produce ENCs compliant to the SOLAS V carriage requirement. This standard is based on the International Organization for Standardization (ISO) 8211. The aim was to provide worldwide coverage with an ENC, which has been successfully achieved.

The use of ISO 8211 has the major advantage to support the incremental update of database content. That reduces the size of updated files significantly. Small data size was a significant feature in 1992 when the provision of big data was an issue. Contrarily, the disadvantage of ISO 8211 is that this standard will more or less exclusively be used for ENC production. The data model has limited flexibility because it is embedded in the encapsulation. The acceptance of ISO 8211 outside the ENC world is rather limited. In addition, the standard was not able to support

*Refer to appropriate resolutions and recommendations of the International Hydrographic Organization concerning the authority and responsibilities of coastal States in the provision of charting in accordance with regulation 9.

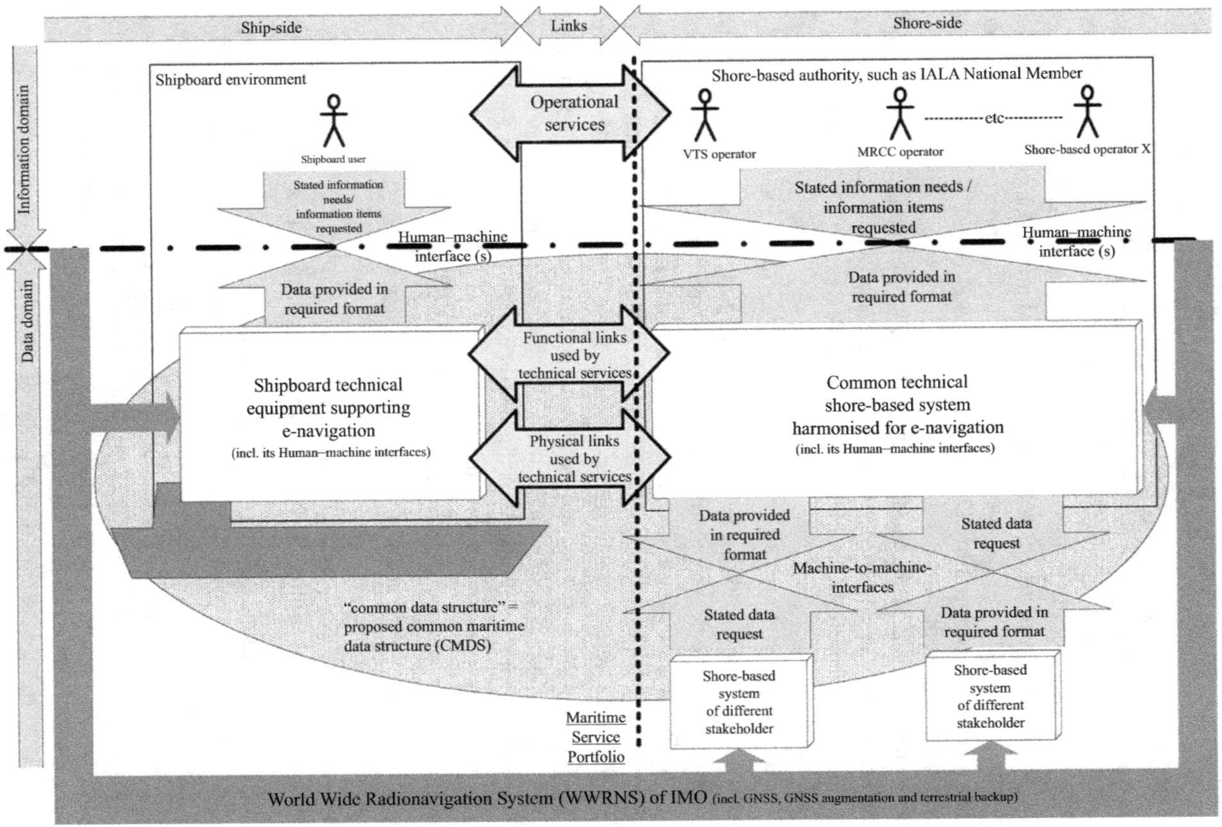

Figure 3.3 IMO e-navigation concept. IMO, e-Navigation Strategy Implementation Plan (SIP) NCSR 1/28

real-time data or gridded bathymetry provision. A further disadvantage is that portrayal instructions have to be provided separately by the S-52 standard. This standard is an integrated part of any ECDIS device.

The emerging request to provide new data to compliment ENC information unveils a further significant disadvantage; S-52 updates have to be done individually on every single on-board system. The limited flexibilities of the standards itself, the necessary workload to manage the S-52 updates, and the fact that older ECDIS exist on-board with no manufacturer being able to manage the up-to-dateness of ECIDS, stressed IHO to reconsider the future usefulness of S-57 and S-52.

These disadvantages required freezing of the standards and the end of further developments. IHO started the development of a flexible, extensible and acceptable Universal Hydrographic Data Model named S-100. The first edition of S-100 was released in 2008.

S-100 is able to manage and support a wider variety of data, more exchange formats and new applications. Among these new possibilities in data processing, standards based on S-100 fulfil the SOLAS Chapter V, Regulation 2 requirements.

3.3.2 S-100 as the Universal Hydrographic Data Model

Although the S-100 data model can be employed to support data sources outside the hydrographic domain, the development of the model is based initially on the request to support a wider range of digital data which have a relationship with hydrography. Stakeholder demanded the IHO to provide an extensible standard which supports inter alia gridded data, real-time data, three-dimensional (3D) data and other specific data. Standards based on S-100 should be able to support GIS applications, bathymetric data and information additional to those presented on a navigational chart.

S-100 is not a format solely used for digital hydrographic data; it is a universal data model that defines all components used:

- to develop S-100 compliant product specifications;
- to develop components needed for quick product data updates;
- to develop S-100-based services;
- to develop plug-and-play concepts; and
- to develop S-100-based interoperability between products (see Section 3.6.4).

One of the core functions of S-100 is to acquire, to access, to process, to analyse and to present data from various sources to be useful for various stakeholders. This was the key factor convincing IMO to specify S-100 as one of the basic e-navigation components. The feature that the file content is independent from the file carrier was a further convincing argument. This feature offers a wider flexibility in data provision. A fact, which has an importance, relates to the foreseeable innovations in data transmission methods.

The use of S-100 offers additional benefits of which IMO can benefit for defining shore-based monitoring systems features, such as the:

- capability to further extensions as S-100 is based on ISO 91xxx series;
- greater and more cost-effective use by using ISO 91xxx standards;

- conformance with ISO/TC211 and therefore the possibility to use off-the-shelf GIS software;
- possibility to introduce the interoperability between products and therefore avoidance of redundant data storage; and
- possibility to combine data from various stakeholders.

3.3.3 S-100-based product specifications

According to the current edition 4.0.0 of S-100 (2019) [8], the following components are requested to deliver a complete product specification:

- specification scope;
- dataset identification;
- data content and structure;
- coordinate reference system;
- data quality;
- data capture and encoding;
- maintenance;
- portrayal;
- data product format;
- data product delivery; and
- metadata.

Several stakeholders have developed or are still developing product specifications on the basis of S-100 (Table 3.3).

Although the product specification under the remit of the IHO is essential to provide a well-designed chart background and comprehensive supporting information to the mariner, the most relevant product specification for shore-based monitoring systems is S-211 which coordinates and standardises reports which will be usually exchanged between ships, ports, terminals, agents and ship owners.

Also of relevance for shore-based monitoring systems is S-212 that standardises the reports and messages to be exchanged between a ship and a VTS and between two or more VTSs. The information exchange between VTSs may reduce the administrative burden to the mariners and can also simplify the workload (Figure 3.4).

3.3.4 Tailor-made information provision

The S-100 data model offers the possibility to tailor-made information. The data model consists of following elements which can be optional, conditional or mandatory:

- feature class;
- information feature class;
- simple attribute;
- simple attribute, enumerated types;
- simple attributes, codelist types;
- complex attribute; and
- relationship and associations.

Table 3.3 List and description of product specifications and the responsible bodies

Standard number	Responsibility	Standard name	Status
S-101	IHO	ENC	O
S-102	IHO	Bathymetric Surface	O
S-104	IHO	Water Level Information for Surface Navigation	O
S-111	IHO	Surface Currents	O
S-112	IHO	Dynamic Water Level Data Transfer	P
S-121	IHO	Maritime Limits and Boundaries	R
S-122	IHO	Marine Protected Areas	R
S-123	IHO	Marine Radio Services	R
S-124	IHO/IMO	Navigational Warnings	O
S-125	IHO supported by IALA	Marine Navigational Services	P
S-126	IHO	Marine Physical Environment	O
S-128	IHO	Catalogue of Nautical Products	O
S-129	IHO	Under Keel Clearance Management	O
S-201	IALA	Aids to Navigation Information	R
S-210	IALA	Inter-VTS Exchange Format	P
S-211	IALA	Port Call Message Format	O
S-212	IALA	VTS Digital Information Service	
S-230	IALA	Application Specific Messages	P
S-240	IALA	DGNSS Station Almanac	O
S-245	IALA	eLoran ASF Data	O
S-246	IALA	eLoran Station Almanac	O
S-247	IALA	Differential eLoran Reference Station Almanac	O
S-401	IEHG	Inland ENC	O
S-402	IEHG	Bathymetric Inland ENC	O
S-411	WMO	Ice Information	R
S-412	WMO	Ocean Forecasts	O
S-413	WMO	Weather and Wave Conditions	O
S-414	WMO	Weather and Wave Observations	O
S-421	IEC	Route Exchange Format	O

O, under development, this covers also product specification which has been released for testing purposes; P, planned; R, released.

Based on that data model, information can be assigned to specific vessels by using inclusion-type association. The next example has been copied from the product specification S-122 (see [9]) (Figure 3.5). It shows that a vessel with certain characteristics either needs to provide a report or not. The characteristics are:

- in ballast or not;
- loaded with cargo (type of cargo);
- loaded with dangerous or hazardous cargo (IMDG Class);
- type of vessel;

Figure 3.4 The S-100 concept. Korean Hydrographic and Oceanographic Agency

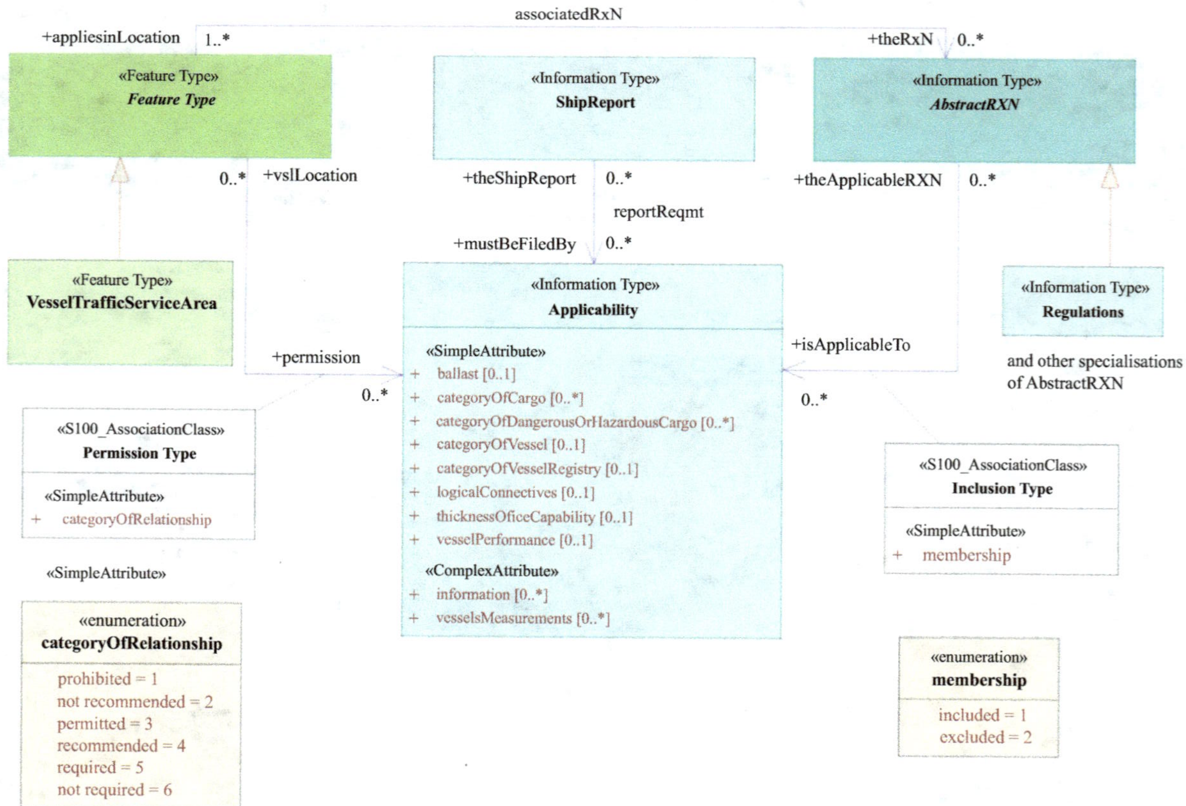

Figure 3.5 Applicability of vessels associated with reports [9]. © 2020 International Hydrographic Organization

- type of vessel's registry;
- thickness of ice capability;
- vessel's performance; and
- vessel's measurements.

The possibility to assign services or information to a particular type of vessel offers a wide range of options for shore-based monitoring systems.

In a future imaginable scenario, an automated process transmits permanently a ship's characteristics, and a shore-based monitoring service receives and assesses this information. Based on the general characteristics or on a particular characteristics value, a ship may receive tailor-made information from the shore-based system. One example could be the provision of information on how to approach a fairway best or the suggestion of an alternative route, which is more suitable for the ship.

3.4 Supporting elements

3.4.1 Why are supporting elements required?

Having e-navigation solutions and S-100 as a framework for hydrographic data can be compared with the building of a skyscraper. Living in such a house would be impossible without lifts, stairs, window cleaning facilities, etc.

The same analogy applies in the e-navigation context. e-Navigation solutions and S-100 do not guarantee that all components work well together. Different feature IDs and feature encoding for the same fact cause problems with data interoperability and data consistency. Ship details can be one example. One e-navigation solution might use ships' details which are different from the details used to operate an ECDIS.

More worse is it if different media are involved. An example is that a vessel completes electronically a reporting form requested by a harbour and a printout is the result. This printout will be submitted as a facsimile that can be considered as e-navigation ready because the transmission is made by electronic means. The harbour copies manually the information from the facsimile into their database system.

The ideal way would be that a vessel's report is compiling the information from an on-board database, the submission will be made by electronic means and that the port integrates the report data electronically in his system.

The necessary exchange of the amount of data between ships and shore or between ships and ships requires a permanent availability of appropriate and possibly redundant telecommunication methods, which have to be permanently synchronised.

3.4.2 Common marine data infrastructure

Considering that several international bodies are involved in the development of e-navigation solutions, and taking further into account that the results of the e-navigation solutions should be interoperable to certain extent; it was decided by IMO to develop or to specify a commonly used marine data structure. An agreed and adopted CMDS is a core element of the e-navigation developments.

IMO was ascertained that the maintenance of such a CMDS goes beyond the original IMO work and that building up expertise within IMO is costly and time-consuming. Therefore, IMO decided to use the IHO S-100 (see Section 3.3.2) as the underlying CMDS. This decision is based on the fact that S-100 is based on ISO 91xxx series and that this offers the most flexible use.

Other common data structures (CDSs) focussing on the provision of a CDS for non-hydrographic information are under consideration. Most of this information is operational.

At some point, both the CMDS and one or more CDS(s) overlap to certain extent. A good example is the different provision of information on dangerous cargo. In the S-100 CMDS, the dangerous cargo information is a feature attribute with specific values. Within another CDS, e.g. dynamic cargo tracking the dangerous cargo, information can be much more specific and may go down to the IMDG-Code number for each piece of cargo.

The challenge for the implementation of the e-navigation concept will be to combine both the CMDS and the CDS with no overlap or reduction of information, or to make sure that one system does not corrupt specific information of the other.

Marine resource names (MRNs) can be a possible solution. IMO and IHO appreciate the idea and advice the further development of this concept for e-navigation purpose.

3.4.3 Marine resource names

An IHO technical working group confronted to compare hydrographic-office-specific light numbering systems of their individual list of lights entries with those provided by the United Kingdom Hydrographic Office (UKHO) in 2009, made the first attempt to develop a unique identifier for hydrographic information. The hydrographic community recognised the UKHO light number as the international light number at that time.

The problem was addressed, and a solution named MRN was developed by the International Association of Lighthouse Authorities (IALA). The MRN concept bases on the use of Uniform Resource Identifiers, a mechanism used in the Internet world. Initially motivated to resolve the light numbering problem, this persistent unique identifier (PUI) could be used to unambiguously name any geographic feature in the e-navigation environment.

Other PUIs are in operation for cargo handling and tracking. One example is the cargo tracking note needed by customs for clearance.

S-100 specifies the MRN concept for creating PUI in S-100-based products.

The use of MRN supports the IMO e-navigation SIP solution S3, which calls for improved reliability, resilience and integrity of bridge equipment and navigation information. Within the e-navigation context, the most predicted use of MRN is the machine-to-machine communication. This underlines the importance of having a strict syntax specification.

Although the strict syntax specification sets limits, the MRN concept is highly flexible. This flexibility, in turn, requires strict governance to avoid the disharmonious use of MRNs, which could diminish the trust in the concept.

The MRN concept is currently hypothetical. The proof-of-concept is a task for IHO and IALA to be completed before the MRN concept becomes operational. It is assumed that most stakeholders have assigned individual IDs to geographical or non-geographical features. In theory, these presently assigned feature IDs can be reused and would be simply extended by additional information in front of each ID. The reuse of existing feature ID concepts and possible interoperability aspects has not yet been assessed and needs to be evaluated.

3.4.4 *Sufficient data bandwidth availability*

Communication is one key element of e-navigation. The different services need a sufficient provision of data, preferably in real-time. That requires a communication service which:

- is permanently available;
- is reliable;
- is flexible;
- can provide the required amount of data;
- can provide sufficient data transfer speed;
- can provide a cybersecure connection; and
- is offered to an appropriate price.

Bandwidth is the most commonly used term to describe the maximum data transfer rate capability of a communication service. The bandwidth of a service is measured in bit per second.

Gigabits of data will be exchanged within an e-navigation environment. This data exchange can be bidirectional between the involved parties. Taking the massive amount of data into account, a sufficient bandwidth is essential (Table 3.4).

Although certain components are still under development and awaiting official IMO adoption afterwards, very high frequency (VHF) data exchange system (VDES) has been widely accepted as one of the most promising data exchange systems for low-volume data between shore and ships. One of the core benefits is the possibility of bidirectional data exchange. Another convincing benefit is that the necessary technical equipment onboard is not costly. The required technical infrastructure is either available or can be installed quickly and no additional costs for the stay in shipyards or so are necessary.

VDES can be provided globally, and the used connection system is based on the vessel's location. It can be either VDE-TER or VDE-SAT. The intended spatial coverages are:

- VDE-TER: ship-to-ship and ship-to-shore communication; and
- VDE-SAT: ship-to-satellite.

That means that VDE-TER can be used for all kind of communication in VHF range and between ship and shore. VDE-SAT will be used in areas with no VHF

Table 3.4 Examples of list of communication systems used or planned in marine environment (2019)

Communication systems	Services
AIS (used)	• Aids to navigation • Collision avoidance • Search and rescue • Fishing fleet control and monitoring • Maritime security • Accident investigation • Ocean currents estimates • Infrastructure protection • Fleet and cargo tracking
UHF, VHF, LF, MF (used)	• Traffic management communication • Broadcast of maritime safety information (MSI) • Search and rescue (SAR) communications • All kinds of voice and digital telecommunication
VDES (planned)	• Search and rescue (SAR) communications • Broadcast of maritime safety information (MSI) • Broadcast of temporary and preliminary notices to mariners • Transmission of facilitation of international maritime traffic (FAL) forms • Traffic management communication • Download of updated digital publications • Download of the Global Maritime Distress and Safety System (GMDSS) Master Plan • Route exchange • Broadcast by maritime infrastructure
G4 (used) and G5 (planned)	• High mobility communication • Real-time data provision

coverage. That includes polar regions and very remote regions such as regions in the south of the Atlantic Ocean or the Pacific Ocean.

VDE also enables data exchange between oceanographic and meteorological sensors and shore-based stations or directly between those sensors and ships.

However, certain circumstances may require the exchange of high-volume data. This can be particularly important if real-time data exchange is necessary. One example can be the provision of high bathymetric data and real-time water level information used to improve the navigational capability of a waterway. In this case, the VDES would use the best available system for high-volume data exchange such as G-4 or G-5 mobile telecommunication, possibly as a Long-Term Evolution

Service, or sufficient satellite services such as ka/ku-Band very small aperture terminal or several Inmarsat services if necessary.

3.5 Shore-based monitoring systems

3.5.1 Why can shore-based monitoring systems benefit from e-navigation?

As mentioned in Section 3.2.1, e-navigation supports the exchange of data and information in a systematic and coherent way. This, in turn, offers the opportunity to develop services which can easily access and process the data or which can provide the information for human-based decision-making processes. Two principle shore-based systems can be distinguished.

The first system controls how a ship operates in a certain area. This can be beneficial in difficult traffic situations. A set of different services has been established to control ships' movements. The most commonly known systems assess the ship's automatic identification system (AIS) data. Other systems rely on radar surveillance or on messages received from ships. The systems are either active or passive. That means they either actively provide advice to ships or simply check passively the ships' movement and provide advice through other communication channels.

The second system has access to the ship's operational data. Shore-based services may support the ship's crew in repair service questions or in urgent medical situations. These services are also able to control certain ship's parameter and could provide service instructions to the crew. Industry has developed a lot of services to support ships, management and insurances in having a comprehensive overview of ship's parameter or different services status. Not being able to provide an overview of all available solutions, the DNV-GL applications ShipManager^{TM} and Navigator^{TM} show in exemplary way possible solutions (Table 3.5).

3.5.2 Shore-based ship operation centres

The established and commonly known ship operation centres are VTSs and ship reporting systems and both are supported by e-navigation MS 1 (VTS Information Systems), MS 2 (VTS Navigational Assistance Service) and MS 3 (VTS Traffic Organisation Service).

These maritime services in the context of e-navigation establish a digital data exchange between ships and shore to reduce reporting management burden to the ship, simplify the work by reusing existing information and reduce the information overload for both ship- and shore-based ship operation centres.

Intensive trials should figure out what e-navigation set up fits the IMO requirements best. These trials will usually be conducted in test beds. Such test beds have been established in many places around the world to test either the shore-based operation or the interaction between ships and shore.

Table 3.5 List of services provided by DNV-GL applications ShipManager™ and Navigator™ and a selection of provided solutions [10]

ShipManager™	**Solutions (selected examples)**
ShipManager Technical	• Prepare and document all planned and unplanned maintenance tasks performed • Automatically update stock counts in consideration of spare parts consumed for maintenance tasks • Keep running hours for each equipment item individually, regardless of where it is installed • Keep a lifecycle record of each equipment and manage defects
ShipManager Procurement	• Purchase any type of goods and services be it spares, stores, repair services, lubes, paints, or provisions • Monitor the consumption of paints, chemicals, lubes or other consumables • Manage re-deliveries
ShipManager QHSE	• Report and follow-up incidents, accidents, injuries and near-misses • Monitor the validity of vessel certificates and surveys
ShipManager Crewing	• Run fleetwide crew planning/scheduling processes and optimise crew deployment in line with safe manning requirements and company policies • Manage certificates, licences, qualifications, experience, competences and contracts
ShipManager Projects	• Manage requests for quotations from yards (locations) and suppliers • Compare prices and select the best yard/supplier combination for the project
ShipManager Hull	• Issue unscheduled inspections from vessel and office • Integrate thickness measurements in the 3D digital twin
ShipManager Survey Simulator	• Simulate a step-by-step inspection, including where to focus your attention, i.e. where hull structural deficiencies are likely to occur • Use a set of real-life tools relevant for documentation and report inspection results
ShipManager Analyzer	• Draw conclusions about fleet and vessel performance even though data are spread across numerous software systems and/or departments • Easily compare vessels, see trends and bring information that was not related together
Navigator™	
Navigator Port	• Prepare for port arrival and departure with the latest arrival/departure checklists for more than 3,500 ports and more than 1,600 different forms that are automatically prefilled with, e.g., ship information and crew data • Perform electronic reporting directly to authorities
Navigator Insight	• Onshore system collects data for the entire fleet in operational reports (log abstracts, noon, departure and arrival, bunker, ballast water operations, disposals and many more) and management reports • Allows comprehensive fleet performance monitoring with in-depth analysis • All information is structured for reuse in any type of analysis or reporting to stakeholders

Shore-based monitoring systems are involved in most of the test beds. These systems either both control and support the traffic on sea by providing ships and shore-based services with relevant traffic situation information or they support the ship-to-shore, ship-to-ship and shore-to-shore data exchange.

Important sea trials have been conducted by the Sea Traffic Management (STM) Validation Project in Europe and the SMART-Navigation Project in the Republic of Korea (SMART) (Figure 3.6).

STM focussed on test different e-navigation components within a SOLAS compliant environment, whereas SMART focussed on the provision of e-navigation services to the non-SOLAS market, which includes inter alia recreation crafts, small coastal vessels and fishery.

The development and intend release of the S-100 compliant S-421 Product Specification (route exchange format) in 2020 is one important result of trials conducted under the e-navigation test scenarios. This standard describes the routeing information exchange between surrounding ships and between ships and shore-based facilities, such as harbours or traffic control offices.

The provision of maritime safety information (MSI) transmitted from shore to ships is also a key element of e-navigation MS 5 (MSI). Various transmission methods were investigated. Tests emphasised the importance of permanently available communication possibility for both voice and data.

It was further emphasised that every e-navigation solution can only be as good as the involved parties provide the necessary information timely and in a high level of accuracy.

The involvement of shore-based ship operation centres in an e-navigation environment has several advantages. The technical infrastructure of such systems is much more advanced than those on-board. They are able to process more data in a very short time and consequently are able to provide advice to ships, which can assist a ship in the reducing of fuel consumption, in optimising the voyage and in reducing collision and grounding risks.

Figure 3.6 The SMART concept [15]

3.5.3 Shore-based ship monitoring centres

The most shipping lines, which operate globally, have deployed vessel monitoring systems (VMSs). A VMS collects data through sensors on-board. These data will be collected and usually be transferred to the ship owner or charterer once a day or on request. The main purposes of these VMSs are the control of different marine operations such as fishing or deep-water exploitation and the ship's operational status.

The VMSs have access to the ship's operation data, including status of the main and auxiliary engines, the fuel consumption, the ship's position, the speed and the course. It is also possible to collect, transmit and provide information on ship's trim and list, the shaft power and both the energy generation and the energy use, which includes the ship's and cargo's energy consumption.

The use of e-navigation can extend the operational service of these monitoring systems and may establish a more efficient bidirectional information exchange between experts ashore and personnel on-board.

Having e-navigation components in operation, the original VMS range of data exchange will be extended. Decisions can now be based on exchanged real-time ship's data. MS 9 (Telemedical Assistance Service) may be used to perform urgent medical treatments. MS 6 (Pilotage Service) and MS 7 (Tug Services) information can be preprocessed ashore and a ship could receive tailor-made information supporting the ship crew in preparation for pilot or tug arrangements. This can prevent the ship or the shipping company from unnecessary claims.

Several shore-based monitoring aspects have been discussed, and trials have been conducted under the e-navigation concept. All trials have extensive data exchange between ship–shore, ship–ship or shore–ship in common. Four concepts developed by the STM and their predecessor projects (https://www.stmvalidation.eu/documents/) are provided next in more detail:

- the 'Enhanced Shore-based Monitoring' service;
- the 'Flow Optimisation' service;
- the 'Area Management Instruction' service; and
- the 'Fleet Management' service.

The 'Enhanced Shore-based Monitoring' service is intended to be provided in critical sea area and during the route monitoring phase. Only the planned route segment relevant for a particular part of the whole route will be exchanged between ship and shore. The transit along the planned route will be monitored by radar and/or AIS data and offers the shore-based service a better image traffic. Decisions can be made considering the ship's intention, and appropriate advice can be given if the ship diverts from the intended route or if critical situations could potentially arise.

The 'Flow Optimisation' service is intended to be provided during the route planning and route monitoring phases. The whole planned route will be exchanged between ship and shore. The service can be used to synchronise traffic movements in a certain area taking into account the route planning of all vessels in this area. It can also be used to provide a ship with MSI relevant for the area of navigation. A

'Flow Optimisation' service may provide the vessel proposals to change the route in critical situations.

The 'Area Management Instruction' service provides all vessels in an area of concern with up-to-date navigational information and MSI in particular. The systems could benefit from a bidirectional data exchange in a way that recognition of information is traceable.

The 'Fleet Management' service is similar to the VMS described earlier. The service tracks all ships of a shipping company and provides well-advanced analysis tools. This can improve the company's fleet performance for single voyages or for voyages on regular services.

3.6 The future

3.6.1 Why can the future of the provision of navigational information be different from today?

The current provision of navigational information is highly regulated. It bases on international regulations, standards, specifications and recommendations defined by IMO, IEC and IHO. It is assumed that the legal and regulatory precondition will be similar in the future.

However, and regardless of the current legal and regulatory limitations, the e-navigation concept offers much wider possibilities of data provisions and computing. It offers operational options far beyond from what are currently used in marine business on-board and ashore.

These operational options open also the window for a new development in shipping which looms – autonomous vessels. Saying that the move from paper to electronic navigation is a paradigm shift, the move towards autonomous shipping is the next radical development in marine business requiring a new approach to the provision of navigational information. IMO has recognised this new development and initiated investigations to collect IMO-affected regulations, guidelines and standards. This goes from the definition of what the term 'ship' means to standards having the portrayal of information incorporated.

Autonomous vessels will not appear immediately in marine business. The operation with autonomous vessels will be gradually implemented starting with automated processes and decision support over remotely controlled vessel with or without mariners on-board and will end with fully automated ships able to make decisions and determine actions by themselves.

The e-navigation solution structure supports this development. Parts irrelevant for autonomous shipping will be separated from the main content and can be removed from regulations, standards and guidelines if necessary or no longer relevant. Autonomous shipping does not need portrayal. It needs data to distinguish between go-areas and no-go-areas. The definition of these areas is relevant for the decision-making process.

The following sub-chapters provide possible future solutions for data provision and supporting elements in ECDIS. The distinction between ECDIS screens used for route monitoring or route planning is not subject of the discussion.

A further sub-chapter discusses a possible routeing optimisation solution by employing shore-based traffic monitoring/control systems as a tool to provide the necessary routeing advice.

3.6.2 Split of data provision responsibilities

The split of a major part of maritime business in various maritime services under the concept of e-navigation offers services the possibility to provide their information independent from other services.

An example is the Nautical Chart as it is now. Several sources provide information collected, assessed and presented in an appropriate way on paper or electronic charts, such as:

- aids to navigation information provided by the responsible IALA administration;
- sea areas defined by administrations responsible for the environmental protected;
- sea areas defined by the Ministry of Defence;
- topographic information provided by land survey administrations;
- information on natural and cultural features by land mapping administrations;
- underwater feature information provided by the hydrographic survey administration;
- tracks and routes defined by the IMO and national shipping administrations; and
- ice, tides and current information provided by the hydrographic offices.

Theoretically, this collection of information on charts can be split in a way that each data owner provides his information independently. That means the data provision is independent, both timely and locally. At the end, the information will be merged directly on-board or in the hydrographic office ashore before providing the data to the ship. The result would be the electronic chart as known today.

The preparation and development of certain S-100-based product specifications support this idea. A sufficient interoperability specification defines the order of the information received and that the more up-to-date and more precise information prevails.

Another option is to stream data on-board with the aim to replace or enhance chart content. Dense bathymetric data useful to provide recent survey results replace original chart bathymetric data. Contemporary data may replace charted information of the height of the tide, the tidal stream or the surface current.

The most benefit for the end user could be gained if all mentioned information will be provided simultaneously and enable a ship to navigate an area with either more possible draught or with a better navigational time frame. Such a replacement of original chart information would be in particular important for river navigation where the navigable area is limited.

An enormous bandwidth and a permanent availability of transmission services is a necessary precondition to provide such an amount of data. A potential solution

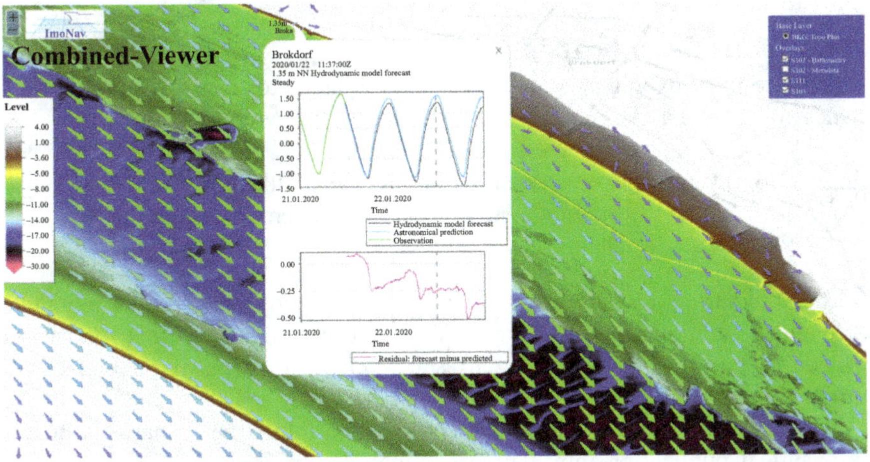

Figure 3.7 ImoNav combined viewer. © 2020 Smile Consult GmbH, Hannover; © 2020 Bundesamt für Seeschifffahrt und Hydrographie, Hamburg/Rostock

to limit the necessary bandwidth and the computing workload on-board could be that a shore-based monitoring system precalculates the data and provides data only for a small spatial around and ahead of the ship.

The split of responsibility of data provision and subsequently, the combination of data from various sources at the end of the process must be quality assured and must consider interoperability aspects (see Section 3.6.4).

The next example shows a portion of the River Elbe fairway. It presents a combined graphical presentation of following data (Figure 3.7):

- a map from the land mapping authority;
- the high density bathymetry combined with water level information from the hydrographic office;
- graphs comparing following height of the tide information:
  - hydrographic model forecast,
  - astronomic prediction,
  - observation, and
  - residual forecast minus predicted.
- the direction and speed of the up-to-date surface current.

3.6.3 Enhanced chart content

In addition to the described replacement of chart content (see Section 3.6.2), a chart display on ECDIS can be enhanced with more information. In theory, the additional presentation of maritime service in context of e-navigation information on an ECDIS screen is possible. However, information overflow is a critical point.

Two thoughts to bear in mind

'Even the most experienced and well trained navigator can make a mistake when forced to handle too much information at the same time'.

'You can argue that the crew member should have sufficient proficiency in ECDIS but what saves lives isn't what people should know, it's what they do know'; adapted from an American Marine Incident Investigation Report.

The integration of nautical publication information and its geo-referenced presentation could easily be achieved. The beauty would be if this information provision is tailor made according to ship's characteristics, cargo and current environmental conditions.

In combination with sophisticated sensors, the following scenario is imaginable. Nautical publication information on tidal stream applies in particular for a specific type of vessel with a certain size. Currently, the information will be provided as follows (strongly simplified):

Physical environment

Tidal stream off harbour entrance

during W wind strong NE current of 3 kn

during N wind strong SE current of 4 kn

Waterway

Restrictions

NE current can be dangerous for sailing vessels with a length of 10 m or less when passing the harbour entrance

The future information provision could be that sensors detect an NE current. This information will be transmitted to all ships in the vicinity by VDES services. According to the ship's parameter, the machine on-board knows that the ship's type is a cargo vessel of 100 m in length. The result will be that this information will not be presented to that ship, whereas a sailing boat with a length of 8 m will receive the information that the 'NE current with a speed of 3 kn has been detected off the entrance. This is dangerous for your ship when passing the harbour entrance'.

The provision of real-time data in addition to the charted content is another option. Weather or ice information is useful to adjust the route accordingly. Information on designated temporary right whale areas is useful to protect the animals present in these areas. Temperature information (water and air) might be helpful to avoid cargo claims.

Underwater temperature and water density information can differ according to the position of the sensors in the water. This information is helpful for survey and submarine activities.

A further option is to put a true-to-scale underwater ship shape on the chart. This offers a vessel a 3D view on the surrounding area and a better indication how

Figure 3.8 Ship shape with high dense bathymetric data. Kongsberg Digital & Electronic Chart Centre

the vessel might react in difficult areas. Combined with the vessel's AIS data, this view is useful for future shore-based monitoring systems to control a vessel's movement in confined areas (Figure 3.8).

3.6.4 Interoperability

The provision of data in an e-navigation environment makes an appropriate specification of interoperability aspects important. This will be different from the current S-57-based ECDIS world, where the avoidance of the information occultation on displays where different information have been loaded and combined is on focus. The current ECDIS portrayal rules define display priorities based on the importance of the information. This applies to chart content as well as to chart overlay content.

Currently, the IMO Performance Standards for ECDIS MSC.232(82) [11] allows the provision of additional information on an ECDIS screen. However, clear instructions on how the integration of this additional information should be done are missing. This lack results in contrary implementation and presentation of this information by ECDIS manufactures.

As mentioned in Sections 3.6.2 and 3.6.3, in a future S-100-based ECDIS being potentially able to provide among chart information also other e-navigation information, a standardised interpretation of interoperability is essential. The IHO is working on such an interoperability catalogue standard. This standard will enable machines to govern the interaction between S-100-based products. Potentially, and considering that S-100 has been accepted as the CMDS for e-navigation services by IMO; it can also be used to define interoperability of information of marine services in context of e-navigation. Noting that no interoperability definition has yet been tested with real data, the developed interoperability definitions are currently conceptual.

The principle idea of interoperability is that all products to be provided on an ECIDS screen will somehow support the mariner in route planning or monitoring.

These products must be part of a machine-readable interoperability catalogue. The S-101 ENC is the underlying basis, and the products enhance/replace the ENC content or provide information in addition to the ENC. The principles of the current S-57 ECDIS to avoid occultation of safety information apply to S-100-based ECDIS too. Furthermore, the amount of data to be displayed should no cause interpretation issues for the mariner.

All products must base on product specifications, which are part of the interoperability catalogue. Data model components of these product specifications must be harmonised. That reduces or eliminates the risk that information of one product will be deleted without having an adequate replacement in a different product, or that less-detailed information of one product replaces detailed information of a different product.

Interoperability rules are not the same for all purposes and interaction between data. The latest draft of the IHO interoperability analyses [12] identifies five levels of interoperability:

- Level 0=overlapping of underlying layer;
- Level 1=simple interleaving;
- Level 2=type-based selectivity and feature class replacement;
- Level 3=feature hybridisation; and
- Level 4=spatial operations to virtually combine data.

The interoperability specification describes Levels 1 and 2 in detail, and the principles have been described for Levels 3 and 4. The detailed description of the latter two is expected in the future and depends on experience gained and stakeholder request.

In Level 0, data will be portrayed as originated and the mariner could turn on or turn off the layer. The current IMO ECDIS Performance Standards apply. Level 0 interoperability is always the fallback if a product will be delivered which is not part of the interoperability catalogue.

In Level 1, data from different products are interleaved. This is specified by the display plan and drawing priority specification of the interoperability catalogue.

In Level 2, data from one product can be replaced by data from another product (as described in Section 3.6.2).

3.6.5 Provision of incremental updates

ISO 8211 is the encoding standard of chart information of future S-100-based ECDIS. Current S-57-based ECDIS use the same standard. The big advantage of using ISO 8211 is that updates to charts are relatively small and that updates are incremental.

However, under the S-100 regime new data formats are possible. These data formats in their current form can either accept incremental updates or not. If a data format does not support incremental updates, two principle solutions are thinkable; split the information and provide it in small products or develop an incremental update mechanism. Currently, the Open Geospatial Consortium is working to make incremental updates possible for Geography-Markup-Language-based products.

3.6.6 Dynamic ships routeing measures

IMO ships' routeing defines routeing measures, which need to be endorsed internationally, see IMO RESOLUTION A.572(14) GENERAL PROVISIONS ON SHIPS' ROUTEING, AS AMENDED.

In addition to these routeing measures, coastal states can define routeing measures within their jurisdiction. It is a legal right of each coastal state to implement further routeing measures within their territorial waters without running the IMO endorsement procedure, see IMO RESOLUTION A.572(14)GENERAL PROVISIONS ON SHIPS' ROUTEING, AS AMENDED, paragraphs 3.12 and 3.13. Whether the IMO will be involved in the designation process or be informed on the existing of such routeing measure belongs to the relevant coastal state. A good example is the routeing measures established in Lake Vänern in Swedish waters. These routeing measures are national, not under IMO regime and therefore, not listed in the IMO Ships' Routeing publication.

However, all routeing measures have this in common that they are static. A vessel must follow the determined routes to avoid all kinds of no-go areas applicable to the vessel.

A possible paradigm shift would be that a vessel receives individual routeing measures provided by a shore-based traffic monitoring/control system based on traffic situation, environmental conditions and ship's characteristics. This routeing measure can overrule the static routeing measures. An example could be that a ship proceeds from a port southwest of Stockholm to a port at the east coast in the Gulf of Bothnia. Under the current legislation, the ship must use east lanes of the traffic separation schemes (TSSs) 'South Aland Sea' and 'North Aland Sea'. According to the current rule, the ship's route crosses the southbound TSS's lanes twice.

Individual routeing measures could direct the vessel through the southbound lanes. That would reduce the transit time, the fuel consumption and possible issues caused by weather or ice conditions (Figure 3.9).

3.7 Conclusion

The full realisation of all e-navigation concept parts is a huge task for the entire marine community. This includes all legislative, regulatory and operational aspects.

The chapter is based on research trying to collect and to describe all necessary components in terms of having a full regulatory environment (by IMO) and the fundamental standards (by IHO) ready. The development of resolutions, regulations, standards and guidelines has either initiated or their necessary revisions have been identified and are in progress to make the e-navigation concept real.

It is commonly recognised that certain supporting elements are needed. The chapter is mentioning what has been considered as essential; the CMDS, the MRN and permanently available communication with sufficient bandwidth.

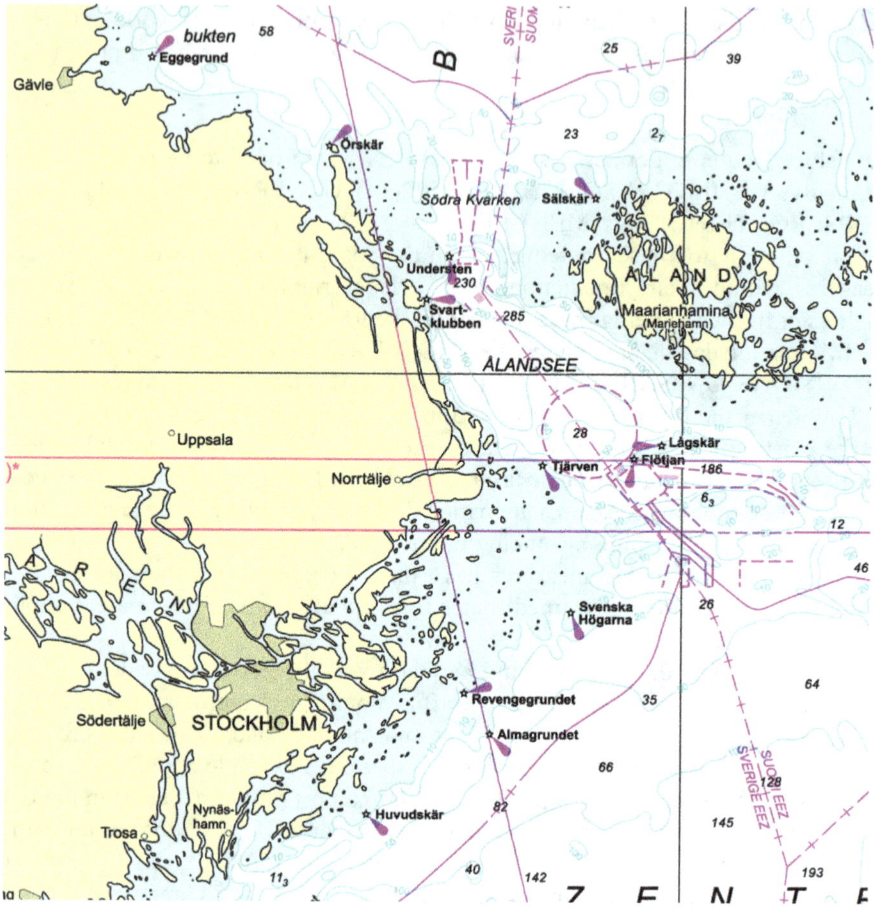

Figure 3.9 Extract from BSH Chart 99. © 2020 Bundesamt für Seeschifffahrt und Hydrographie, Hamburg/Rostock

e-Navigation has impacts on different shore-based systems. This includes both passive and active systems. Systems with different development or implementation status are on the market. The chapter descripted some of those and made no claims of being complete.

The full implementation of e-navigation will have a profound effect on navigation. Considering that ships will navigate autonomously in the future, the provision, assessment and use of information necessary to navigate a ship safely will change the future of navigation significantly. This includes the split of data provision and consequently, the increase of a data vendor's responsibility. Ensuring quality proof mechanism throughout the whole delivery chain becomes vital for building up the confidence of the marine industry in e-navigation.

However, the full implementation of the e-navigation concept will take many years' time. In the meantime, selected e-navigation solutions will be implemented assisting the marine community to reduce the administrative burden for mariners and offices ashore. The implementation of other e-navigation solutions will simplify the navigation and therefore help one to improve the safety of navigation.

List of abbreviations

AIS	automatic identification system
CDS	common data structure
CMDS	common maritime data structure
ECDIS	Electronic Chart Display and Information Systems
ENC	electronic navigational chart
IALA	International Association of Lighthouse Authorities
IEC	International Electrotechnical Commission
IHO	International Hydrographic Organization
IMHA	International Maritime Health Association
IMO FAL	International Maritime Organization, The Facilitation Committee
IMO	International Maritime Organization
IMPA	International Maritime Pilots' Association
INS	inertial navigation system
ISO	International Organization for Standardization
MRN	Marine resource name
MSC	Maritime Safety Committee
MSI	maritime safety information
PUI	persistent unique identifier
QHSE	quality, health, safety and environment
SDs	Sailing Directions
SIP	strategic implementation plan
SMART	SMART-Navigation Project
STM	Sea Traffic Management
TSS	traffic separation scheme
UHF	ultra high frequency
UKHO	UK Hydrographic Office
VDES	VHF data exchange system
VHF	very high frequency
VMS	vessel monitoring system
VTS	Vessel Traffic Service
WMO	World Meteorological Organization

Appendix A

List of tasks and intended outputs defined by IMO to fulfil e-Navigation Strategic Implementation Plan Solutions S2, S4 and S9 (adapted from [6]).

Number	Task	Output
T6	Develop a methodology of how accuracy and reliability of navigation equipment may be displayed. This includes a harmonised display system	Guidelines on the display of accuracy and reliability of navigation equipment
T7	Investigate if an INS, as defined by Resolution MSC.252(83), is the right integrator and display of navigation information for e-navigation and identify the modifications it will need, including a communications port and a PNT module. If necessary, prepare a draft-revised performance standard. Refer to Resolution MSC.191(79) and SN/Circ.243	1. Report on the suitability of INS 2. New or additional modules for the Performance Standards for INS
T8	Member states to agree on standardised format guideline for ship reporting so as to enable 'single window' worldwide (SOLAS regulation V/28, Resolution A.851(20) and SN.1/Circ.289)	Updated guidelines on single window reporting
T9	Investigate the best way to automate the collection of internal ship data for reporting, including static and dynamic information	Technical report on the automated collection of internal ship data for reporting
T11	Development of draft guidelines for software quality assurance (SQA) in e-navigation. This task should include an investigation into the type approval process to ensure that software lifetime assurance (software updates) can be carried out without major re-approval and consequential additional costs. Refer to SN/Circ/266/Rev.1 and MSC.1/Circ.1389	Guidelines for software quality assurance (SQA) in e-navigation
T13	Develop guidelines showing how navigation information received by communications equipment can be displayed in a harmonised way and what equipment functionality is necessary	Guidelines on the harmonised display of navigation information received from communications equipment
T14	Develop a common maritime data structure and include parameters for priority, source and ownership of information based on the IHO S-100 data model. Harmonisation will be required for both use on shore and use on the ship and the two must be coordinated (two domains). Develop further the standardised interfaces for data exchange used on-board (IEC 61162 series) to support the transfer of information from communication equipment to navigational systems (INS), including appropriate firewalls (IEC 61162-450 and 460)	1. Guidelines on a common maritime data structure 2. Further develop the IEC standards for data exchange used on-board, including firewalls

(Continues)

(*Continued*)

Number	Task	Output
T15	Identify and draft guidelines on seamless integration of all currently available communication infrastructure and how they can be used (e.g. range, bandwidth) and what systems are being developed (e.g. maritime cloud) and could be used for e-navigation The task should look at short range systems such as VHF, 4G and 5G as well as HF and satellite systems taking into account the six areas defined for the MSPs	Guidelines on seamless integration of all currently available communication infrastructure and how they can be used and what future systems are being developed along with the revised GMDSS
T16	Investigate how the harmonisation of conventions and regulations for navigation and communication equipment would be best carried out. Consideration should be given to an all-encompassing e-navigation performance standard containing all the changes necessary rather than revising over 30 existing performance standards	Report on the harmonisation of conventions and regulations for navigation and communication equipment would be best carried out
T17	Further develop the MSPs to refine services and responsibilities ahead of implementing transition arrangements	Resolution on Maritime Service Portfolios

References

[1] Augustus. The Deeds of the Divine Augustus. *The Odyssey: Book XII.* [Online] [Cited: 09 12 2019.] http://classics.mit.edu/Homer/odyssey.mb.txt.

[2] Lipiński, Edward. *Orientalia Lovaniensia Analecta, 127*. Leuven, Belgium: Peeters Publishers & Department of Orienta Studies, 2004.

[3] Wikipedia. *Wikipedia.* [Online] [Cited: 18 02 2020.] https://en.wikipedia.org/wiki/Portuguese_discoveries#Sailing_techniques.

[4] IHO Publications. *IHO Publications.* IHO. [Online] [Cited: 19 02 2020.] https://iho.int/en/iho-publications.

[5] IMO Assembly. www.imo.org. [Online] [Cited: 07 02 2020.] https://www.google.com/url?sa=t&rct=j&q=&esrc=s&source=web&cd=1&cad=rja&uact=8&ved=2ahUKEwidzvzr_77nAhVSIlAKHXKFBhgQFjAAegQIBBAB&url=http%3A%2F%2Fwww.imo.org%2Fen%2FKnowledgeCentre%2FIndexofIMOResolutions%2FAssembly%2FDocuments%2FA.893(21).pdf&usg=AOvVaw0. A.893(21).

[6] IHO Secretariat. *E-Navigation*. International Maritime Organization. [Online] [Cited: 10 12 2019.] http://www.imo.org/en/OurWork/Safety/Navigation/Documents/enavigation/SIP.pdf.

[7] IHO Secretariat. *HGDM 2/10 – Report of the Second Meeting of the IMO/IHO Harmonization Group on Data Modelling*. International Maritime Organization, 2018. [Online] [Cited: 12 12 2019.] https://docs.imo.org/Category.aspx?cid=786 (protected access, registration requested). HGDM2/10.

[8] IHO Secretariat. *S-10n-Product Specification Template 4.0.0*. 2019. [Online] [Cited: 17 12 2019.] https://www.iho.int/mtg_docs/com_wg/S-100WG/MISC/Part%2011%20Appendix%20D_S-10n-ProductSpecificationTemplate_4.0.0.docx.

[9] IHO Secretariat. *S-100 Geospatial Information Registry*. [Online] [Cited: 17 12 2019.] http://registry.iho.int/beta/productspec/view.do?idx=73&product_ID=S-122&statusS=5&domainS=ALL&category=product_ID&searchValue=.

[10] DNV-GL. *DNV-GL Marine Fleet Management Software and Ship Management Systems, ShipManager 114260*. [Online] [Cited: 20 01 2020.] https://brand-central.dnvgl.com/download/DownloadGateway.dll?h=BE1B38BB718539CC0AB58A5FF2EA7A832F16E63D7ED855158DE3FDEE70B7B960E00484241EBFDFAA9AEF8F1E2C5C5095.

[11] Maritime Safety Committee. www.imo.org. [Online] [Cited: 07 02 2020.] https://www.google.com/url?sa=t&rct=j&q=&esrc=s&source=web&cd=1&cad=rja&uact=8&ved=2ahUKEwiCi_6x8r7nAhVNhqQKHQXOCyEQFjAAegQIBBAB&url=http%3A%2F%2Fwww.imo.org%2Fen%2FKnowledgeCentre%2FIndexofIMOResolutions%2FMaritime-Safety-Committee-(MSC)%2FDocuments%2FMS. MSC.232(82).

[12] IHO S-100 Test Strategy Meeting. www.iho.int. 2019. [Online]. [Cited: 08 02 2020.] https://legacy.iho.int/mtg_docs/com_wg/S-100WG/TSM7/TSM7_2019_4.9_Interoperability_functional_principles_overview-white%20paper-20180707.docx. S-98.

[13] Homer Odyssee. Available from: https://upload.wikimedia.org/wikipedia/commons/c/c7/Homer%2C_Odyssey%2C_Vaticanus_Palatinus_graecus_7.jpg [Accessed 23 March 2021].

[14] Historic chart Piri Reis. Available from: https://de.wikipedia.org/wiki/Karte_des_Piri_Reis#/media/Datei:Piri_reis_world_map_01.jpg [Accessed 23 March 2021].

[15] The SMART navigation project. Available from: http://smartnav.org/eng/html/SMART-Navigation/about_smart_navigation.php [Accessed 23 March 2021].

Chapter 4

Maritime transportation along the Northern Sea Route

Alexandra Middleton[1]

During the Soviet Union times, the Northern Sea Route (NSR) served an essential part in the Arctic resources' exploration. Climate change and consequent melting of the Arctic Ocean create preconditions for the development of the NSR. This chapter addresses the increasingly important role of global maritime transportation. While the NSR offers the shortest transit time between Europe and Asia, the geopolitical and legal status of the territory, the NSR belongs to, is of crucial importance, taken that it is operated within Russia's exclusive economic zone (EEZ). The main drivers for the NSR development discussed are increased demand for natural resources globally, the availability of them in the Arctic, the growing demand for extreme tourism products and the strategic interest shared by the several Arctic and non-Arctic stakeholders to prepare for future container shipping. Maritime transportation along the NSR today is dominated by destination shipping and by transporting of natural resources.

This chapter provides a comprehensive review of shipping activity along the NSR by type of shipping and type of cargo, including container shipping. Additionally, a case study of how global demand for liquefied natural gas (LNG) is affecting the development of the NSR is highlighted. Finally, the future of the NSR development by looking at sustainability requirements in the Arctic and worldwide that shapes the future of the NSR is discussed. Smart shipping is likely to affect global shipping, thus several initiatives for the development of smart shipping for the NSR are discussed too. In conclusion, this chapter provides an outlook for the development of shipping along the NSR in the future.

4.1 Northern Sea Route in a global maritime transportation

Maritime transport is fundamental for world trade accounting for about 80% of all goods' volume transported [1]. Most of the maritime transportation from East to

[1]Oulu Business School, University of Oulu, Oulu, Finland

West is done via the Suez Canal shipping lane that corresponded to 12% of the world trade volume [2]. At the same time, the world shipping is entering a new era with new shipping routes in the Arctic becoming increasingly available. With the melting of the Arctic ice, sea routes in the Arctic stay navigable for longer periods. In the long term, Arctic shipping routes have the potential to change international trade and shipping. Figure 4.1 demonstrates three alternative Arctic routes: Northwest Passage, Central Arctic Ocean Route and Northeast Passage/NSR. Northeast Passage is a water route along the northern coast of Europe and Asia, between the Atlantic and Pacific Oceans.

The NSR is a defined part of the Northeast Passage. The borders of the NSR are marked by black color (see Figure 4.2). The NSR encompasses the waters adjacent to the northern coast of the Russian Federation, including the inland and territorial seas, the contiguous zone and the EEZ of the Russian Federation. The borders of the NSR are defined by the Russian Federal law of 2012 [3]. The NSR borders are limited in the East by the line delimitating the sea areas with the United States of America and

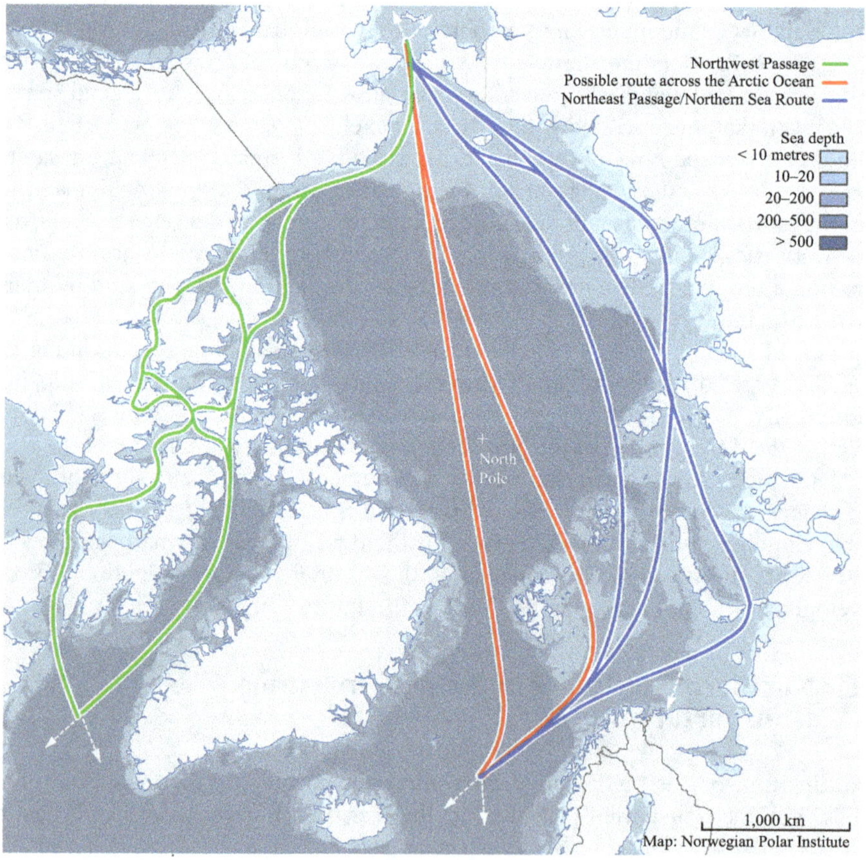

Figure 4.1 Arctic shipping routes. Source: Norwegian Polar Institute

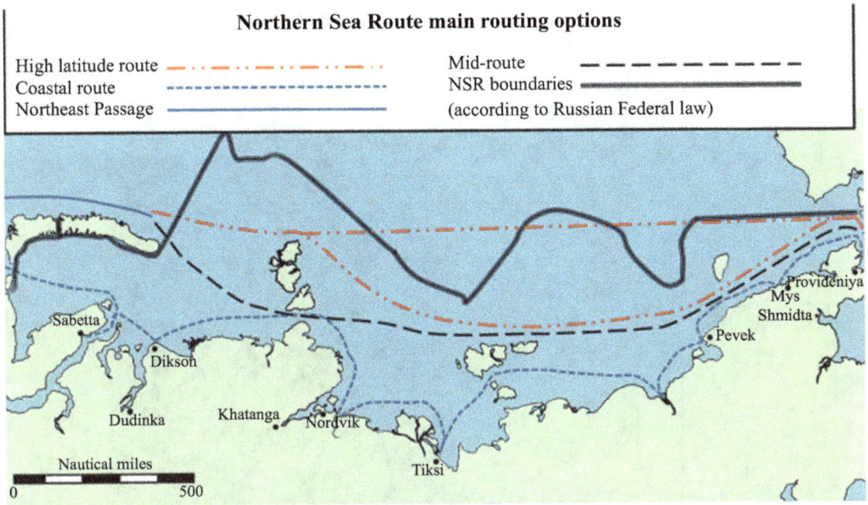

Figure 4.2 The borders of NSR according to the Russian Federal law 132-FZ [4]

by the parallel of the Dezhnev Cape in the Bering Strait; in the West, by the meridian of the Cape Zhelanie to the Novaya Zemlya archipelago, by the east coastal line of the Novaya Zemlya archipelago and the western limits of the Matochkin Shar, Kara Gates and Yugorski Shar Straits [4]. It is important to note that the terms the "NSR" and the "Northeast Passage" are often used interchangeably; still from the legal standpoint, the boundaries of the NSR are delineated in Figure 4.2.

The NSR lies in Arctic waters within Russia's EEZ. EEZ is an area beyond and adjacent to the territorial sea: it can extend to a maximum of 200 nautical miles from the baselines (see Figure 4.3). A coastal state enjoys sovereign rights over its natural resources, within the EEZ. Rights and freedoms of other states are governed by the relevant provisions contained under Chapter V of the United Nations Convention on the Law of the Sea (UNCLOS) (art. 55–75) [5,6].

Out of the three Arctic routes (see Figure 4.1), the NSR is the most widely used and most operationally viable at the moment. The usage of the NSR results in distance reduction and 10–14 days faster journey times, i.e., shipping from Japan to Europe via NSR results in 40% distance saving, from South Korea to Europe is 30% shorter and from China 20% [7].

The NSR is considered one of the feasible shipping routes yet to be fully commercialized. It may bring positive economic effects to come as savings in shipping distance and time [8], reductions in emissions [9] and resulting in energy efficiency [10]. Time-saving effects are linked to the duration of ice-free conditions. Arctic Marine Shipping Assessment report by the Arctic Council forecasted that the NSR would be open 90–100 days by 2080 [11]. Currently, all year shipping is possible only in the west segment of the NSR. Moderately, ice-strengthened vessels are likely to use Arctic transits for 10–12 months by the late century [12].

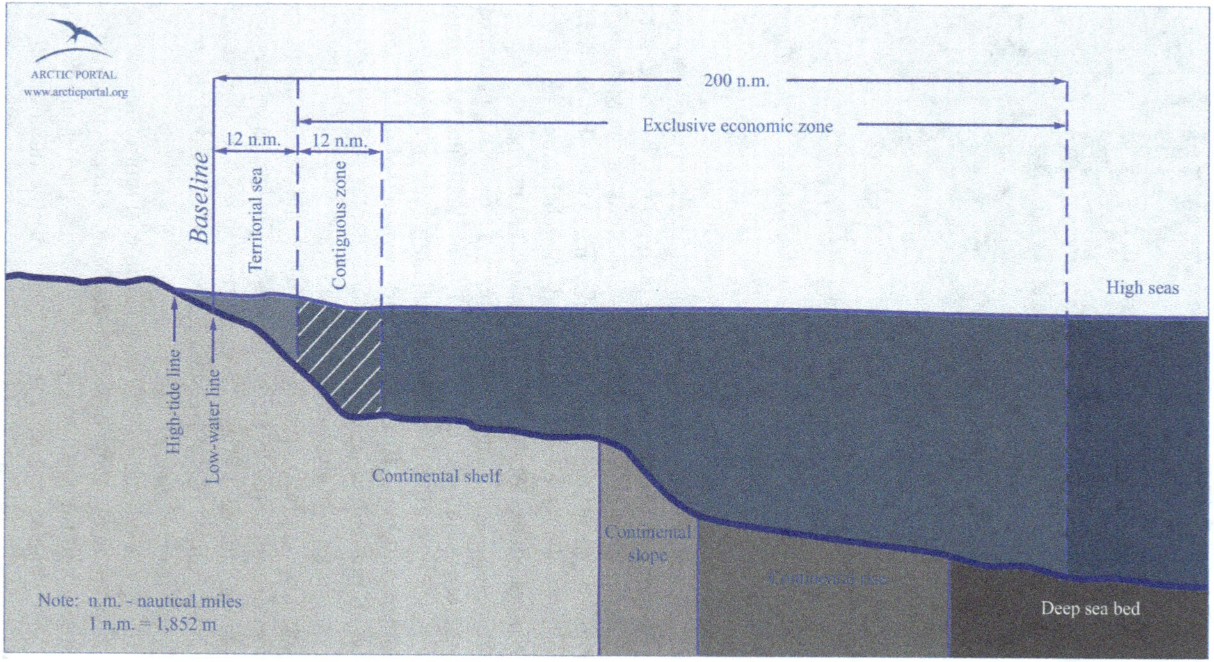

Figure 4.3 Exclusive economic zone. Source: Arctic Portal

Russia's Arctic coastline stretches more than 14,000 km. The NSR lies between the Kara Gate, at the western entry of the Novaya Zemlya straits, and the Provideniya Bay, at the southern opening of the Bering Strait, for a total length of 5,600 km. The Barents Sea does not fall under the NSR's legal regime [3] but is often included as a part of the NSR (see Figure 4.4). The NSR represents a shipping route that has multiple lanes such as coastal and mid-route that cross through waters of varying status: internal, territorial and adjacent waters [13]. According to the Russian position, the NSR is a historically existing national unified transport route of the Russian Federation in the Arctic and therefore is considered to be under its exclusive jurisdiction [13]. However, some countries contest Russian exclusive control over the NSR, which is later discussed in the section focusing on legal analysis of the NSR.

The route consists of five individual seas: the Kara, the Laptev, the East Siberian, the Chukchi and the Barents Seas. The route spans through three archipelagoes: the Novaya Zemlya, the Severnaya Zemlya and the East Siberian Islands [14]. Notably, all the five seas lie in the continental shelf zone with an average depth of just above 200 m (see Table 4.1). Additionally, bathymetry in some of the straights, e.g., Dmitry Laptev Strait, with the depth of 8–9 m may create challenges for safe passage of the ships [15]. The northern sea route (NRS) is linked to the largest rivers in Siberia and is part of a larger Russian internal shipping network. Ships may operate in the western NSR for substantially longer seasons, including year-round navigation when using appropriate ice-class vessels, compared with more limited accessibility remaining in the eastern part of the NSR (Laptev, East Siberian and Chukchi seas) [16]. Ports along the NSR are marked as circles (see Figure 4.4). In the west segment, the port of Murmansk is the only ice-free deep port with dedicated shipyards and maritime services [17]. Many ports are considered shallow and either do not have facilities, or sufficient wharf space [18]. Future shipping along the NSR would require substantial investment into port infrastructure that has been already undertaken in the port of Sabetta as a result of the Yamal-LNG project. Additional port facilities and developments are discussed further in this chapter.

4.1.1 Role of NSR in a historical context

4.1.1.1 Explorations in sixteenth to nineteenth centuries

The knowledge of the Arctic was still imprecise in the second half of the sixteenth century. The Dutch and English sent expeditions sailing eastward along the coast of Russia, culminating in expeditions (1594, 1595 and 1596) by the Dutch explorer William Barents reaching Novaya Zemlya. Since then, Russia took a leading part in the exploration of the Arctic lands and sea routes. In the seventeenth century, further advances along the coast of Russia were made by the Russian explorers, e.g., in 1648 Russian seafarer Semyon Dezhnev sailed from Kolyma to the Pacific Ocean and discovered the Bering strait 80 years ahead of the Bering expedition [19].

During the Great Northern expedition (1733–43), Danish cartographer and explorer in Russian service Vitus Bering mapped almost the entire coast of the Russian Arctic and discovered Alaska. Regular trade shipping along the NSR began

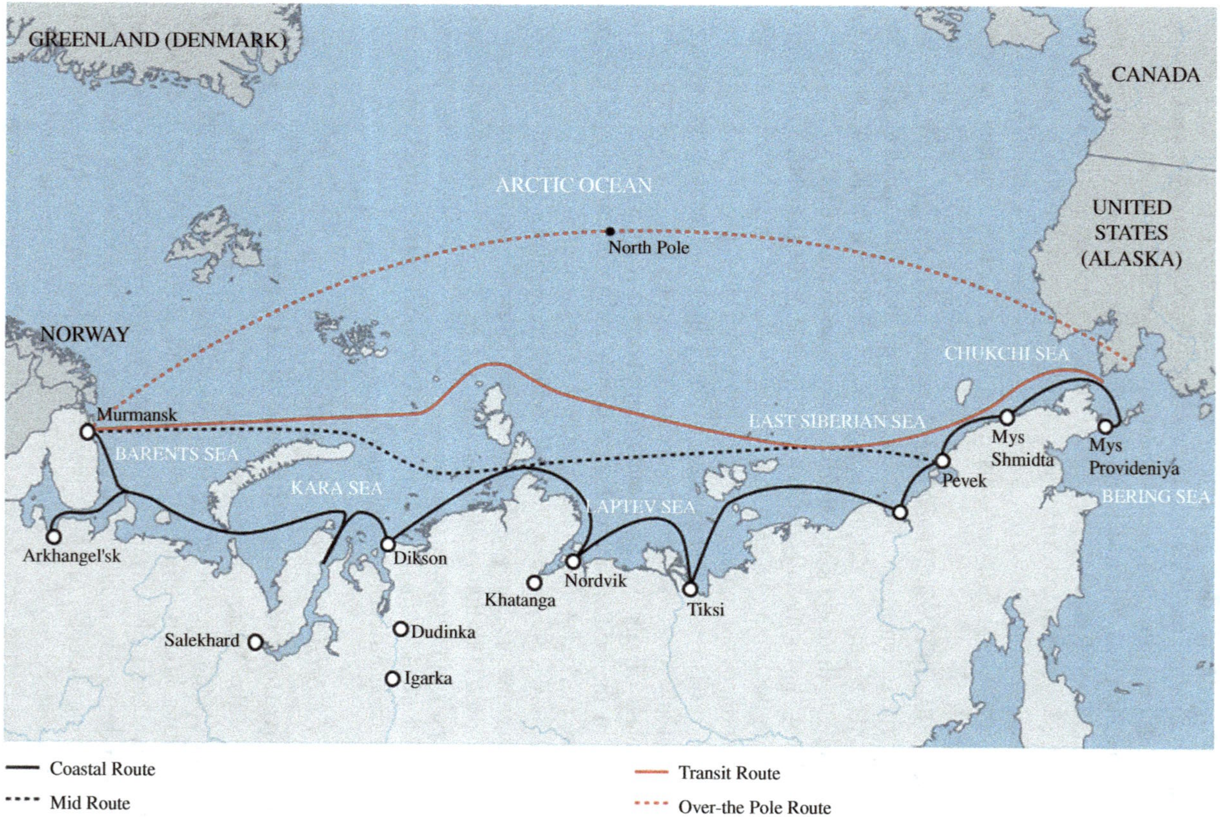

Figure 4.4 Map of the Northern Sea Route [13]

Table 4.1 Areal extent and average depth of seas along the Northern Sea Route

Sea	Areal extent in sq. km	Average depth in meters
Kara Sea	900,000	90
Laptev Sea	650,000	578
East Siberian Sea	900,000	58
Chukchi Sea	600,000	200
Barents Sea	1,400,400	200

Source: Adapted from [14].

in the nineteenth century, exporting Siberian mineral resources and importing industrial goods; during 1876–1919, a total of 86 successful sailings shipped a total of 55.2 tons of goods [19]. Already then, the NSR proved to require finance and special arrangements for infrastructure, including the need for technical capabilities, icebreaker fleet, ice-strengthened ships, supplying radio and meteorological stations along the route [19].

4.1.1.2 Soviet period 1919–91

The Soviet period brought a new wave in the Arctic exploration of natural resources and resulted in the growth of extractive industries in the Arctic. Several parts of the NSR served economic development and provided means to ship supplies to and from remote parts of the country. The first full NSR trip within one navigational season from Arkhangelsk to Petropavlovsk-Kamchatsky was completed in 1932 [20]. During the same year, the NSR Administration (NSRA) was established. Development of the ports, infrastructure and settlements along the NSR corresponded with the exploration of mineral resources, e.g., apatite and timber via Murmansk port, transporting gold and timber from the Kolyma region via Pevek port, transporting coal from Dixon, Tiksi, Khatanga and Providenie ports.

The NSR played an important strategic role during WWII in providing supplies and serving military purposes. The Soviet period can be characterized by investments into port infrastructure, connecting it to the railway networks, the building of icebreaker fleet, aviation bases and airports along the coast of the NSR, which was rather advanced already in the 1980s to serve sea-borne traffic along NSR [20]. Due to military and political reasons, the NSR was factually closed to foreign vessels during the Soviet period [7].

4.1.1.3 Post-Soviet period and current state

The Soviet Union collapse and economic crisis of the 1990s greatly affected shipping along the NSR with sometimes just a few ships a year. The usage of the NSR took up starting from 2006 with cargo shipping reaching 2 million tons and continued to grow since. Figure 4.5 summarizes the NSR development in terms of cargo volume shipped. In 2016, 7.5 million tons of cargo were shipped that exceeded the 1987 record of 6.6 million tons. The growth had been very fast in

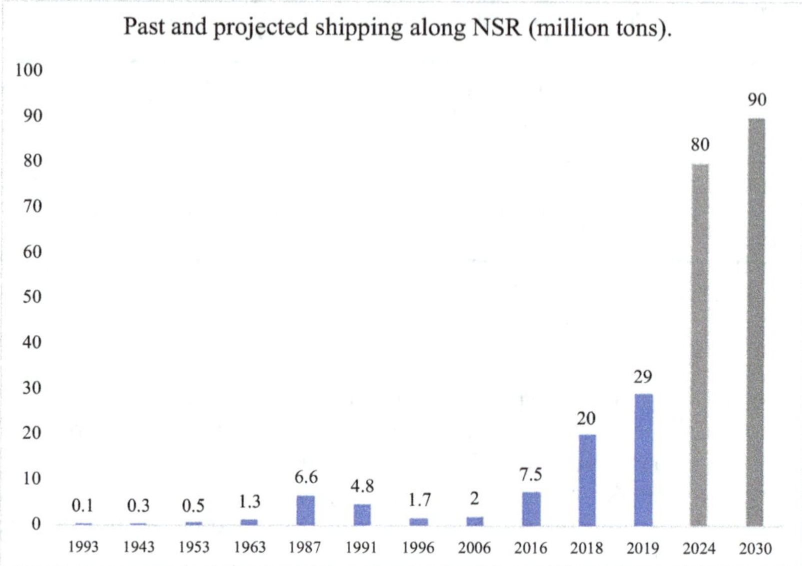

Figure 4.5 Past and projected shipping along NSR, 1933–2030, million tons. Source: Compiled by author CHNL, Rosatom

2016–19, reaching 289% more cargo shipped in 2019 as compared to 2016. Russian Government expects the volume rising fourfold compared to 2016 and reaching 80 million tons in 2024 and 90 million tons in 2030.

The growth in the NRS usage during 2016–19 is due to the Arctic development projects, such as shipping of 7.7 million tons of oil via Arctic Gates oil-loading terminal in 2019 [21] and shipping of 20.5 million tons of LNG gas from the Yamal LNG project since 2017 [22]. New projects aimed at the exploration of hydrocarbons in Russia will result in an extra 50 million tons shipping. Russia is investing in its icebreaker capabilities adding three new icebreakers by 2022 to four existing ones. Furthermore, two new Lieder-type icebreakers are expected to be built by 2027 [23].

4.1.2 Effect of climate change on NSR

4.1.2.1 Diminishing sea ice

Climate change affects the Arctic to a larger extent, with the Arctic warming twice as rapidly compared to the rest of the planet. Figure 4.6 illustrates the Arctic in September when there is the least sea ice before water starts freezing again. Satellite records show that the Arctic lost 13% of its ice each decade since 1979 [24]. The latest study predicts the September sea ice disappearance in the Arctic already in 2044–67 [24]. The decline is coincident with the abrupt global and Arctic warming over the last 30 years [25].

Figure 4.6 Sea ice extent in 2019 in September compared to the corresponding average between 1981 and 2010. Source: UCLA

With the loss of ice, the Arctic Ocean becomes accessible for oil and natural gas exploration and the development of new transport routes and Arctic tourism. The economic viability of the NSR development is taking into consideration the melting of the Arctic sea ice. At the same time, current models poorly represent spatial patterns of sea ice thickness [26]. In order to develop economic activities along the NRS, there is a need for improved predictions of sea ice thickness, sea ice retreat dates and open water duration.

4.1.2.2 Navigation along NSR

Despite ice loss, navigation along the NSR remains challenging (Table 4.2). A report by the American Bureau of Shipping lists major risks that need to be considered when using NSR [15]. Enormous variability is inherent in the prevailing conditions of the NSR, coupled with limitations of prediction models. During the period in which the route is open, vessels can expect to encounter moderate and strong winds, low air temperatures, a high number of days with fog and, late in the season, long polar nights, snowstorms and possible blizzards, air temperature, visibility.

Sufficient preparation of ship crews to be able to operate in icy waters has been identified as very important. Due to efficient organization and control measures, an estimated possibility to get heavy water leaking ice damage of the ship corpus does not exceed 2% from the number of vessels operating on the NSR, which corresponds to the probability of collisions in coastal waters of the World Ocean [27].

Table 4.2 Navigational risks along NSR

Risk	Manifestation
Air temperature	Summer temperatures remaining close to 0 °C
Visibility	Frequent fogs during the summer months, blowing snow, optical haze
Wind	Blizzards may be encountered early (June) and late (October)
Sea level variations	Pronounced seasonal level variations
Waves	Wave heights of 4–5 m in early autumn (September and October)
Ice	Ice may be encountered on the NSR at any time. In years with heavy ice conditions, Arctic seas are almost completely covered with drifting ice throughout the summer

Source: Adapted from ABS Report.

4.1.3 Frameworks governing the NSR

Maritime transportation along the NRS depends on many factors, including geopolitical forces, environmental conditions, commodity prices and legal regulation. The main frameworks governing the NSR are summarized later. The UNCLOS governs the human use of the sea, territorial claims, economic rights and broad environmental responsibilities [28]. International Maritime Organization (IMO) regulates commercial shipping by setting regulations on safety, security and environmental concerns. IMO developed the Polar Code that came into force in 2017. The Polar Code covers the full range of design, construction, equipment, operational, training, search and rescue (SAR) and environmental protection matters relevant to ships operating in the waters surrounding the two poles [29].

The Arctic Council formed in 1996 represents an intergovernmental forum for Arctic states, indigenous peoples and non-Arctic observer states. Concerning the NSR, Arctic-Council-produced Arctic Maritime Shipping Assessment [11] focused on the shipping in the Arctic Ocean, its potential impacts on humans and the Arctic marine environment and the Arctic shipping infrastructure requirements. Furthermore, the Arctic Council facilitated two binding treaties: Agreement on Cooperation on Aeronautical and Maritime Search and Rescue in the Arctic (2011) [30] and Agreement on Cooperation on Marine Oil Pollution Preparedness and Response (2013) [31]. On the national level, Russian Federation sets as its priority development of the NSR and required infrastructure in its Arctic Strategy until 2020 [32] and in the plan for infrastructure development of the NSR for the period 2020–35 [33]. The Russian Federation regulates the legal status of the NSR by law on the NSR [3]. Additionally, there are other Federal laws and rules of navigations along the NSR that apply to the NSR are out of the scope of this analysis.

Major frameworks governing the Northern Sea Route (NSR)

The United Nations Convention on the Law of the Sea:

Article 234

International Maritime Organization:

Polar Code (2017)

Arctic Council:

Arctic Maritime Shipping Assessment (2009)
Cooperation on Aeronautical and Maritime Search and Rescue in the Arctic (2011)
Agreement on Cooperation on Marine Oil Pollution Preparedness and Response (2013)

National Laws and Regulations:

Arctic Strategy of the Russian Federation until 2020
Law on the NSR (2012)
Plan for infrastructure development of the NSR for the period 2020–35 (2020)

4.1.3.1 Non-Arctic states in the Arctic

Non-Arctic countries emerge as Arctic stakeholders by producing their own national Arctic policies and strategies. Currently, Germany, France, Spain, the United Kingdom, Scotland within the United Kingdom, the Netherlands, Switzerland as well as Japan, South Korea and China have produced strategic or policy papers on the Arctic that among other issues address maritime transportation and opening of the Arctic Ocean [34].

Russia has close cooperation with China on its Arctic energy development projects and the development of the NSR infrastructure, whereby China provides capital investments needed [35]. China exhibits interest in the Arctic due to the internal growing energy demands and recurrent problems with its maritime transport in the Indian Ocean [36].

4.1.3.2 Legal status of the NSR

Russia defines the NSR as a historically existing national unified transport route of the Russian Federation in the Arctic and therefore considers it to be under its exclusive jurisdiction [13]. At the beginning of 2013, the law on the NSR entered into force, for the first time establishing the boundaries of its water area [3].

Article 234 of UNCLOS provides coastal States with special rights to adopt and enforce nondiscriminatory laws and regulations to prevent, reduce and control marine pollution in ice-covered areas with particularly severe climatic conditions within their EEZ [37]. Article 234 provides Russia and Canada legal grounds for establishing control over shipping in the NSR and Northwest Passage, respectively. Canada included Northwest Passage into its internal waters in 1985 establishing a notification process for crossing the passage [38]. The United States contests Russia's and Canada's position regarding Article 234 due to restrictions for other states to exercise their right to the freedom of navigation in the Arctic [39].

In 2013, Russia established rules of navigation in the water area of the NSR which is an authorization-based procedure for passing through the NSR by domestic and foreign vessels, including portions of the Russian EEZ that are part of the NSR water area [40]. The permit–granting process is open to ships of all flags [41]. The authority to organize the navigation of vessels in the waters of the NSR is vested in the administration of the NSR, which was formed in March 2013 by Order of the Government of the Russian Federation. The NRS does not have a predetermined single channel for ships to follow, rather each time the route is chosen on the basis of current conditions. Ice conditions are the determining factors that affect the chosen route on every occasion, passing along ice-covered waters from 2,200 to 2,900 nautical miles [14]. The functions of the NSRA Office include [42]

- receiving applications;
- examining these applications and issuing permits for sailing in the NSR;
- monitoring ice and navigation conditions;
- coordinating the installation of navigation equipment in areas of hydrographic work, providing ships with information services;
- search and rescue (SAR) operations and
- issuing certificates for ice pilotage of vessels.

In practice, the admittance to the NSR presents a flexible system where a permit is granted depending on ice conditions, vessel ice-class and intended passage in the NSR area. The application with all required documents has to be sent to NSRA via e-mail not earlier than 120 and no later than 15 calendar days of the intended date of entry into the NSR. Icebreaker assistance can be either mandatory or optional depending on ice conditions. The tariff of icebreaker assistance depends on the following factors: number of navigational zones along the NSR for which icebreaker assistance is needed, number of icebreakers needed, the season of navigation and vessel's ice-class and gross tonnage [43]. The latest addition to the management of the NSR is the formation of the NSR Directorate that will oversee implementation of the state policy and development strategy for the NSR, interaction with regions and authorities, management of development projects, including infrastructure, as well as monitoring the achievement of state targets. In addition, the Directorate will be in charge of managing the nuclear icebreaker fleet [44].

Introduced in 2017, the Polar Code developed by the IMO stipulates a series of additional requirements for vessels and crews performing Arctic navigation in order to increase the safety of shipping in Polar waters [29]. The Polar Code does not affect the special rights of the Arctic States to unilaterally establish national standards for vessel activity in their EEZs in the Arctic Ocean [40].

4.2 Drivers for NSR development

4.2.1 Increased demand for natural resources and Arctic development

The Arctic region is home to vast natural and mineral resources that make this area highly attractive on a global scale. According to the US Geological Survey (USGS), it is estimated that the Arctic contains up to 30% of the world's undiscovered gas and 13% of the world's undiscovered oil resources [45]. The largest deposits of gas are found in the Russian Arctic offshore territory (see Figure 4.7). Growing demand for natural resources in China and other Asian countries serves as an accelerator for the Arctic development and the use of NSR in its turn. Russia, the largest grain exporter, announced that it is planning to use the NSR to ship the grain that will be produced in the south Siberian regions to the north on the rivers of Irtysh and Ob to the seaport of Sabetta in Yamal and later delivered to the Asian markets [46].

The NRS is the most convenient route for transporting vast amounts of natural resources, such as oil and gas and mineral resources in the Russian Arctic [41]. The growth of the Arctic resources' exploration contributes to increased destination bulk cargo transportation on NSR. According to the Ministry of Natural Resources and Environment of the Russian Federation estimates, the volume of exported oil might amount to 21 million tons, gas condensate up to 2.5 million tons, coal up to 23 million tons and the volume of exported timber and cargo supplies up to 5 million tons [47]. Adding the growth in LNG the scale of cargo transportation along NSR is likely to exceed 80 million tons by 2024 [47]. Russia is preparing an initiative to create the Arctic cluster, which can provide by 2030 production of

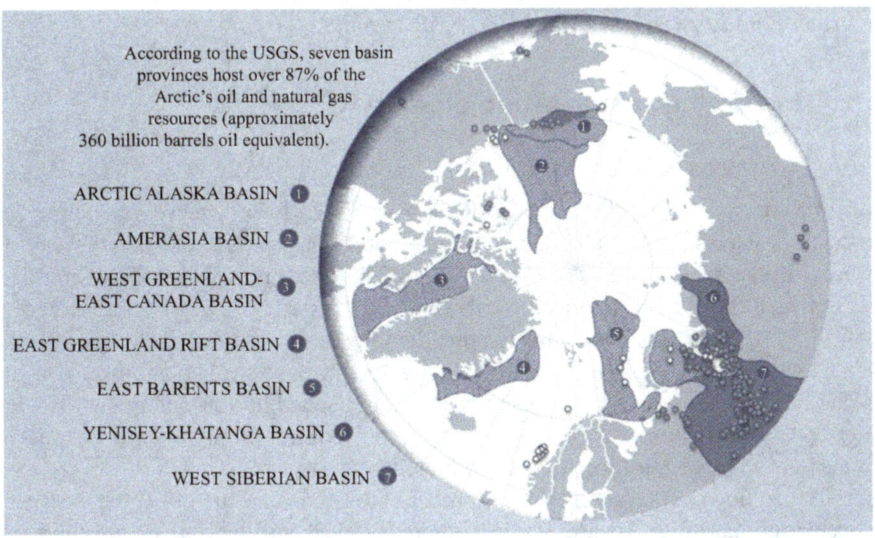

Figure 4.7 Arctic oil and natural gas resources. Source: Visualcapitalist

100 million tons of oil per year [48]. China is the largest market for Russian hydrocarbons, in 2014 Russia and China signed a $400 billion deal on the supply of 38 billion cubic meters of natural gas for 30 years [49].

4.2.2 Tourism in the Arctic

Extreme tourism is growing in the world with tourists buying trips to the Arctic and Antarctic. The main Arctic cruise tourism markets are Greenland and Svalbard [50]. During the season 2019, Svalbard was visited by cruise ships 300 times, with nearly 20,000 tourists. Association of Arctic Expedition Cruise Operators (AECO) was founded in 2003 and has since become an important organization representing the concerns and views of AECO with 29 full members in 2019, comprising in total 60 cruise ships and 10 yachts [51]. AECO developed guidelines on sustainable tourism in the Arctic [52].

In the Russian Arctic, Franz Josef Land archipelago was one of the most visited cruises operator destinations with over 1,000 tourist visits in 2018 [53]. While some cruise operators had offered cruises along the Northwest Passage, tourist cruises were not available on the NSR. The first opportunity for tourists to explore the NSR would have been in Summer 2020 when two companies were to launch NSR cruises offer. Silversea Cruises is a luxury cruise line with its headquarters in Monaco offering a 25 days cruise along the NSR on a luxury ice-strengthened ship beginning in Anadyr (Russia) or Nome (Alaska) [54]. The Russian company would have offered cruises on an ice-class vessel "Akademik Shokalskiy" starting in Anadyr or Murmansk [55]. However, the start of the cruising on NSR was delayed until 2021 due to COVID-19 outbreak. It has been estimated that at least 32 polar expedition-style cruise ships with high Polar Class are currently being built, most are targeted at an ultra-luxury experience [56]. Hence, tourism is expected to affect the growth in the Arctic traffic, on the NSR included.

4.3 Maritime transportation along NRS

4.3.1 Cost considerations

Maritime transportation along the NSR can significantly reduce voyage time, but the carriers need to take into consideration additional costs (cost of icebreaking services, pilotage, additional insurance, etc.). The cost breakdown of the comparison between the Suez Canal and the NSR for a single leg voyage is presented in Table 4.3 [10]. The cost structure is considerably different when compared to the Suez Canal. In the example (used for illustrative purposes only), shipping via the NSR is fuel efficient due to decreased mileage and speed limitations in ice-covered waters. At the same time, passage costs that require pilotage and icebreaker services are higher. Additional costs of added insurance premium need to be considered. It is, however, argued that with more ships choosing the NSR, the expenses for icebreaker cargo escort are likely to decrease [57]. Still, variances in speed and fuel efficiency due to uncertainty in ice and weather conditions prevail on the NSR.

Table 4.3 Cost structure comparison in % for single-leg voyage via Suez and NSR

Cost type	Via Suez (%)	Via NSR (%)
Vessel charter rate	35	43
Fuel cost for propulsion	54	28
Passage cost	11	22
Added insurance premium	–	7
Total costs	100	100

Source: Adapted from [10] (compiled by author).

The average speed recorded in 2010–16 equaled 10 knots but can be reduced to 5–6 knots depending on ice and weather conditions [58]. The extensive literature review identified capital costs and icebreaker fees as the primary cost factors on the NSR [58]. Furthermore, studies report increased crew costs on the NSR in the range from 10% to 28% [58,59].

Most insurance firms refer to the unified International Association of Classification Societies (IACS) requirements concerning Polar Class (2016) [60], a document, established by the IACS and drawing from the recommendation of the IMO's Polar Code guidelines for ships operating in Arctic ice-covered waters [61]. Evidence suggests that insurance firms consider Baltic class 1A to be a minimum for shipping along the NSR, but often insurance companies would evaluate each voyage individually [58], and for commercial reasons, insurance firms do not disclose their policies and tariffs [61]. Several studies indicate that sailing season time affects the cost of using NSR. Furthermore, capital cost premium for ice-class vessels ranges from 20% to 30% and crew costs are assumed on average 15% higher [61].

Due to the variabilities of methods, models and assumptions, the studies on the cost-efficiency of the NSR produce conflicting results. In general, Arctic routes are considered to be competitive for single-leg voyages, and less cost-competitive for year-round operations due to icy conditions and capital cost requirements. However, studies indicate the cost efficiency of the NSR for year-round operation in the long term [58].

4.3.2 Ship classes

The Polar Code [29] defines operational limitations in the polar waters that are affected by ice conditions, temperature and latitude. Ice correspondences are not easy to compile as they depend on ship size, hull form and structural organization. Several attempts have been made to establish correspondences between 13 largest marine classification societies in the world (e.g., Det Norske Veritas, Lloyd's Register) that are the members of the IACS [60]. The Polar Code uses ship categories as a marker in the code to identify the application of certain requirements based on anticipated hazards encountered for the expected operating environment. Three ship categories A, B, and C are defined (see Table 4.4) [29,60,62,63].

Table 4.4 Correspondence of Polar Code ship categories vs. IACS requirements concerning Polar Class and the Finnish–Swedish Ice Class Rules (FSICR) ice-classes

Polar code category	IACS/FSICR ice-class correspondence	Operating capability	Ice thickness for machinery strength design (m)
Category A: a ship designed to operate in polar waters in at least medium first-year ice, which may include old ice inclusions	PC1	Year-round in all polar waters	4.0
	PC2	Year-round in moderate multiyear ice	3.5
	PC3	Year-round in second-year ice, which may include multiyear inclusions	3.0
	PC4	Year-round operation in thick first-year ice, which may include old ice inclusions	2.5
	PC5	Year-round operation in medium first-year ice, which may include old ice inclusions	2.0
Category B: a ship designed to operate in polar waters in at least thin first-year ice, which may include old ice inclusions	PC6	Summer/autumn operation in medium first-year ice, which may include old ice inclusions	1.75
	PC7	Summer/autumn operation in thin first-year ice, which may include old ice inclusions	1.5
Category C: a ship designed to operate in open water or ice conditions less severe than those included in Categories A and B	1AS	Capable of navigating in difficult ice conditions, without the assistance of icebreakers	1.0
	1A	Capable of navigating in difficult ice conditions, with the assistance of icebreakers	0.8
	1B	Capable of navigating in moderate ice conditions, with the assistance of icebreakers when necessary	0.6
	1C	Capable of navigating in light ice conditions, with the assistance of icebreakers when necessary	0.4

Source: Polar Code, IACS Requirements Concerning Polar Class, DNV GL, The Finnish–Swedish Ice Class Rules (FSICR) (compiled by author).

4.3.2.1 Seasonal variations

Annual navigation on the NSR has two main seasons: summer season (typically July–November) and extended season of the rest of the year [15]. Ice thinness and concentration directly affect the navigability of the ships along the NSR. Duration

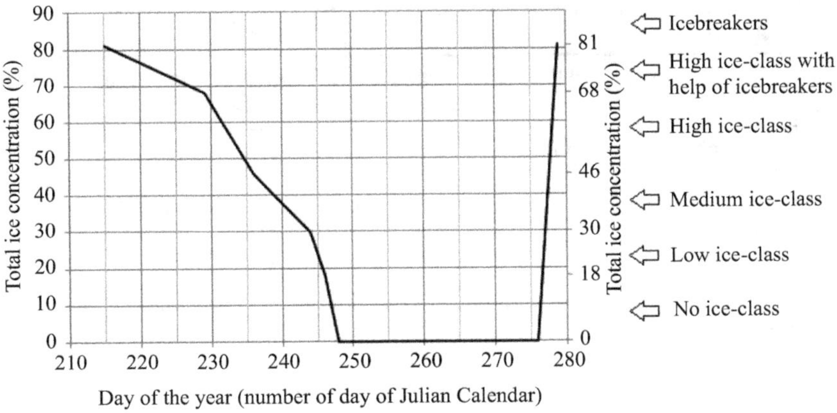

Figure 4.8 Ice thickness and time frame of navigation season for vessels with specific ice-breaking capabilities (ice-classes) [64]

of navigation season corresponds to approximately 28 days for a ship without ice-class, while for vessels of high class sailing on the NSR with icebreaker support is estimated twice as along. Ice growth from 0% to 81% can occur rapidly within 3 days, meaning that all commercial vessels, including high class, should leave that part of the NSR (see Figure 4.8) [64].

4.3.2.2 Ship sizes

Logistics operators consider different ship sizes and ice-classes in their calculations in routes viability. There are two potential restrictions for vessels using the NSR: a draft of 12.5 m due to shallow water between the Russian Arctic islands and a width restriction of 30.0 m, which is the width of the current icebreakers and the consequent ice corridor width provided by them [65]. The thinning of the ice has opened a deep-draft route along the NSR, hence the restrictions can be relaxed to some degree [4].

Size restrictions, extra costs of constructing ice-class vessels and icebreakers are considered as the major obstacles for economically successful exploitation of the NSR for transits [65]. The larger icebreakers are planned to be constructed to change the width limitation. An icebreaker with a projected maximum ice-breaking capacity of 4 m will be able to create a channel with a width of more than 50 m. For the remaining icebreakers being designed or under construction, the maximum hull width does not exceed 32 m [66].

4.3.3 Volumes and types of shipping

Shipping along the NSR is dominated by cargo shipped from the Russian Arctic to domestic and international markets. The NRS is the first choice for shipping between Russian Arctic ports; it is extensively used for shipping natural resources from the Arctic and resupplying local cities and ports [41]. While the growth in total cargo shipped grew nearly 10-fold, the volumes of transit cargo remain a small fraction of all cargo averaging just at 2% during 2016–19. Out of 29 transits in

Table 4.5 Volumes and transits on NSR

Volume and type of shipping	2010	2011	2013	2014	2015	2016	2017	2018	2019
Total cargo shipped (in 1,000 t)	n/a	3,111	3,930	3,982	5,392	7,479	10,729	20,180	30,280
Transit cargo (in 1,000 t)	111	821	1,356	274	39	214	194	491	697
Number of transit vessels	13	38	71	31	18	19	27	27	29

Source: CHNL, [4,50].

Figure 4.9 Shipping lanes of vessels on the Northern Sea Route in 2018 mapped from satellite Automatic Identification System (AIS) data. Source: Business Index North Report

2019, 16 voyages were made between Russian ports (see Table 4.5) [4,50,67]. The most popular destination ports were Murmansk and Arkhangelsk. Ships departed most often from Petropavlovsk-Kamchatsky and Arkhangelsk. The small number of transits indicate that so far NSR is not considered as an alternative to Suez Canal that had 18,880 ship transits carrying 1,031 million tons in 2019 [68].

Most of the shipping activity took place during July–November. Figure 4.9 demonstrates that the most year-round operated segment of the NSR is within the Kara Sea, where shipping is driven by LNG projects in Sabetta and the Arctic Gate Terminal [67,69].

4.3.4 Container shipping

Container shipping relies on a "just-in-time" system, including precise scheduling for efficient loading, shipping and unloading for minimizing push costs [70]. The NSR can be unpredictable due to seasonal variations in ice conditions and local ice drift [28]. Hence, due to limitations in seasonal use, the NRS is not currently used for container shipping that relies on schedule reliability. Furthermore, ice conditions remain challenging for containerships [71].

In summer 2018, the biggest logistics company in the world Maersk performed a first trial commercial container shipping along the NSR. A brand new 1AS ice-class vessel "Venta Maersk" (width: 36 m, length: 200 m. TEU: 3,600) was used for the shipping. The sailing took 37 days from Vladivostok to Saint Petersburg [72]. The twenty-foot equivalent unit (TEU) used to measure a containership's cargo carrying capacity and port throughput capacity. The dimensions of one TEU are equal to that of a standard 20′ shipping container (20×8×8 ft) [73].

An example of Venta Maersk proved that container shipping is possible along the NSR. The trial demonstrated that even a class 1AS vessel required icebreaker assistance on the route [72]. The crew was prepared for sailing in polar waters according to the Polar Code. Maersk announced that the use of the NSR for shipping is not economically viable [72]. However, in 2019, Maersk indicated that together with Atomflot it explores the possibility of offering transport of goods from East to West Russia [74]. Still, the main factors limiting the use of container shipping on the NSR are cost considerations, seasonality, type and size of the vessels optimal for use on the NSR. The average container vessel size on the Asia–Europe trade in 2020 is estimated to be around 16,730 TEU [75], while on the NSR restrictions for up to 4,000 TEU may be applied [76]. The situation is likely to change with the uptake of a new class of icebreakers and the usage of a deep-draft route along the NSR [4,76].

4.3.5 A case study of liquefied natural gas (LNG)

Yamal LNG is an integrated project for natural gas production, liquefaction and marketing launched in 2013. The project is owned by Novatek (Russia; 50.1% shares), Total (France; 20%), CNPC (China; 20%) and the Silk Road Fund (China; 9.9%). The project infrastructure includes the LNG plant and export terminal at the port of Sabetta (Yamalo-Nenets Autonomous District) [77]. The NSR serves as the primary route for the transport of LNG to Asia which takes 15 days compared to 30 days via the Suez Canal. The project initiated the construction of port infrastructure on Yamal and the order of ice-class LNG carriers. A new model of LNG icebreaker tanker with technical characteristics to sail in the ice of up to 2.1 m thick without the assistance of ice breakers was designed for Yamal LNG. Yamalmax is the type of icebreaking LNG carriers designed specifically for the Yamal LNG project. The vessels are based on the Double Acting Ship concept and capable of operating independently without icebreaker escort along the NSR [78]. Altogether 15 LNG carriers were constructed and commissioned in 2019.

The development of Yamal LNG was a facilitating factor for the construction of the fleet of new Russian nuclear icebreakers. Novatek has far-reaching plans for expansion of its operations; by 2025, the output of Novatek's LNG projects is set to reach a level of 40 m, and by 2030 may be as high as 70 m [79]. The company plans to create a cluster in the Yamal–Gydan region to produce as much as 140 m by 2035 [79]. Russia is becoming a major player in the LNG market, and the role of the NSR is expected to grow. The expansion of LNG development projects will bring increasing volumes shipped along the NSR and would require additional investments in the port infrastructure and the fleet of vessels operating along the NSR.

4.3.6 Shipping sustainability requirements worldwide and in the Arctic

The international shipping industry accounts for 3% of global emissions that is under pressure to transform to become more sustainable and environment friendly [80]. In 2020, IMO Sulphur cap entered force setting the limit for the sulfur content of ships' fuel oil to 0.50% m/m (from 3.5% m/m) [81]. Compliance can be achieved by switching to using ships powered by cleaner fuels or by installing "scrubbers," which "clean" the emissions before they are released into the atmosphere [81].

IMO Polar Code that is operational since 2017 sets certain requirements for environmental protection, especially regarding fuel used, sewage, garbage, chemicals and protection against invasive species. The Polar Code prohibits discharge into the sea of oil or oily mixtures from any ship [29]. The usage of heavy fuel oil (HFO) is prohibited in Antarctica under MARPOL but is allowed in the Arctic; however, ships are encouraged not to use or carry HFO in the Arctic [82].

Nongovernmental organizations and indigenous communities campaign for the ban of HFO use in the Arctic. During the IMO Arctic summit 2020, the draft of HFO regulation was agreed upon [83]. The draft proposes that there will be no change in the use and carriage of HFO in the Arctic before the middle of 2024 when the regulation takes effect. However, the Arctic countries receive the permit for the continued use of HFO by their vessels by 2029 [83]. Apart from legislative reform, some players voluntarily impose bans on the use of HFO in the Arctic. For instance, in 2019, Association of AECO announced a ban on the use and carriage of HFO in the Arctic for all its members [84].

Other recent initiatives include the Agreement to Prevent Unregulated High Seas Fisheries in the Central Arctic Ocean that was signed in 2018 by the EU and nine countries with potential interest in sustainable fisheries [85]. The viability of the Arctic shipping needs to be accessed in the light of environmental concerns, e.g., oil and chemical spills, emissions, chemical and noise pollution [86]. Following environmental concerns, Ocean Conservancy introduced Arctic Shipping Corporate Pledge in 2019 to avoid Arctic transhipment routes and promote precautionary Arctic shipping practices [87]. The Arctic Shipping Corporate Pledge is supported by fashion-related companies and some big ocean carriers. However, fashion companies relying on the container shipment were not using the

NSR in the first place, and carrier companies' signees do not have an ice-strengthened fleet to operate in the Arctic waters [88]. It is believed that the development of the shipping in the Arctic waters will be following sustainability requirements worldwide with potential new regulations being designed specifically for shipping in the Arctic waters that take into consideration the vulnerability of the Arctic ecosystems.

4.3.7 Smart shipping along NSR

The advent of digitalization, smart censoring and the usage of big data creates preconditions for smart shipping. Smart shipping that is often understood as autonomous shipping is a much broader concept. Smart shipping includes the following [89, p. 1–2]:

- Smart vessel
  These are the vessels that are highly automated and are therefore equipped with automated systems using (external) data to optimize the key functions of the vessel.
- Smart traffic management and infrastructure
  Management is based on real-time data.
- Smart travel and transport
  The interaction between ships and third logistic parties.
- Smart regulation and facilitation
  Interaction between the ship and third (government) parties for regulation or inspection.

  The Russian Federation is investing in the autonomous management of commercial vessels of marine fleets along the NSR. The Ministry of Industry and Trade of Russian Federation posted on the procurement website an application for the development of a unified technological platform for crewless management of commercial vessels of marine fleets. In the proposal, 311 million rubles were allocated for the development of a unified platform for the use of artificial intelligence in navigation [90]. The results are expected to be used in the modernization of commercial vessels starting from 2020. The contractor will be required to create a technical platform for the following three types of vessels [90]:

- shuttle tankers, a series of Arctic tankers for operation along the NSR;
- self-propelled cargo vessels; and
- self-propelled dry-cargo vessel of the sea and mixed "river/sea" class navigation.

Autonomous shipping technologies are developed by research centers and companies included in the industry center "Marinet" of the National Technological Initiative that receives allocated state funding. As part of the National Technology Initiative, in 2015, a road map for autonomous maritime transport system was developed. Furthermore, the work is currently underway in Russia to equip existing vessels with artificial intelligence. Marinet intended to test crewless technologies in practice in the Arctic, Baltic, Black and Azov Seas. The total budget of the project "Creating a platform for the development of autonomous navigation technologies using computer simulation in a virtual environment" is 439 million rubles of state support contributing to 60% of the project budget [91]. Additionally, the Rosatom state corporation is in the process of developing its digital model of an autonomous vessel, which will be designed to operate on the NSR [92]. The plan for the NSR development until 2035 addresses the creation of digital systems for logistics along the NSR.

4.3.8 Challenges and recommendations for sustainable shipping along NSR

Analysis of the NSR demonstrates that the NSR plays an important role in the Russian internal and international shipping, especially due to LNG transport from the Yamal LNG project. However, the viability of commercial use of the NSR for international transit shipping remains questionable. In sum, identified advantages of the NSR are reduced journey time, lower emissions and safety due to the lack of piracy threats. The challenges of shipping along the NSR include political climate (NSR as part of Russian territorial waters), environmental concerns, operational conditions (weather, predictability), capital expenditure (ice-class vessels and ice-breaker charges), contractual issues and uncertainty of ice-free seasons [93].

In 2019, Russia published a most comprehensive plan for infrastructure development of the NSR for the period 2020–35 that aims to make maritime transportation along the NSR internationally competitive [23,33]. As specified in the Arctic Council's 2011 agreement on Cooperation on Aeronautical and Maritime Search and Rescue in the Arctic (2011) [30], the Russian Federation is responsible for SAR along the NSR. Currently, along the NSR there are the Maritime Rescue Coordination Center located in Dikson and two Maritime Rescue subcenters in Pevek and Tiksi. Dikson operates year-round, whereas Tiksi and Pevek are open from July to October [42]. The plan proposes the construction of rescue coordinating centers in ports Pevek and Sabetta by 2020 and in Tiksi by 2022. Altogether, 11 new SAR vessels serving various purposes are expected to be constructed and in operation by 2024 [23,33].

The plan includes increased navigational and hydrographic support by modernizing three existing vessels and having 13 new navigational and hydrographic vessels built by 2022–23. Icebreaking capabilities will be strengthened by introducing five new class icebreakers from 2022 to 2024 and additional three super-powerful Lieder class icebreakers scheduled to be ready by 2027, 2030 and 2032. Ports of Murmansk and Petropavlovsk-Kamchatsky are the candidates for transhipment hubs. Murmansk is already the main transhipment hub for the exit of the Arctic resources. In the future, Sabetta port supported by Yamalmax LNG carriers fleet will be exporting production directly to the customer [93]. The necessity of transhipment hubs is stipulated by ports limiting vessel sizes and climate conditions. Furthermore, Russia plans to establish a domestic containership operator using ice-class vessels to increase containership traffic.

The future of the NSR will depend on the holistic approach that covers both investment and political will to proceed with infrastructure and safety improvements. The plan covers creating stable and continuous satellite communication for the users of the NSR by launching four geostationary satellites by 2024. The launch of six space modules by 2024 will secure a high-speed automatic identification system on the NSR. The density of meteorological stations is expected to increase to provide more precise weather forecasting; moreover, by 2025, the hydrometeorological data would be available via geosynchronous (GEO) satellites. The safety is addressed by measures to improve the forecasting of ice conditions. Moreover, year-round shipping along NSR would require the creation of a unifying single operational control center expected to be launched by 2021. Regarding ecological measures, the plan states that the mandatory application of the best available technologies aimed at pollution reduction in the seas will be applied [23,33].

Overall, maritime transportation on the NSR is expected to increase, including tourism. Improved safety, SAR and planned dredging works lifting vessel size restrictions will influence the accessibility of the NSR for international shipping. The prospects of shipping along the NSR shall be evaluated continuously taking into account changing geopolitical forces, legal aspects and environmental concerns over the Arctic shipping in general.

References

[1] Brooks MR and Faust P. 50 Years of review of maritime transport, 1968–2018: Reflecting on the past, exploring the future (no. UNCTAD/DTL/2018/1). Available from: https://unctad.org/en/PublicationsLibrary/dtl2018d1_en.pdf [Accessed 15 February 2020].

[2] Hellenic Shipping News. [Online]. Available from: https://www.hellenicshippingnews.com/suez-canal-makes-all-time-traffic-record/ [Accessed 5 March 2020].

[3] Russian Federation. Federal Law 132-FZ: On amendments to certain legislation of the Russian Federation regarding state regulation of merchant shipping in the waters of the Northern Sea Route. 2012. Available from: http://base.garant.ru/70207760/ [Accessed 17 February 2020].

[4] Kiiski T. Feasibility of commercial cargo shipping along the Northern Sea Route. 2017. Available from: https://www.utupub.fi/handle/10024/130546 [Accessed 15 February 2020].

[5] The United Nations Convention on Law of Sea. Available from: https://www.un.org/depts/los/convention_agreements/texts/unclos/part5.htm [Accessed 17 March 2020].

[6] Arctic Portal. Exclusive economic zone. Available from: https://portlets.arcticportal.org/exclusive-economic-zones [Accessed 17 March 2020].

[7] Didenko NI and Cherenkov VI. Economic and geopolitical aspects of developing the Northern Sea Route. In: IOP conference series: Earth and environmental science (vol. 180, no. 1, p. 012012). IOP Publishing. 2018.

[8] Lee SW and Song JM. Economic possibilities of shipping though Northern Sea route. The Asian Journal of Shipping and Logistics. 2014;30(3):415–30.

[9] Furuichi M and Otsuka N. Proposing a common platform of shipping cost analysis of the Northern Sea Route and the Suez Canal Route. Maritime Economics & Logistics. 2015;17(1):9–31.

[10] Schøyen H and Bråthen S. The Northern Sea Route versus the Suez Canal: Cases from bulk shipping. Journal of Transport Geography. 2011;19(4): 977–83.

[11] Arctic Council. Arctic Marine Shipping Assessment Report (AMSA). 2009. Available from: https://www.pmel.noaa.gov/arctic-zone/detect/documents/AMSA_2009_Report_2nd_print.pdf [Accessed 7 March 2020].

[12] Melia N, Haines K, and Hawkins E. Sea ice decline and 21st century trans-Arctic shipping routes. Geophysical Research Letters. 2016;43(18):9720–8.

[13] Heininen L, Sergunin A, and Yarovoy G. Russian strategies in the Arctic: Avoiding a new Cold War.Valdai Club. 2014. Available from: https://valdaiclub.com/files/11482/ [Accessed 2 January 2020].

[14] Brubaker RD and Ostreng W. The Northern Sea Route Regime: Exquisite superpower subterfuge? Ocean Development & International Law. 1999;30 (4):299–331.

[15] American Bureau of Shipping (ABS). Navigating the Northern Sea Route: Status and guidance. 2014. Available from: https://www.wilsoncenter.org/sites/default/files/media/documents/event/navigating_the_northern_sea_route_status_and_guidance.pdf [Accessed 10 December 2019].

[16] Stephenson SR, Brigham LW, and Smith LC. Marine accessibility along Russia's Northern Sea route. Polar Geography. 2014;37(2):111–33.

[17] Baudu H. La route maritime du Nord, réalité et perspectives. Available from: https://cqegheiulaval.com/la-route-maritime-du-nord-realite-et-perspectives/ [Accessed 3 March 2020]. French.

[18] Nordregio. Sea Routes and ports in the Arctic. Available from: https://nordregio.org/maps/sea-routes-and-ports-in-the-arctic/ [Accessed 13 March 2020].

[19] Borodachev VY and Alexandrov VY. In: Johannessen OM, editor. History of the Northern Sea Route. Remote sensing of sea ice in the Northern Sea Route. New York: Springer Berlin Heidelberg. 2007, p.1–23.

[20] Gavrilova K, Vakhtin N, and Vasilyeva V. Anthropology of the Northern Sea Route: Introducing the topic. The Polar Journal. 2017;7(1):46–57.

[21] Gazprom. The Novy Port project. Available from: https://www.gazprom-neft.com/company/major-projects/new-port/ [Accessed 20 March 2020].

[22] The Barents Observer. Russian Arctic shipping up 430% in 3 years. 2020. Available from: https://www.themoscowtimes.com/2020/02/28/russian-arctic-shipping-up-430-in-3-years-a69454 [Accessed 20 March 2020].

[23] Middleton A. Northern Sea Route: From speculations to reality by 2035. The High North News. 2020 January 7. Available from: https://www.highnorthnews.com/en/northern-sea-route-speculations-reality-2035 [Accessed 25 March 2020].

[24] Thackeray CW and Hall A. An emergent constraint on future Arctic sea-ice albedo feedback. Nature Climate Change. 2019;9(12):972–8.

[25] Hartmann DL and Ceppi P. Trends in the CERES dataset, 2000–13: The effects of sea ice and jet shifts and comparison to climate models. Journal of Climate. 2014;27(6):2444–56.

[26] Stroeve JC, Markus T, Boisvert L, Miller J, and Barrett A. Changes in Arctic melt season and implications for sea ice loss. Geophysical Research Letters. 2014;41(4):1216–25.

[27] Pazovsky VM. Problems of navigation safety provisions on the Northern Sea Route. Asia-Pacific Journal of Marine Science & Education. 2012;2(2):17–24.

[28] Buixadé Farré A, Stephenson SR, Chen L, *et al.* Commercial Arctic shipping through the Northeast Passage: Routes, resources, governance, technology, and infrastructure. Polar Geography. 2014;37(4):298–324.

[29] International Maritime Organization. Polar Code. 2017. Available from: http://www.imo.org/en/MediaCentre/HotTopics/polar/Documents/POLAR%20CODE%20TEXT%20AS%20ADOPTED.pdf [Accessed 25 March 2020].

[30] Arctic Council. Agreement on cooperation on aeronautical and maritime search and rescue in the Arctic. 2011. Available from: https://oaarchive.arctic-council.org/handle/11374/531 [Accessed 30 March 2020].

[31] Arctic Council. Agreement on cooperation on marine oil pollution preparedness and response in the Arctic. 2013. Available from: https://oaarchive.arctic-council.org/handle/11374/529 [Accessed 30 March 2020].

[32] Arctic strategy of the Russian Federation until 2020. 2013. Available from: http://government.ru/info/18360/ [Accessed 29 March 2020]. Russian.

[33] Plan for infrastructure development of the Northern Sea Route for the Period 2020–2035. 2019. Available from: http://government.ru/docs/38714/ [Accessed 29 March 2020]. Russian.

[34] Heininen L, Everett K, Padrtova B, and Reissell A. Arctic policies and strategies—Analysis, synthesis, and trends. International Institute for Applied Systems Analysis 2020. Available from: http://pure.iiasa.ac.at/id/eprint/16175/1/ArticReport_WEB_new.pdf [Accessed 1 April 2020].

[35] Hsiung CW and Røseth T. The Arctic dimension in Sino-Russian relations. In: Bekkevold JI and Lo B, editors. Sino-Russian relations in the 21st century. Cham: Palgrave Macmillan. 2019, p. 167–87. Available from: https://www.palgrave.com/gp/book/9783319925158

[36] Li X and Peng B. The Rise of China in the emergence of a new Arctic order. In: Finger M and Heininen L, editors. The Global Arctic handbook. Springer. 2019, p.197–213. Available from: https://www.springer.com/gp/book/9783319919942

[37] United Nations Convention on the Law of the Sea. 1982. Available from: https://www.un.org/depts/los/convention_agreements/texts/unclos/unclos_e.pdf [Accessed 19 March 2020].

[38] Sharapova AA. Legal regulation of the Northern Sea Route: Current state of the problem. Asia-Pacific Journal of Marine Science and Education. 2017:52–61.

[39] Franckx E, Cambou D, Benatar M, *et al.* In: Proelß A, editor. The United Nations Convention on the law of the sea: A commentary. München: C.H. BECK. 2017, p.2512–53.

[40] Gavrilov V, Dremliuga R, and Nurimbetov R. Article 234 of the 1982 United Nations Convention on the law of the sea and reduction of ice cover in the Arctic Ocean. Marine Policy. 2019;106:103518.
[41] Hill E, LaNore M, and Véronneau S. Northern sea route: An overview of transportation risks, safety, and security. Journal of Transportation Security. 2015;8(3–4):69–78.
[42] The Northern Sea Route Administration Office (NSRA). Available from: http://www.nsra.ru/en/home.html [Accessed 2 April 2020].
[43] Milaković AS, Gunnarsson B, Balmasov S, *et al.* Current status and future operational models for transit shipping along the Northern Sea Route. Marine Policy. 2018;94:53–60.
[44] Kommersant. Severniy Morskoy Trest. 2018. Available from: https://www.kommersant.ru/doc/3695428 [Accessed 30 November 2019]. Russian.
[45] Gautier DL, Bird KJ, Charpentier RR, *et al.* Assessment of undiscovered oil and gas in the Arctic. Science. 2009;324(5931):1175–9.
[46] Staalesen A. Rosatom eyes shipments of grain on Northern Sea Route. The Barents Observer. 2019 December 30. Available from: https://thebarentsobserver.com/en/industry-and-energy/2019/12/rosatom-eyes-shipments-grain-along-northern-sea-route#.Xgm9C1ZSG6o [Accessed 30 January 2020].
[47] Doklad Ministra Alexandra Novaka na Soveschanii s Chlenami Gosudarstvennoy Komissii po Voprosam Razvitiya Arktiki. Available from: https://minenergo.gov.ru/node/13453 [Accessed 10 February 2020]. Russian.
[48] Starinskaya A and Chervonnaya A. Rosneft mozhet sozdat Arkticheskiy kluster. Vedomosti. 2019 April 1. Available from: https://www.vedomosti.ru/business/articles/2019/04/01/797956-sevmorput [Accessed 30 January 2020]. Russian.
[49] Anishchuk A. As Putin looks east, China and Russia sign $400-billion gas deal. Reuters. 2014 May 7. Available from: https://www.reuters.com/article/us-china-russia-gas/as-putin-looks-east-china-and-russia-sign-400-billion-gas-deal-idUSBREA4K07K20140521 [Accessed 18 February 2020].
[50] Lasserre F. Arctic shipping: A contrasted expansion of a largely destinational market. In: Finger M and Heininen L, editors. The Global Arctic handbook. Springer. 2019, p.83–100. Available from: https://www.springer.com/gp/book/9783319919942
[51] Association of Arctic Expedition Cruise Operators (AECO). Available from: https://www.aeco.no/ [Accessed 8 April 2020].
[52] Association of Arctic Expedition Cruise Operators (AECO). Guidelines by Association of Arctic Expedition Cruise Operators (AECO). Available from: https://www.aeco.no/guidelines/ [Accessed 8 April 2020].
[53] Rivadeneira K. Russia is betting on Arctic tourism. Arctic Today. 2018 November 8. Available from: https://www.arctictoday.com/17947-2/ [Accessed 1 April 2020].
[54] Silversea. Available from: https://www.silversea.com/destinations/cruises-arctic-greenland/nome-to-tromso-e1200822025.html [Accessed 1 April 2020].

[55] Heritage Expeditions. Available from: https://www.heritage-expeditions.com/destinations/russian-far-east-travel/north-east-passage-westbound-cruise/ [Accessed 1 April 2020].

[56] Wright C. The rise and rise and rise of Polar Cruising. The Maritime Executive. 2018 May 30. Available from: https://www.maritime-executive.com/features/the-rise-and-rise-and-rise-of-polar-cruising [Accessed 7 April 2020].

[57] Stephenson SR, Smith LC, Brigham LW, and Agnew JA. Projected 21st-century changes to Arctic marine access. Climatic Change. 2013;118(3):885–99.

[58] Theocharis D, Rodrigues VS, Pettit S, and Haider J. Feasibility of the Northern Sea Route: The role of distance, fuel prices, ice breaking fees and ship size for the product tanker market. Transportation Research Part E: Logistics and Transportation Review. 2019;129:111–35.

[59] Lasserre F. Case studies of shipping along Arctic routes. Analysis and profitability perspectives for the container sector. Transportation Research Part A: Policy and Practice. 2014;66:144–61.

[60] International Association of Classification Societies. Requirements concerning Polar Class. Available from: http://www.iacs.org.uk/download/1803 [Accessed 10 April 2020].

[61] Sarrabezoles A, Lasserre F, and Hagouagn'rin Z. Arctic shipping insurance: Towards a harmonisation of practices and costs? Polar Record. 2016;52(4):393–8.

[62] DNV GL. Polar Ship categories. Available from: https://www.dnvgl.com/maritime/polar/requirements.html [Accessed 10 April 2020].

[63] Finnish Transport Communications Agency TRAFI. The Finnish-Swedish Ice Class Rules (FSICR). Available from: https://www.traficom.fi/en/transport/maritime/ice-classes-ships [Accessed 2 April 2020].

[64] Pastusiak T. Scheduling transit voyages of vessels of various ice classes across the Northern Sea Route. Annual of Navigation. 2019;26(1):114–26.

[65] Pruyn JF. Will the Northern Sea Route ever be a viable alternative? Maritime Policy & Management. 2016;43(6):661–75.

[66] Dobrodeev A and Sazonov K. Challenges of speedy icebreaker-assisted operation of heavy-tonnage vessels in ice. In: ASME 2018 37th international conference on ocean, offshore and Arctic engineering. American Society of Mechanical Engineers Digital Collection. 2018. Available from: https://asmedigitalcollection.asme.org/OMAE/proceedings-abstract/OMAE2018/51296/V008T07A007/275770 [Accessed 12 April 2020].

[67] Center for High North Logistics (CHNL) Office. Available from: https://arctic-lio.com/ [Accessed 4 April 2020].

[68] Suez Canal Administration. Available from: https://www.suezcanal.gov.eg/English/Navigation/Pages/NavigationStatistics.aspx [Accessed 6 April 2020].

[69] Business Index North. Maritime traffic and transportation infrastructure along the Northern Sea Route. 2019. Available from: https://businessindexnorth.com/sites/b/businessindexnorth.com/files/binrapport2019_small.pdf [Accessed 6 April 2020].

[70] Lasserre F and Pelletier S. Polar super seaways? Maritime transport in the Arctic: An analysis of Shipowners' intentions. Journal of Transport Geography. 2011;19(6):1465–73.

[71] Zhang Y, Meng Q, and Zhang L. Is the Northern Sea Route attractive to shipping companies? Some insights from recent ship traffic data. Marine Policy. 2016;73:53–60.

[72] The Maritime Executive. Venta Maersk completes Northern Sea Route passage. 2018. Available from: https://www.maritime-executive.com/article/venta-maersk-completes-northern-sea-route-passage [Accessed 2 April 2020].

[73] Pappas A. What is a TEU? Supply chain dictionary. Available from: https://dedola.com/2011/10/what-is-a-teu/ [Accessed 12 March 2020].

[74] Humpert M. Shipping company Maersk to return to Arctic Northern Sea Route this Summer. The High North News. 2019 June 14. Available from: https://www.highnorthnews.com/en/shipping-company-maersk-return-arctic-northern-sea-route-summer [Accessed 12 March 2020].

[75] Port Technology. Average containership sizes on global routes. Available from: https://www.porttechnology.org/news/average_containership_sizes_on_global_routes/ [Accessed 10 January 2020].

[76] PwC. Northern Sea Route: Development prospects and uncertainties. 2020. Available from: https://www.pwc.dk/da/publikationer/2020/pwc-severnyy-morskoy-put-perspektivy-i-neopredelennosti-razvitiya-en.pdf [Accessed 22 March 2020].

[77] Total. Yamal LNG: The gas that came in from the cold. Available from: https://www.total.com/energy-expertise/projects/oil-gas/lng/yamal-lng-cold-environment-gas [Accessed 23 March 2020].

[78] Aker Arctic. Christophe de Margerie. Available from: https://akerarctic.fi/en/reference/yamalmax/ [Accessed 26 March 2020].

[79] Henderson J and Yermakov V. Russian LNG: Becoming a global force. The Oxford Institute for Energy Studies. Available from: https://www.oxfordenergy.org/publications/russian-lng-becoming-a-global-force/?v=f0aa03aaca95 [Accessed 27 March 2020].

[80] CDP. Shipping heavyweights at risk of missing climate targets. 2019. Available from: https://www.cdp.net/en/articles/media/shipping-heavyweights-at-risk-of-missing-climate-targets [Accessed 28 March 2020].

[81] IMO. Sulphur 2020 – Cutting sulphur oxide emissions. Available from: http://www.imo.org/en/MediaCentre/HotTopics/Pages/Sulphur-2020.aspx [Accessed 28 March 2020].

[82] UN. New regulations to reduce marine pollution in Antarctic enter into force – UN. Available from: https://news.un.org/en/story/2011/08/383212-new-regulations-reduce-marine-pollution-antarctic-enter-force-un [Accessed 29 March 2020].

[83] IMO. IMO Arctic Summit 2020. Available from: https://imoarcticsummit.org/ [Accessed 8 April 2020].

[84] AECO. AECO self-imposed ban on heavy fuel oil. Available from: https://www.aeco.no/2019/11/expedition-cruise-industry-charts-course-for-sustainable-arctic-tourism/ [Accessed 9 April 2020].

[85] EU. Agreement to prevent unregulated high seas fisheries in the Central Arctic Ocean. Available from: http://publications.europa.eu/resource/cellar/f075cd03-46f3-11e9-a8ed-01aa75ed71a1.0006.03/DOC_1 [Accessed 29 March 2020].

[86] Østreng W, editor. National Security and International Environmental Cooperation in the Arctic—The case of the Northern Sea Route. Springer Science & Business Media. 2012.

[87] Ocean Conservancy. The Arctic Corporate Shipping Pledge. Available from: https://oceanconservancy.org/protecting-the-arctic/take-the-pledge/ [Accessed 30 March 2020].

[88] Middleton A. In the name of the Arctic. The High North News. 2019 November 6. Available from: https://www.highnorthnews.com/en/name-arctic [Accessed 30 March 2020].

[89] Pauwelyn AS, De Waterweg V, Kuiters L, and Schreuder M. Smart shipping on Inland Waterways. Available from: https://www.pianc.org/uploads/files/InCom/ToR/ToR-InCom-WG-210-Smart-shipping-on-inland-waterways.pdf [Accessed 30 March 2020].

[90] Regnum. Minpromtorg nameren osvaivat Arktiku pri pomoschi sudov-bespilotnikov. Available from: https://regnum.ru/news/polit/2697413.html [Accessed 10 April 2020]. Russian.

[91] Government will support computer simulation platform for unmanned navigation development. Available from: https://marinet.org/governmet-will-support-computer-simulation-platform-for-unmanned-navigation-development/ [Accessed 11 April 2020].

[92] Rspectr. Korably bez ekipazhey vyidyt v more letom. Available from: https://rspectr.com/articles/495/korabli-bez-ekipazhej-vyjdut-v-more-letom [Accessed 13 April 2020]. Russian.

[93] Pierre C and Olivier F. Relevance of the Northern Sea Route (NSR) for bulk shipping. Transportation Research Part A: Policy and Practice. 2015;78:337–46.

Chapter 5
Smart shipping beyond e-navigation

Yung-Ho Yu[1]

5.1 Introduction

This chapter discusses the purpose of smart and autonomous ships as well as the technologies and services necessary for efficiency in the context of the recent investments in this field by the South Korean Government and compliance requirements with International Maritime Organization (IMO) environmental regulation. South Korea is one of several nations that have embarked on an autonomous ship project seeking to boost the domestic eco-friendly and smart shipping industry. Already being home to one of the largest ship-building industries worldwide, one of the key objectives for the nation is to achieve a 50% global market share by 2030. In this context, the chapter looks at how shipbuilders and manufacturers are seeking to innovate by enabling newbuilds to comply with IMO environmental regulation drawing using information and communication technology (ICT), and more advanced services and shore support.

To do this, the chapter begins by briefly reviewing the projects being spearheaded by the South Korean Government to develop technologies to tackle the challenge of increasingly stringent IMO environmental regulation. The e-navigation project will be described in this context. It proceeds to provide an overview of the IMO's environmental requirements and how these impact newbuilds.

The author then moves on to look in detail at developments in the field of smart and autonomous ships. He first introduces the topic and provides a brief overview of the key projects underway around the world. Then he proceeds to explain some of the core principles behind the technological components of these projects. The focus then moves on to discussing the advanced shore services and e-navigation, which forms the second part of the overall puzzle of the e-navigation project. The section explains the technologies necessary to achieve advanced shore services and fluent information exchange between ship and shore. The final part of the text explains the key role of cybersecurity in the projects under consideration. The perception and countermeasures for overcoming cyberthreats are relatively

[1]Marine Electronics Industry Promotion Association (MEIPA), Busan, Republic of Korea

weak in comparison to the actual vulnerabilities. This chapter will detail technologies required for a safe network, namely, authentication, encryption, and decryption, integrity used in network, and maritime domain.

5.2 The South Korean vision

The shipbuilding industry has been mired by slow economic growth in the last decade, similarly to the shipping industry in the last few years. South Korea is one of the countries with the largest share in the sector and therefore has an important stake in ensuring that the industry grows faster than it has in the past. In view of the increased focus on climate change mitigation policies and IMO regulation in respect to emission targets, an important challenge as well as opportunity emerges for the shipbuilding sector to ensure that newbuilds conform and if possible outpace the regulatory requirements that will be imposed on the shipping industry in the short-, medium-, and longer term.

The South Korean Government is well aware of this and being so has recently announced the allocation of funds that will be invested over 6 years from 2020 to 2025 [1]. Financing will be provided both by the government and private shipbuilders, including Hyundai Heavy Industries, Daewoo Shipbuilding & Marine Engineering, and Samsung Heavy industries. There are reportedly four main goals of the project: (1) develop an intelligent navigation system, (2) engine automation system, (3) construction of autonomous ship performance center, and (4) development of operation technology and standardization. The overall outcome will be to develop an intelligent system capable of autonomous navigation, integrated platform to control engine room, collision, and accident prevention, situational awareness, and decision-making. By combining future technologies, the project intends to develop a level 3 (L3) autonomous operating vessel by 2025. According to the IMO, L3 ships are remotely controlled without a crew, while level 4 is a fully autonomous ship [2].

Another related initiative is the smart e-navigation project promoted by a number of partners, including academia and the private sector (Tables 5.1 and 5.2).* The three areas of activity under the scope of this project include the development of core technologies, smart navigation operating systems and digital maritime infrastructure, and international standardization.

This project responds directly to the challenges discussed by the IMO Maritime Safety Committee (MSC). During the 101st session held June 5–14, 2019, the MSC approved a number of circulars related to the development of e-navigation. The e-navigation is defined as "the harmonized collection, integration, exchange, presentation and analysis of marine information on board and ashore by electronic means to enhance berth to berth navigation and related

*This includes the Ministry of Ocean and Fisheries, The Korean Institute of Marine Science and Technology Promotion, KRISO (Korean Research Institute of Ships and Ocean Engineering), Korean Register.

Table 5.1 Smart project activities—e-navigation project

Activity 1	Core technology development for e-navigation service	Enabling comprehensive situational awareness and responding service	WP1	SV1-Navigation Monitoring & Assistance Service (NAMAS)
		e-Navigation service development covering Korean maritime traffic	WP2	SV2-Ship-Borne System Monitoring Service (SBSMS)
			WP3	SV3-Safe & Optimal Route Planning Service (SORPS)
			WP4	SV4-Real-Time Electronic Navigational Chart Distribution & Streaming Service (REDSS)
		Developing essential services required by IMO e-navigation	WP5	SV5-Pilot & Tugs Assistance Service (PITAS)
			WP6	SV5-2-Maritime Environment and Safety Information Service (MESIS)
Activity 2	Developing e-navigation operating system and digital maritime communication		WP7	Integrated e-Navigation System
			WP8	Establishing high-speed wireless maritime network (LTE-Maritime)
			WP9	Establishing digital system for maritime wireless communication (VDES/D-HF)
Activity 3	Harmonization with e-navigation international standards		WP10	Developing maritime data standards (S-10X)
			WP11	Developing maritime connectivity platform
			WP12	Developing maritime wireless communication technology
			WP13	S-mode

Source: Website (http://www.smartnav.org/eng/html/Index/).

services for safety and security at sea and protection of the marine environment."[†] The MSC adopted items focusing on standards and harmonization, which seems to be the priority at present. This includes the circular on guidelines for the standardization of user interface design for navigation equipment, amendments to the

[†]http://www.smartnav.org/eng/html/Community/overseas_news.php?mode=v&bbs_data=aWR4PTI1Njkmc3RhcnRQYWdlPSZsaXN0Tm89MTMmdGFibGU9aWRfYmJzX2RhdGEmY29kZT1vdmVyc2Vhc19uZXdzX2VuJnNlYXJjaF9pdGVtPSZzZWFyY2hfb3JkZXI9||

Table 5.2 Navigation services—e-navigation project

ID	Service	Target vessels	Communication method
SV1	SV1-Navigation Monitoring & Assistance Service (NAMAS)	Korean passenger ship (domestic/international), Korean dangerous cargo vessel, and Korean international ship	LTE-Maritime VDES
SV2	SV2-Ship-Borne System Monitoring Service (SBSMS)	Korean passenger ship (domestic/international) upon request	LTE-Maritime VDES/SAT
SV3	SV3-Safe & Optimal Route Planning Service (SORPS)	Korean passenger ship (domestic/international) upon request	LTE-Maritime VDES/SAT
SV4	SV4-Real-Time Electronic Navigational Chart Distribution & Streaming Service (REDSS)	Domestic costal vessel	LTE-Maritime
SV5-1	SV5-Pilot & Tugs Assistance Service (PITAS)	Pilots and tugs	LTE-Maritime
SV5-2	SV5-2-Maritime Environment and Safety Information Service (MESIS)	Upon request	LTE-Maritime VDES/SAT

Source: Website (http://www.smartnav.org/eng/html/Index/).

performance standards for the presentation of navigation-related information on ship-borne navigational displays (resolution MSC.191(79)). In addition, the MSC resolution on guidance on the definition and harmonization of the format and structure of maritime services (MS) in the context of e-navigation aims to ensure that maritime-related information and data exchanged as part of different MS are implemented internationally in a harmonized, standardized, and unified format.

The links of this project with the environmental sustainability of shipping and ports can be summarized in a few key points. First, increased operational efficiency will reduce emissions by decreasing time delays in various aspects of the marine side maneuvering in ports. Second, monitoring is a crucial aspect of environmental accident prevention and rapid response. In addition, the collection of information through these systems allow for better decision-making in the context of port planning and operations.

5.3 Approaches to compliance with IMO regulation

One of the key drivers of innovation in the shipbuilding industry today is regulatory compliance, which, in turn, has the overriding objective to reduce emissions from shipping, including ports. The 74th session of the IMO's Marine Environment Protection Committee (MEPC 74) was held to support the achievement of the initial objectives set out the IMO to reduce greenhouse gas emissions from ships [3]. The MEPC approved the amendments to MARPOL Annex VI, which significantly enhanced the phase 3 requirements of the energy efficiency design index (EEDI). According to the draft amendment, the phase 3 requirements of the EEDI will become effective as of January 1, 2022, which is 3 years earlier than originally planned.

The EEDI defines that the amount of CO_2 generated by a ship while doing a 1 t mi (gCO_2/t nm) can be calculated by the following equation:

$$EEDI = \frac{\text{Actual } CO_2 \text{ emission}}{\text{Transport work}} = \frac{\text{Engind power} \times SFOC \times CF}{DWT \times \text{speed}} \tag{5.1}$$

where *SFOC* is specific fuel oil consumption and *CF* is conversion factor depending on fuel kinds showed in Table 5.3.

The following shows the 2022 EEDI (classified by DeadWeight Tonnage, DWT) reduction rates relative to the size of a container ship (Figure 5.1):

- for container ships $\geq$ 200,000 DWT, 50% reduction rate;
- for container ships $\geq$ 120,000 DWT, but <200,000 DWT, 45% reduction rate;
- for container ships $\geq$ 80,000 DWT, but <120,000 DWT, 40% reduction rate;
- for container ships $\geq$ 40,000 DWT, but <80,000 DWT, 35% reduction rate; and
- for container ships $\geq$ 15,000 DWT, but <40,000 DWT, 30% reduction rate.

The MEPC also agreed to the terms of reference for a correspondence group to look into the introduction of a potential phase 4 requirement for the EEDI.

5.4 IMO regulation and approaches to compliance for newbuilds

Shipping and shipbuilding companies are researching efficient methods to comply with the EEDI requirements. As shown in Figure 5.2, emission requirements are

Table 5.3 Conversion factor for fuel mass to CO_2 mass

Type of fuel	Reference	Carbon content	*CF* (t-CO_2/t-fuel)
Diesel/gas oil	ISO 8217 Grades DMX to DMC	0.875	3.206000
Light fuel oil (LFO)	ISO 8217 Grades RMA to RMD	0.26	3.151040
Heavy fuel oil (HFO)	ISO 8217 Grades RME to RMK	0.85	3.114400
Liquefied petroleum gas (LPG)	Propane Butane	0.819 0.827	3.000000 3.030000
Liquefied natural gas (LNG)		0.75	2.750000

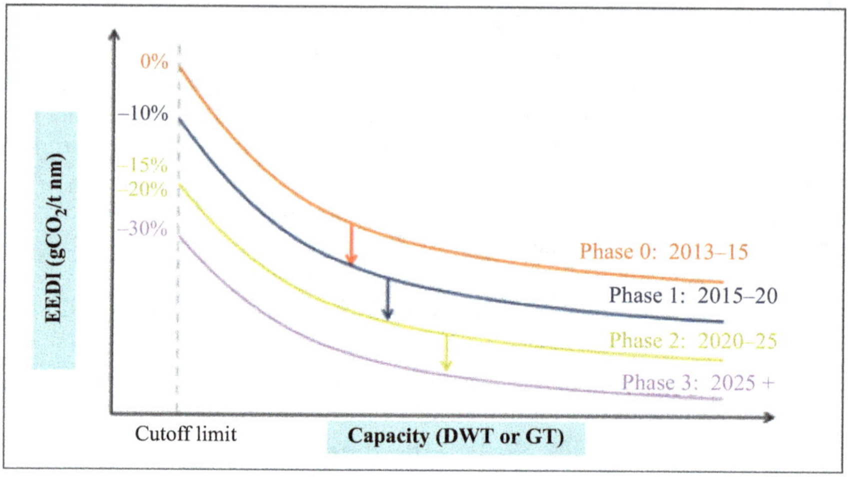

Figure 5.1 EEDI curves relative to vessel size. Initial plans of the EEDI curves depended on phase in relation to vessel size

continuously becoming more and more stringent. Different remedies are being explored in correlation to emission items. For example, selective catalytic reduction (SCR) systems are being considered for coping with NO_x fuel, while scrubbers are being considered for SO_x. Tables 5.4 and 5.5 show economic evaluations of application methods that rely on NO_x and SO_x remedies from the view point of capital expenditures (CAPEX), operating expenditure (OPEX), system installation, and weight and space. Table 5.4 shows that HP and LP SCR are more economical. However, the system installation process is complex and requires more volume in comparison to exhaust gas recirculation (EGR).

Table 5.5 indicates that the CAPEX of low-sulphur fuel (LS fuel) is low and can be primarily attributed to the lower requirements of facilitating additional equipment. On the other hand, the OPEX is high because of the high prices of LS fuel as compared to the use of LNG.

Shipbuilding companies believe that compliance with EEDI requirements can be achieved through a combination of these various methods.

Some of the effective methods that reduce emissions to comply with EEDI requirements are shipbuilder studies in advanced hull design, attaching energy saving devices (ESD) to the hull, changing the shape of the rudder, altering waste heat recovery systems, and using gas fuel. Table 5.6 shows the effectiveness of EEDI reduction strategies based on various researches. The most compelling approach would be using an liquefied natural gas (LNG) fuel system, but the CAPEX for constructing LNG bunkering facilities and revising engine systems should be considered synthetically [4].

When comparing the LNG dual fuel (DF) and legacy engine systems, which would be a better method for complying with EEDI requirements of reducing emissions? Figure 5.3 illustrates the emissions of LNG DF and legacy diesel engines.

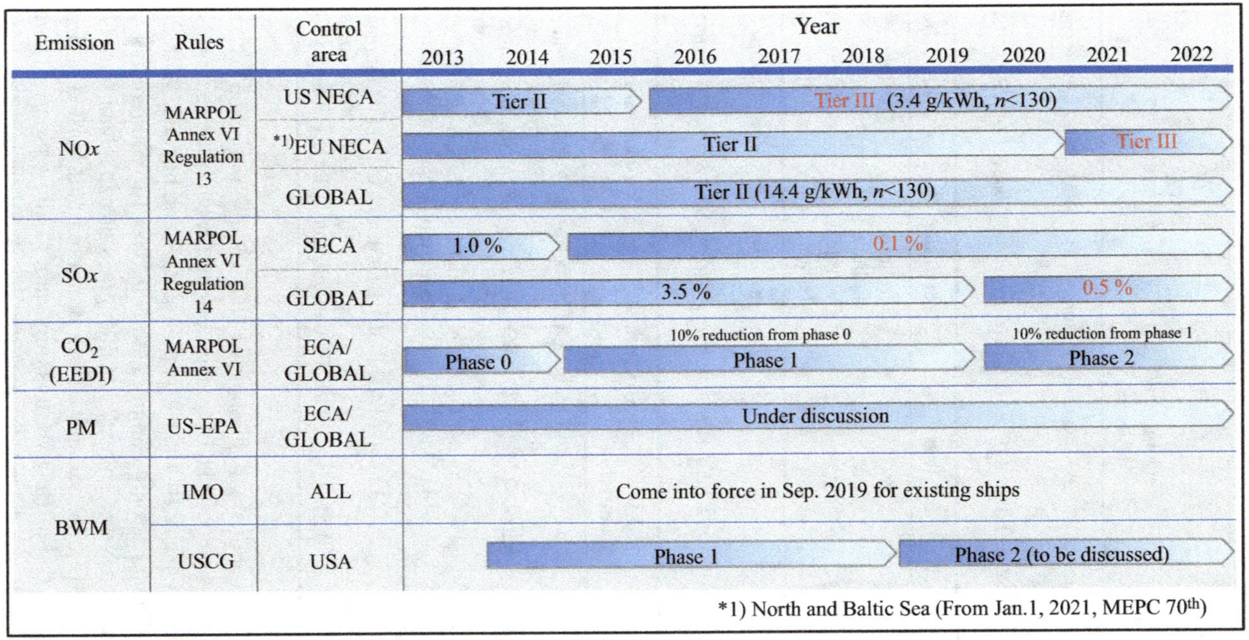

Figure 5.2 Emission plan of MEPC 70 according to various sea areas

Table 5.4 Economic evaluation of application method based on NO_x therapy

Item	HP SCR	LP SCR	EGR
CAPEX	Low	Low	High
OPEX	Low	Low	High
System installation	Complex	Complex	Simple
Weight and space	Medium	High	Low

Note: HP SCR, High Pressure SCR; LP SCR, Low Pressure SCR; EGR, Exhaust Gas Recirculation.

Table 5.5 Economic evaluation of application method based on SO_x therapy

Item	LS fuel	Scrubber	LNG fuelled ship
CAPEX	Low	Medium	High
OPEX	High	Medium	Low
System installation	Simple	Complex	Complex
Weight and space	Low	Medium	High

Table 5.6 Effectiveness of EEDI reduction according to various research

No.	Development item	EEDI reduction	Advantage	Disadvantage
1	Hull lines	0.3%–1%	M/E FOC reduction	Additional reduction is difficult
2	ESD, rudder			
3	LNG fuel system	20%–22%	NO_x, SO_x, PM reduction	Insufficient infra for LNG bunkering
4	Shaft Gen.	3%–3.5%	G/E FOC reduction	E/R length increase
5	Waste heat recovery system	4%–4.5%	G/E FOC reduction	E/R casing expansion
6	Air Lub. System	2%–3.5%	M/E FOC reduction	G/E FOC increase, limited condition for using
7	Solar panel	0.5%	G/E FOC reduction	Less efficiency

LNG DF engines only emit small amounts of NO_x-, SO_x-, and other IMO-regulated particulates. SO_x emissions of LNG DF engines comply with the requirement of all sea areas, including NO_x Tier III and EEDI phases 2, 3, and beyond. With the current technologies, X-DF engines only comply with NO_x Tier III requirements.

So then, what is an X-DF engine? X-DF engines utilize a lean-burn Otto cycle combustion principle [5,6]. Fuel and air are premixed and burned at a relatively high air-to-fuel ratio. To be effective, it is important to allow plenty of time for gas to mix in order to control the rate of combustion. X-DF engines utilize uniflow scavenging processing and the stratification of an air and gas mixture. Winterthur Gas & Diesel began developing the X-DF technology in 2011 and tested successfully in 2013 (Figure 5.4).

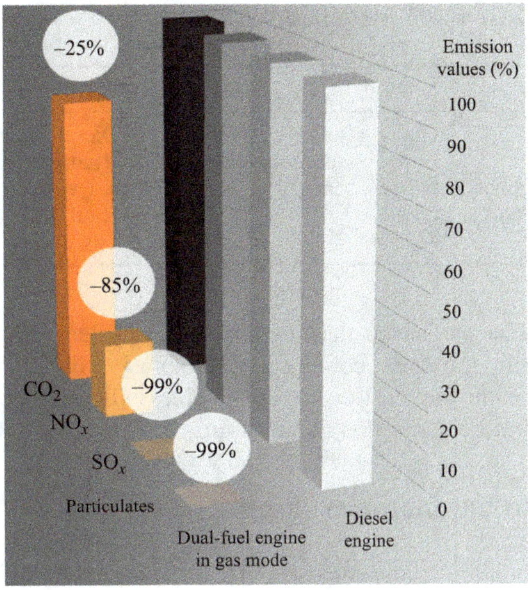

Figure 5.3 Comparison with emission of LNG DF and legacy engine

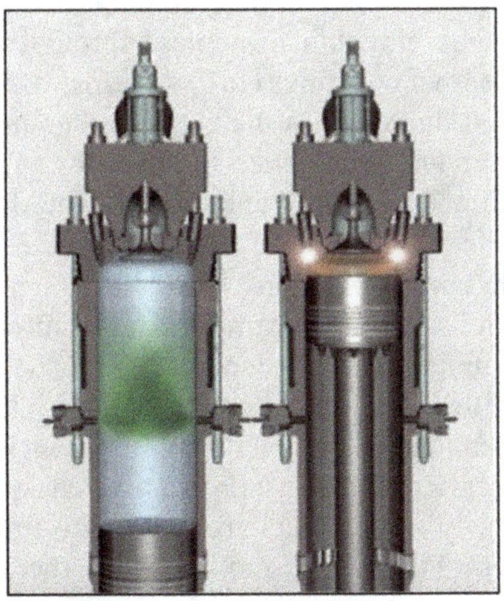

Figure 5.4 X-DF principle with gas admission switching to gas and ignition. Source: WinGD [5]

Wärtsilä implemented this technology to the larger bore engine types X62DF and X72DF. In 2015, a full-scale demonstration of the 6X72DF engine's technology was set up at Wärtsilä's licensee Diesel United in Aioi, Japan. The most significant differences between the X-DF and ME-GI engines are the underlying

Table 5.7 Comparison with X-DF and ME-GI engine

	X-DF	ME-GI
Principle	Otto cycle	Diesel cycle
FGSS	Low pressure (abt. 12 bar g)	High pressure (abt. 300 bar g)
NO_x Tier III Compliance	O (gas mode)	X (requires SCR or EGR)
CAPEX (M/E+FGSS)	Low	High
OPEX	Moderate	Low

combustion principles as well as the fuel gas supply system (FGSS), as shown in Table 5.7. Although Otto cycle combustion is explosive in comparison to Diesel cycle's slow and calm combustion, X-DF engines still manage to supply low-pressure gas and a more controlled combustion.

5.5 Smart and autonomous ships

Shipping and shipbuilding companies are perpetually trying to find ways to comply with IMO emission requirements through the use of new technologies, ICT systems, and equipment. Moreover, ICT systems communicate with shore systems to receive shore services that support optimal vessel maneuvering and safe navigation from the e-navigation center. Smart ships have over 5,000 data points and 7,000 channels to reinforce situation awareness through things such as closed-circuit monitoring and Internet of Things (IoT) systems, while 200 sensors from the engine room monitor fault diagnosis in the cargo and engine rooms. Some of this data are analyzed on-board and transmitted to shore for an enhanced data analysis to optimize control and improve ship maneuvering. The data include images, but sometimes vocal audio is transmitted to the shore center or shipping company headquarters via satellite communication. This communication system must have the ability to transmit data safely and seamlessly even through bad sea conditions that degrade satellite communication system capabilities (Figure 5.5).

It is important that the ship status and navigation situation of the ship are included in the data transmitted to the e-navigation center so that e-navigation can provide safe navigation services through a proper awareness and understating of the ship's situation. This is why shipbuilders are looking to implement the smart ship technologies displayed in Figure 5.6 and branding them with names such as SVESSEL,[†] SMARTSHIP[§] 3.0, and DS4 [7], just to name a few. Smart ship technologies can be categorized into safe navigation, green eco ship, connecting ship and shore, and IT technologies to support these systems. Some of the correlating subject matter for each technological category are shown in Figure 5.6. To summarize, therefore, the safe navigation category covers e-navigation, the green

[†]cf. Samsung Heavy Industries webpage, https://shi.svessel.com/ (accessed 21.01.21).

[§]cf. Moxa website, "Smart Ships for Smooth Sailing", webpage https://www.moxa.com/en/spotlight/integrated-solutions/marine/smart-ship/index.htm (accessed 21.01.21).

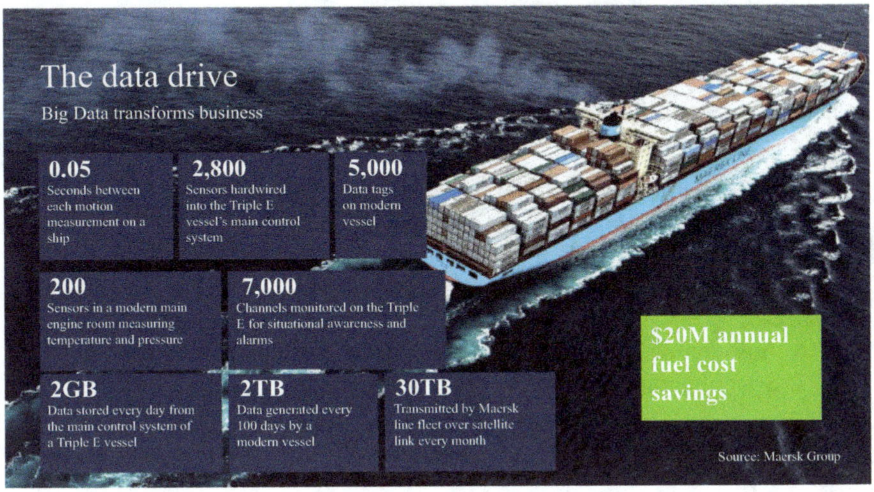

Figure 5.5 Data aggregation from smart ship

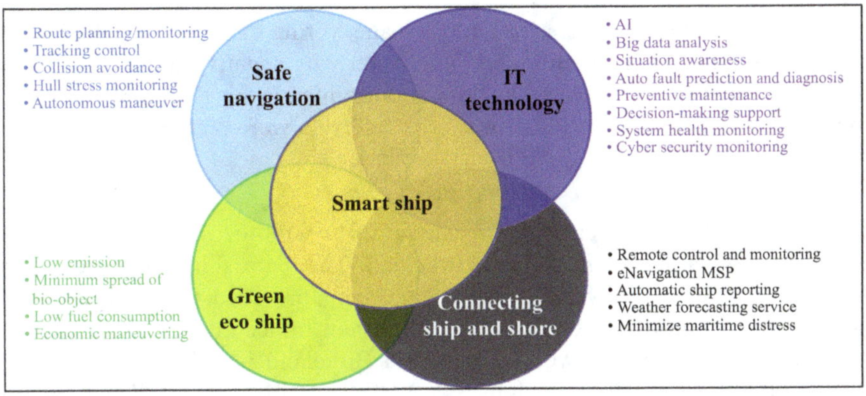

Figure 5.6 Technology categorized for smart ships (cf. [8])

eco ship category is relative to shipping companies and compliance with IMO environmental requirements, while the connecting ship and shore and IT technology categories cover technologies that support the other technologies [8].

Several autonomous ship projects on the worldwide stage that already took place are currently under way. The next section provides a brief review of each [9].

5.5.1 Project MUNIN (Maritime Unmanned Navigation through Intelligence in Networks) [10]

This project researched navigation technologies, economic assessments, and legal systems of autonomous ships and was launched in the EU in 2012 involving two research institutes, three universities, and three enterprises, spanning over a 3-year time span.

At the time of its activity, project Maritime Unmanned Navigation through Intelligence in Networks was composed of ten work packages (WPs) that focused on different aspects.

- WP1–WP3: Administration and management;
- WP4: Architecture for on-board data management and reliable communications between ship and shore;
- WP5: Autonomous bridge for analysis of bridge work and shore-side remote controls;
- WP6: Autonomous engine room for autonomous engine operation;
- WP7: Shore operation center software tools and processing shore-side remote controls;
- WP8: Proof of concept for proving and verifying developed concept and technology;
- WP9: Assessments of cost benefits, safety, legality, and research results of each WP; and
- WP10: Future concepts for analyzing future effects of autonomous vessels.

The project found that the autonomous maneuvering of ships controlled by the shore control center via satellite communication was negative due to high communication expenses. Therefore, autonomous navigation by the ship herself and remote control by the shore center in shorter intervals were recommended.

5.5.2 The Yara-Kongsberg project [11]

The Yara-Kongsberg project began in May of 2017 with the intent of building an electric propulsion ship named Yara Birkeland. The Yara shipbuilding company built the Yara Birkeland vessel, while Kongsberg provided the navigation, the communication, and the electric propulsion system in accordance to the memorandum of understanding between them. Yara Birkeland is a 120-TEU container ship that is 79.5 m in length, weighs 3,200 DWT, and has a 7–9 MW h battery, designed for zero emissions.

This project had a plan to be operated by crew as of the end of 2018 and full autonomous operation starting in 2020 but was delayed due to a late contract to build at VARD shipbuilding. This ship is estimated to be delivered by Q1 2020 and enter into autonomous operation in 2022. Although this project is encountering setbacks, the intrigue is high due to the technological contributions of Kongsberg in developing this autonomous ship.

5.5.3 The ReVolt project by DNV-GL [12]

The ReVolt is a 100-TEU, 60-m long, short sea-operated cargo ship that is propelled by the power supplied through a 2.3–5.5 MW h battery and does not require a single crew member to be on-board. The ReVolt will navigate at a 6-kn speed within a range of 100 nautical miles.

The project research started in 2013 and was announced later in 2014. Since the ship was designed to be crewless, the accommodation area was used to add to its cargo area. The battery of the ship will be charged during its 4 h at berth and is estimated to be used for 15 years due to its low charging rate. This ship is in the concept design stage and will not build until all of the autonomous technologies are mature enough. A downsized model at a 1:20 scale was developed by the Norway Technology National University (NTNU) to test and verify sensors, collision avoidance, and autonomous technologies.

5.5.4 SIMAROS (Safe Implementation of Autonomous and Remote Operation of Ships) project [13]

The Safe Implementation of Autonomous and Remote Operation of Ships Project was initiated in 2016 to develop the world's first autonomous offshore support vessel (OSV) by Kongsberg, Automated Ship, and Bourbon. This OSV was planned to be built at Fjellstrand shipbuilding in Norway. The goal of this project was to develop commercial autonomous and remote control technologies composed with 10 WPs. The Norwegian Maritime Authority (NMA) and Det Norske Veritas-Germanischer Lloyd (DNV-GL) are involved in this project, and the original plans to build the ship by 2018 have been delayed.

5.5.5 ROMAS (Remote Operation of Machinery and Automation System) project [14]

From 2017 to 2019, this project developed shore-side engine control systems to control engine control system on-board supported by nonfunctional requirements (NFR). DNV-GL, Fjord1 Ferry Company, Høglund Marine Solutions, and NMA are working together to develop regulations and rules to verify the effectiveness and robustness of the system.

5.5.6 The AAWA (Advanced Autonomous Waterborne Applications) project [15]

This project was carried out from 2015 to 2018 to advance the technology of remote maneuvering of ships and commercial validity. Rolls Royce, DNV-GL, universities in EU, shipping companies, and several equipment manufacturers were involved in verifying developed on-board systems since 2017, but the results have not been reported yet. Finnish shipping company, FinFerries, demoed this system on their Ro–Ro ships to substantiate the system.

5.5.7 The Autosea project [16]

The purpose of the Autosea project is to develop collision avoidance technologies for autonomous vessels. The NFR (from 2015 to 2019), the NTNU, Kongsberg, DNV-GL, and Maritime Robotics (a zero-man system solution company) took part in this project. This project was divided into four areas: (1) sensor convergence of radars, CCTV cameras, infrared cameras, and light detection and ranging sensors;

(2) collision avoidance through active and passive response methods; (3) system architecture that survey the strategies of cars and airplanes; and (4) developing various methodologies and tests, in addition to experimenting and verification.

5.5.8 The One Sea project [17]

Finland's One Sea project set out with the hopes of constructing an autonomous maritime ecosystem in the Baltic Sea by 2025. An autonomous maritime ecosystem requires efficiencies in minimizing harmful discharge fuels, the prevention of safety accidents, and overall economic improvements for autonomous vessels. This project started in 2016 and includes Rolls Royce, Wärtsilä, DIMECC (which includes an alliance of over 400 organizations), ABB, Ericsson, Business Finland, Finnpilot Pilotage, TIETO, and many more. Not only does this project examine technical and ethical issues but various other parts such as rules and regulations. The One Sea project plans to open to the segment of the public sector looking to test autonomous vessel maneuvering post construction (Figure 5.7).

5.5.9 SSAP (Smart Ship Application Platform) project [18]

Smart Ship Application Platform (SSAP) set out to develop a smart ship open platform by involving shipping companies, ship classification societies, universities, and research institutes. The goal of this project was to standardize data transmissions and communications between functional modules and machinery. A total of 27 organizations and 9 observers were involved during the first leg of this project which lasted from 2012 to 2015.

The SSAP project was to develop specifications for on-board data servers, examine the on-board data servers, and propose standards to the International Organization for Standardization (ISO). A group of 34 organizations, in addition to 9 observers, were involved in the second stage of this project that began in 2015 and lasted until 2017. The focus of the second stage of the project was to practice the proposed international standard and further develop a testing system. Stage three of this project commenced in 2018.

5.5.10 Green Dolphin [19]

Finally, CSSC, which is a Chinese national shipbuilding group, built a smart ship and entered into maneuvering test. This project started in 2015 and resulted in the creation of a green smart bulk ship known as Great Intelligence in 2017. Great Intelligence, a 38,000 DWT ship which has been classified by Lloyd's Register (LR), is able to collect operational data while maneuvering.

5.6 Core technologies of smart and autonomous ships

Smart ships are defined by their ability to communicate the ship's real-time maneuvering data, navigation, and situation awareness with shore-based service centers. Alternatively, shore-side service centers take the information received to

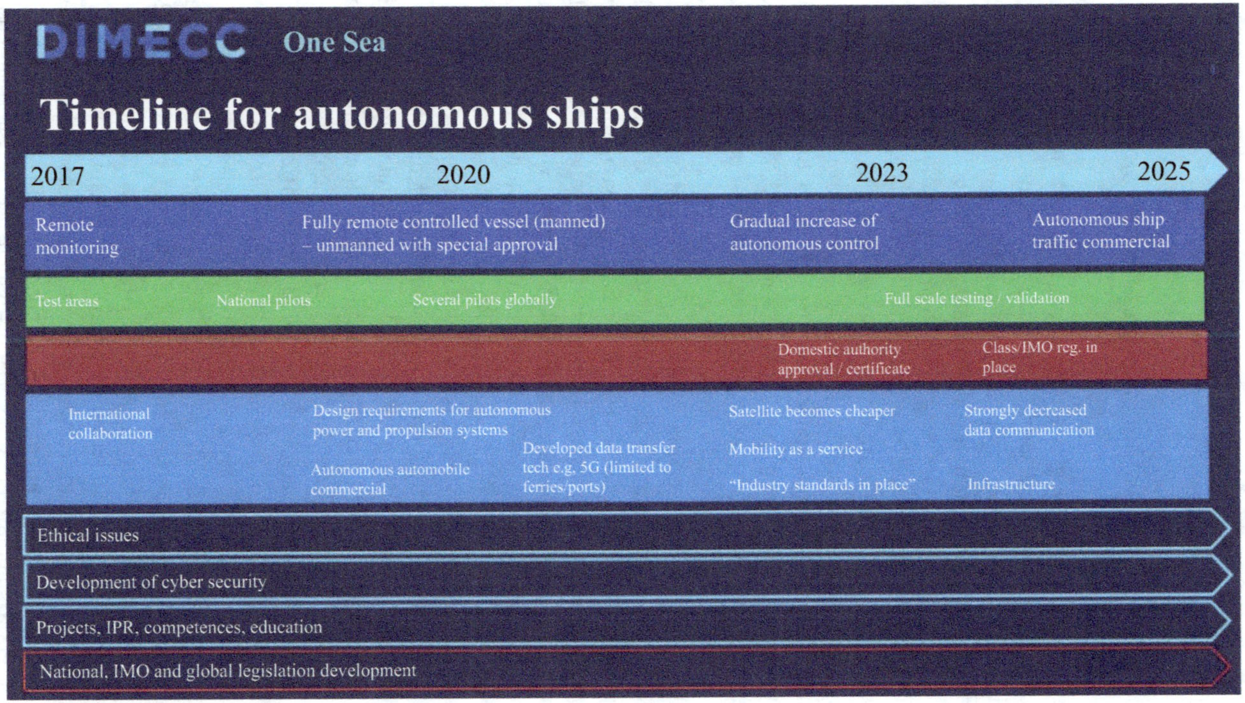

Figure 5.7 Roadmap of the One Sea project

provide the appropriate services to the ship. Shore-based service centers may be an e-navigation service center or even a shipping company that owned service center. This section will describe the technologies that are necessary to provide these services to the ship, particularly information-related ship and shore and information and the environment (e.g., weather).

5.6.1 Technology of connecting ship and shore—on-board data collection

On-board data collection technologies are principal requirements that cover all of the functions of a smart ship. Vital services for safe navigation and green eco ship technologies depend on connecting ship and shore technologies even in bad communication situations resulting from a bad sea state. Moreover, the worst sea state is more reliant on communication for safe navigation. Connection technology itself can also be divided into two parts. The aggregation of ship data, which deals with the ship's network, and the actual connection between ship and shore, which deals with the aspect of wireless communication.

Various ship network systems applied on-board bring a bad interoperability between the system and equipment, resulting in a hindrance of information integration. There are two kinds of networks in large: the navigation and communication network, and the control and monitoring of the engine and cargo systems.

In navigation and communication systems, the standards set forth in IEC 61162 are developed by International Electrotechnical Commission (IEC) Technical Committee (TC) 80 (IEC TC 80—Maritime navigation and radiocommunication equipment and systems). However, in engine and cargo control and monitoring, lots of industrial or private standards are applied, which not only are a different physical layer but also a different protocol (i.e., Modbus, Control Area Network (CAN) Bus, Ethernet, RS-232, RS-422, RS-485). These different physical layers and data protocols lead to interoperability defects between the system and equipment, causing problems such as hindering the efficiency of control and monitoring by delaying conversion by the gateway, and that is why international standards depend on the characteristics and layer requirements of real-time control and monitoring systems. Figure 5.8 represents a navigation and communication system network of a ship governed by IEC 61162-1, -2, -3, -450. IEC 61162-1 and 61162-2 are also known as National Marine Electronics Association (NMEA) 0183. IEC 61162-1 is slow in speed that uses RS-232 with a 4,800 baud rate transmission, while IEC 61162-2 is high in speed using RS-422 with a 38,400 baud rate. Both the standards transmit information as a sentence using NMEA 0183 data protocol. IEC 61162-3 is known as NMEA 2000 and uses a CAN Bus. The CAN Bus was developed by Robert Bosch GmbH in 1983 and is used in real-time control areas such as automobiles. The Society of Automotive Engineers developed a standard for automotive vehicles, or J1939 that the ISO adopted as an international standard, ISO 11783 [20–23].

The NMEA 2000 standard employs J1939, which uses a Parameter Group Number (PGN) format composed of informational fields [24]. Figure 5.9 shows how NMEA 2000 standard is composed with other standard.

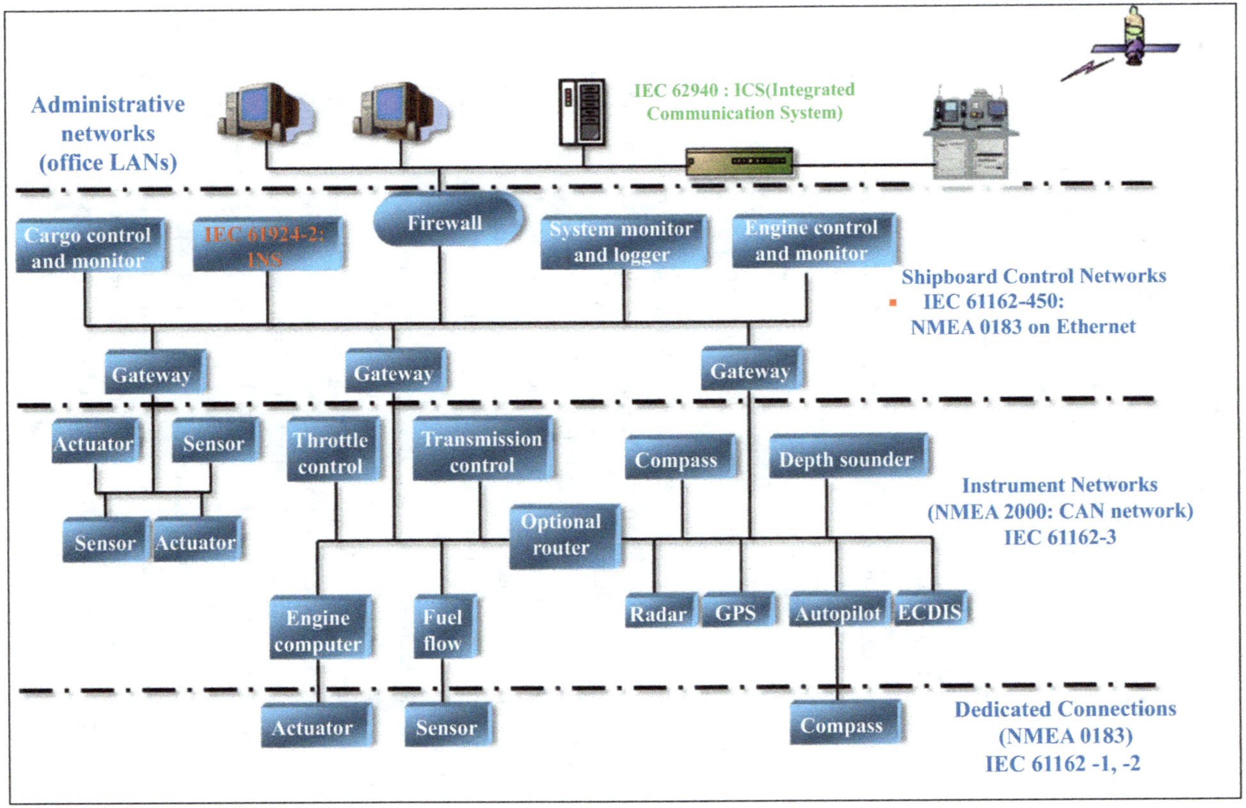

Figure 5.8 IEC 61162 Series ship network [8]

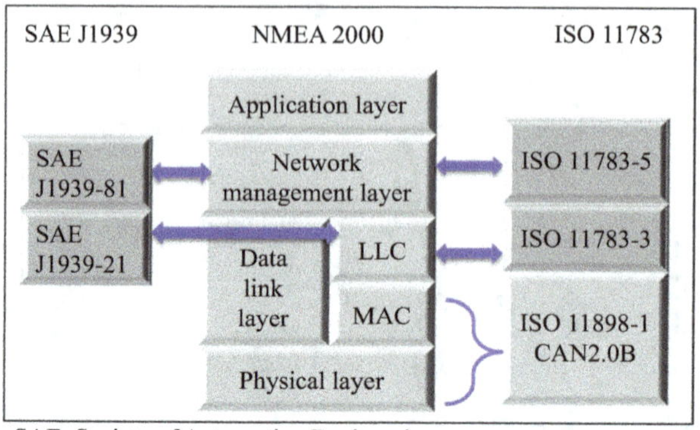

SAE: Society of Automotive Engineering
LLC: Logical Link Control
MAC: Media Access Control

Figure 5.9 Relationship of NMEA 2000 with other standards

The need for ships to gather more information has led to the requirement to base ship networks on Ethernet. IEC TC 80 developed IEC 61162-450 that standardizes the transmission of the information defined in IEC 61162-1 by Ethernet via IEEE 802.3.

Smart ships not only need on-board data but also data from neighboring navigation systems for safe navigation. Smart ships need to collect data from the engine room and cargo space. To collect engine room and cargo data, IoT and machine-to-machine (M2M) technologies are utilized. IoT and M2M sensors must have their own IP addresses that transmit data directly to the server. Sensor data include noise and may be faulty at times. However, sensors now have the ability to remove noise, calculate, and optimize the data using edge computing technology, transmitting reliable data back to the server. Since the need to recalculate and filter the noise in the server system is essentially cut out, it reduces the computing load of server system. There are many IoT sensors installed in ports and straits to monitor traffic and increase navigational situational awareness. Because the address space of IPv4 have been exhausted already, IPv6 is applied for IoT and M2M technologies.

Ship networks also use IPv6 for security and to interoperate with IoT and M2M systems. Recently network and data security are hot topics in the maritime domain. There are lots of reported hacks that have resulted in the loss of large amounts of money. The International Marine Electronic Alliance (IMEA)[¶] and NMEA[||] have started to develop a new ship network called OneNet that uses IPv6 in addition to network security technologies that use Transport Layer Security (TLS) and Datagram TLS (DTLS) defined with the Internet Engineering Task Force (IETF)[**] RFC 5246 and 6347.

[¶] https://www.imea-marine.org/
[||] https://www.nmea.org/
[**] https://www.ietf.org/

Simply put, OneNet will transfer two kinds of PGN, one that aligns with NMEA 2000 and the other a dedicated OneNet PGN that only uses OneNet on an IPv6 Ethernet network. The PGN contains information on engine and deck machinery such as windlass, mooring winch, thrusters, and cargo. OneNet can interoperate with other industrial standards not only IEC 61162-450 standard but also Modbus and Fieldbus using an extension header type defined for specific standards.

The following illustration shows a ship network using OneNet covering the engine room and cargo, interoperating through an IEC 61162-450 network for navigational equipment (Figure 5.10).

5.6.2 Advanced shore service and e-navigation

The IMO identifies 17 MS as advanced shore services. These include but are not limited to vessel traffic services, electronic navigational charts, tug and pilotage, maritime safety information, vessel shore reporting, remote medical, nautical publication, meteorological, search and rescue, and aid to navigation. All e-navigation services are provided from shore to ship. Radio communication systems are needed to provide some of these services and exchange information between ship and shore. The Waterborne Vision 2030[††] among others suggests keeping up with technology to continue the fluent information transfer between ship and shore. To achieve this, the following technologies are necessary to achieve advanced shore services and fluent information exchange between ship and shore:

- Open and integrated data networks protected from cybersecurity risks enabling new innovative ship functions and easy integration with shore services.
- New sensors and real-time collection of ship parameters with automated information management and surveillance, enabling big data analysis in shipping.
- Improved integration with shore support centers for technical operation and remote maintenance. Improved maintenance systems and processes for zero defects during voyage.
- The development of a decision support tool for e-performance, incorporating artificial intelligence (AI) and an energy efficient operation index, adjusted for the actual vessel and voyage particulars.
- The development of decision support for safer and more automated nautical operation of ships at sea and in port, including fully or periodically remote control from shore.

5.6.3 Environmental information (e.g., weather)

Today, many ships are operated with no men attending the engine room at night. A charge engineer watches the status of the engine room remotely through a monitoring system installed in his room. When his watch is over, the monitoring system is changed over to the next engineer's room. This shows that the machinery is

[††] www.waterborne.eu

reliable enough to vacate the engine room, but at the same time, it indicates that it may be necessary to monitor the machinery of the cargo area as well.

Recently, Japanese shipping company Nippon Yusen Kabushiki (NYK) conducted the world's first maritime autonomous surface ship (MASS) trial by manning the autonomous ships for safer operations, aligning with the first stage of the IMO's autonomous ship plan as a result of the SSAP project. Media reports suggest that Iris Leader, an NYK-operated pure car truck carrier, was navigating day and night using the Sherpa System for Real ship navigation system.

When examining the first trials of MASS operations, we can identify the importance of navigation systems that automatically assess the ever-changing environmental situation around the ship. These navigational systems must have the functionality to constantly stay aware of the situation to control the ship safely. In order to achieve this, these systems must calculate collision risk, determine the most optimal and safest route, and adjust the speed while maneuvering the ship. Through this study, the four technological categories for smart ships as referenced as safe navigation, green eco ship, connecting ship and shore, and IT technology are proven to be pillars for developing smart and autonomous ships (refer to Figure 5.6).

5.7 Standards of ship network (standards of IEC, ISO)

5.7.1 IEC TC 80 standard—IEC 61162-1 and IEC 61162-2 [20,21]

The IEC 61162 series standards are for ship networks and are composed of IEC 61162-1, -2, -3, 450, and 460. IEC 61162-1 is known as NMEA 0183 low speed, which is the first network standard to transfer information of autopilot by serial communication using RS-232, with a baud rate of 4,800. IEC 61162-2 is known as NMEA 0183 high speed, using RS-422 at a 38,400 baud rate. These two standards exchange information in the following general sentence format:

$ttsss,d1,d2,....*hh⟨CR⟩⟨LF⟩

The first two letters following the "$" are the talker identifier. The next three characters (sss) are the sentence identifier, followed by a number of data fields separated by commas, followed by a checksum hh after *, and terminated by carriage return/line feed. The data fields are uniquely defined for each sentence type. An example of talker sentence is

$HCHDM,238,M,*hh⟨CR⟩⟨LF⟩

where "HC" specifies the talker as being a magnetic compass, the "HDM" specifies the magnetic heading message follows. The "238" is the heading value, and "M" designates the heading value as magnetic. If data for a field are not available, the field is omitted, but the delimiting commas are still sent, with no space between them. The checksum field consists of a "*" and two hex digits representing the exclusive OR of all characters between, but not including, the "$" and "*."

Proprietary sentence is allowed for individual manufacturers to define proprietary sentence formats. These sentences start with "$P," then a three-letter manufacturer ID, followed by whatever data the manufacturer wishes, following the general format of the standard sentences. The following information is an example of Garmin's manufacturer sensor configuration.

$PGRMC, ⟨1⟩,⟨2⟩,⟨3⟩,⟨4⟩,⟨5⟩,⟨6⟩,⟨7⟩,⟨8⟩,⟨9⟩,⟨10⟩,⟨11⟩,⟨12⟩,⟨13⟩,⟨14⟩*hh⟨CR⟩⟨LF⟩

1. Fix mode, A=automatic (only option).
2. Altitude above/below mean sea level, −1,500.0 to 18,000.0 m.
3. Earth datum index. If the user datum index (96) is specified, fields 5–8 must contain valid values. Otherwise, fields 4–8 must be null.
4. User earth datum semimajor axis, 6,360,000.0–6,380,000.0 m (0.001 m resolution).
5. User earth datum inverse flattening factor, 285.0–310.0 (10–9 resolution).
6. User earth datum delta *x* earth centered coordinate, −5,000.0 to 5,000.0 m (1 m resolution).
7. User earth datum delta *y* earth centered coordinate, −5,000.0 to 5,000.0 m (1 m resolution).
8. User earth datum delta *z* earth centered coordinate, −5,000.0 to 5,000.0 m (1 m resolution).
9. Differential mode, A=automatic (output differential global positioning system (DGPS) data when available, non-DGPs otherwise), D=differential exclusively (output only differential fixes).
10. NMEA Baud rate, 1=1,200; 2=2,400; 3=4,800; 4=9,600.
11. Velocity filter, 0=no filter; 1=automatic filter; 2–255=filter time constant (e.g., 10=10 s filter).
12. Not used.
13. Not used.
14. Dead reckoning valid time 1–30 (s).

A query sentence is a means for a listener to request a particular sentence from a talker as following general format:

$ttllQ,sss,*hh⟨CR⟩⟨LF⟩

The first two characters of the address field are the talker identifier of the requester and the next two characters are the talker identifier of the device being queried (listener). The fifth character is always a "Q" defining the message as a query. The next field (sss) contains the three-letter mnemonic of the sentence being requested. An example query sentence is

$CCGPQ,GGA,*hh⟨CR⟩⟨LF⟩,

where the "CC" device (computer) is requesting from the "GP" device (a GPS unit) the "GGA" sentence. The GPS will then transmit this sentence once per second until a different query is requested.

Talker ID is defined by standard.

5.7.2 IEC TC 80 standard—IEC 61162-3 [22]

IEC 61162-3 is known as NMEA 2000 that transfers information as a PGN. The NMEA 2000 standard is composed of a combination of existing standards and its own application layer as shown in Figure 5.10. Characteristics of NMEA 2000 is CAN Bus, 250 kbps, maximum 200 m in length, 50 physical node, and 252 logical node. NMEA 2000 has three kinds of packet frames: single frame, fast packet frame, and transport protocol frame. Specifications of those frames are shown in Table 5.8.

CAN Bus has two kinds of identifiers that are an 11-bit ID and a 29-bit ID. The NMEA 2000 uses a 29-bit ID. Figure 5.11 shows the structure of a 29-bit ID CAN Bus composed of the former 11-bit ID and the latter 18-bit ID.

In the following illustration (Figure 5.12), a 29-bit ID is composed of a priority 3-bit, an extended data page (EDP) 1-bit, a data page (DP) 1-bit, a protocol data unit (PDU) format (PF) 8-bit, a PDU specific (PS) 8-bit, and a source address 8-bit. If the PDU is separated by PF value, anything with a PF field less than 240 would fall under PDU1, while a PF field greater than 240 would be considered PDU2. PS 8-bit means that the destination address is in PDU1, while PS is included in PGN as a GE (group extended) in PDU2. So PGN is defined with EDP, DP, PF, and PS depending on PS. PGN is expressed by 3-bit and high 6-bit filled with zero (0). If PF is PDU1, PS expresses destination address, but PGN is expressed with PS zero (0).

The PGN information is interpreted on the basis of the PGN number in the PGN database. The PGN database is composed of PGN usage explanations, characteristics of PGN, the default priority of the PGN, the default update rate, the frequency, the destination address that is transmitted to specific devices or global destinations that transmit to all devices, and an explanation of each PGN field. The update rate of each PGN depends on the number of the PGN. Figure 5.13 shows the PGN database information for PGN 59392. Because each field of the PGN database is not separated by a byte boundary and transmitted with binary data, the PGN cannot be interpreted easily without the information found in the PGN database.

5.7.3 IEC TC 80 standard—IEC 61162-450 [23]

IEC 61162-450 relies on the use of an Ethernet network, of which contents are IEC 61162-1, or NMEA 0183 for transmission to many devices with fast speed and large quantity of data.

Figure 5.14 shows how several devices combined with all function blocks defined by IEC 61162-450 standards are connected. SF is a system function block identified by a unique ID and transfers information with the User Datagram Protocol (UDP). NF is a network function block connected to a network defined by this standard as physical node with an IP. SNGF is serial network gateway function block which is an interface with a serial device, such as using the IEC 61162-1 standard. ONF, meaning other NF block, is connected to the network of this standard but does not use this standard requirement and can use other protocol using the same network infrastructure of this standard. This standard defines the following three types of messages:

- Sensor broadcasting message (SBM)
  - composed with one sentence that is defined in IEC 61162-1,
  - transmitted with high update rate and defined in Annex A.

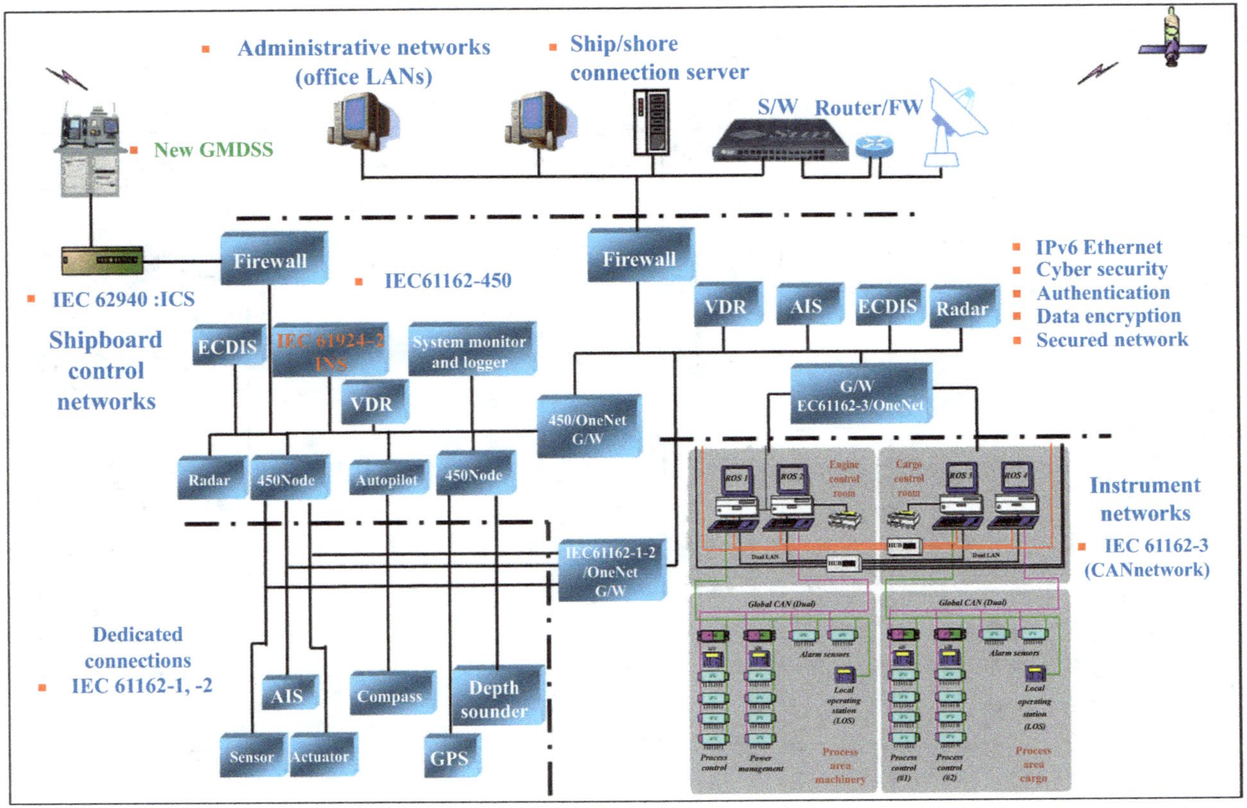

Figure 5.10 OneNet interoperating with IEC 61162-450 network covering all areas of the ship [25]. INS, integrated of navigation system; ICS, integrated communication system; G/W, gate way; VDR, voyage data recorder; ECDIS, electronica chart display and information system; AIS, automatic identification system

Table 5.8 Three types of packet frames of NMEA 2000

Single frame	Transport protocol frame	NMEA 2000 fast packet frame
Up to 8 bytes of data	Can convoy up to 1,785 bytes of data	Limited to 223 bytes of data
Destination only as defined by PGN	Can be used to send any broadcast PGN (greater than 8 bytes) to a specific device	Destination only as defined by PGN
No handshaking	Contains handshaking, i.e., smart transfer method	No handshaking
No transfer protocol delays	Takes long to send the same amount of data as fast packet	Takes less time to send up to 223 bytes. No transport protocol delays
Implemented by all industries	Implemented by all industries	NMEA 2000-specific protocol

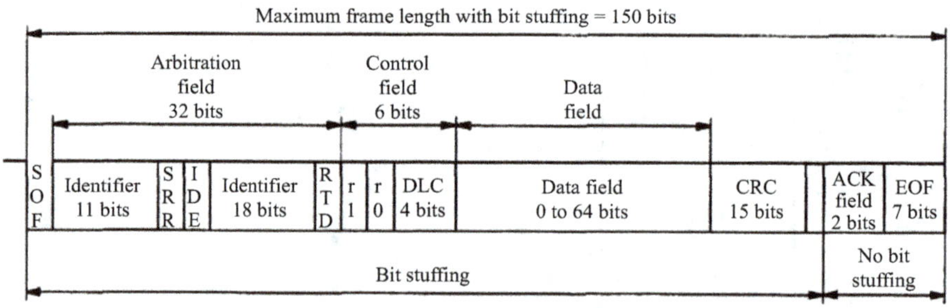

Figure 5.11 CAN frame of NMEA 2000

PDU format	29 bit identifier: Priority 3 bit	EDP 1 bit	DP 1 bit	PF 8 bit	PS 8 bit	SA 8 bit	PGN Nos
		PGN	PGN	PGN	PGN		
PDU1	0–7	0	0	0–239	DA 0–255	SA 0–255	240
			1	0–239	DA 0–255	SA 0–255	240
PDU2	0–7	0	0	240–255	GE 0–255	SA 0–255	16×256 = 4,096
			1	240–255	GE 0–255	SA 0–255	16×256 = 4,096

Figure 5.12 Composition of PGN from 29-bit ID

Example Parameter Group

ISO Acknowledgment **PGN: 059392**

This message is provided by ISO 11783 for a handshake mechanism between transmitting and receiving devices. This message is the possible response to acknowledge the reception of a "normal broadcast" message or the response to a specific command to indicate compliance or failure. The application layer is responsible for determining when this message is desired outside of network management requirements specified by this standard (e.g. response to ISO Request message). This message will always be sent with a destination address of 255.

Single Frame: Yes *Priority Default:* 6 *Default Update Rate:* NA milliseconds *Frequency:* NA cycles per second

Destination: Address *Query Support* *ACK Rqmnts:*

Original Reference ID # 64

Field #	*Field Name*				
1	Control Byte 0x00 = Positive Acknowledgment; 0x01 = Negative Acknowledgment; 0x02 = PGN supported but access denied; 0x03 to 0xFF = Reserved	*Byte Field Size:*	*Bit Field Size:* 8	*Request Parameter*	NO
2	Group Function Value Group Function of PGN being acknowledged. This field identifies for a device the specific group function of a PGN being acknowledged or declined. This field is not used if the PGN being acknowledged or declined is not a group function PGN.	*Byte Field Size:*	*Bit Field Size:* 8	*Request Parameter*	NO
3	Reserved Bits Variable number of reserved bits, all set to logic "1"	*Byte Field Size:*	*Bit Field Size:* resv 24	*Request Parameter*	NO
4	PGN of Requested Information 24-bit Parameter Group Number (PGN) expressed in binary, LSB is transmitted first	*Byte Field Size:*	*Bit Field Size:* 24	*Request Parameter*	NO

Figure 5.13 Example of PGN database

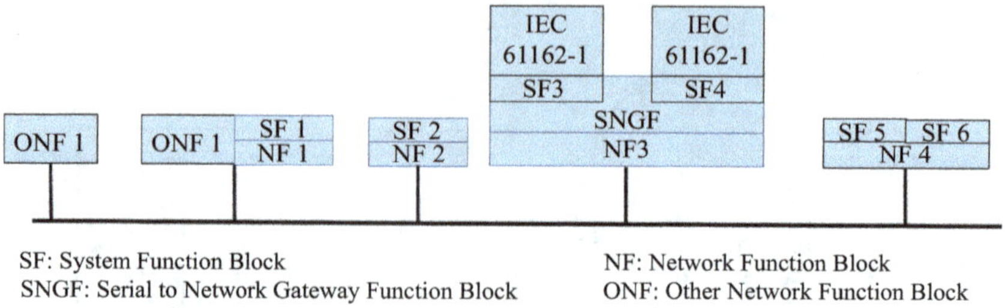

Figure 5.14 Example of IEC 61162-450 network

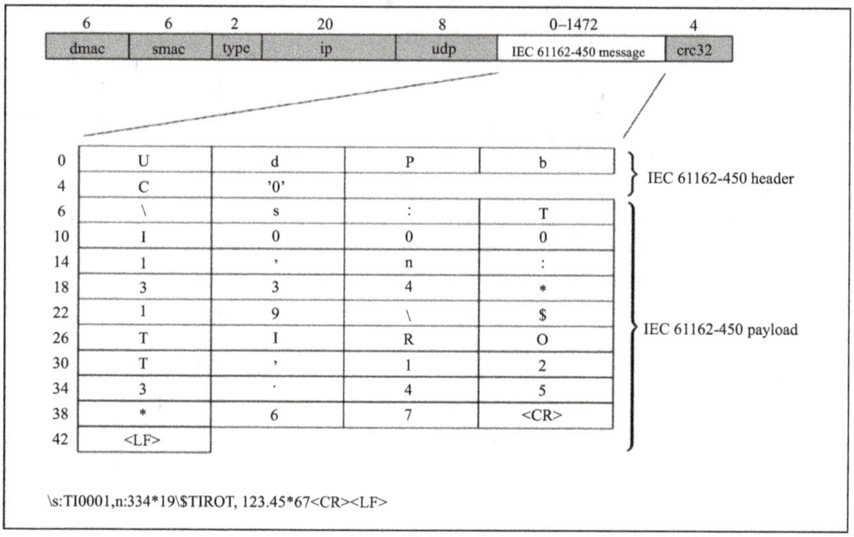

Figure 5.15 Example of data transmission in IEC 61162-450

- Multi-sentence message (MSM)
  - MSM composed of several sentence by tag block,
  - defined in Annex A.
- Command-response pair
  - exchange message to synchronize between command and response device,
  - can be utilized with SBM in some cases,
  - defined in Annex A.

The range of IP addresses defined in this standard is from 172.16.0.1 to 172.31.255.254 which are considered private IP addresses as defined in RFC 1918. Multicast IP addresses range from 239.192.0.1 to 239.192.0.64 and the ports can be allocated from 49152 to 65535.

Figure 5.15 shows data transmission in this standard. DMAC (destination MAC address), SMAC (source MAC address), type, IP, and UDP are included in the IPv4 header, while the payload of this standard is from next part of the

IPv4 header. This example shows transmission of data for rate of turn sensor as in the following sentence:

\s:TI001,n:334*19\$TIROT,123.45*67⟨CR⟩⟨LF⟩

UdPbC is the header of this standard, including NULL character. The next data are the payload of IEC 61162-450, the same as the sentence to be transmitted. In the transmitted sentence s:TI001 means source SFI TI001, n:334 is the serial number of transmitted information, *19 means checksum from the beginning. Sentence is starting from $, TI is the talker ID of ROT, ROT is the mnemonic of rate of turn sensor, 123.45 is the sensor value, *67 is checksum. Talker ID is defined in Annex A. There are three standard header types, UdPbC is a sentence header, RaUdP means binary image transmission, and RrUdP means transmission of re-transmittable binary image.

5.7.4 IEC TC 80 standard—IEC 61162-460 [26]

The IEC 61162-460 standard is the first ship network that took cybersecurity into consideration. Data protocol, electrical characteristics, and others of this standards are the same of IEC 61162-450 except for the network topology. Figure 5.16 shows the network structure of this standard [26].

The concept of the IEC 61162-460 standard is to separate from uncontrolled networks and allow a limited connection through a gateway. Uncontrolled networks mean that the security and reliability of network have not been proven. As shown in Figure 5.16, the 460 network is composed of 460 gateways, 460 switches, 460 nodes, a 460 forwarder, and network status monitoring.

- 460 Gateways separate uncontrolled networks from the 460 network. It has two firewalls, one on the uncontrolled network side and the other on the 460 side as shown in Figure 5.17. Various kinds of servers can be located between firewalls, with some servers being able to connect with the company by using a VPN.
- 460 Nodes are devices that comply with the 460 standard.
- 460 Switches are network switches that connect devices on the 460 network and comply with the 460 standard.
- 460 Forwarders are network devices that connect between controlled networks, including 460 network.
- 450 Nodes are devices that comply with 450 standard.
- 450 Nodes with SNGF are a kind of 450 device with an SNGF which are used on serial network devices such as GPS and other sensors using IEC 61162-1 data protocol.
- Network monitoring function is a monitoring system that is in compliance with network monitoring requirements of the 460 standard.
- System management function is a system management device that complies with system management requirements of the 460 standard.

5.7.5 ISO TC 8 WG 10 smart shipping

The ISO TC 8 established WG 10 smart shipping with the first meeting of WG 10 being held in Qingdao in June of 2017. The scope of WG 10 in Resolution 344 of

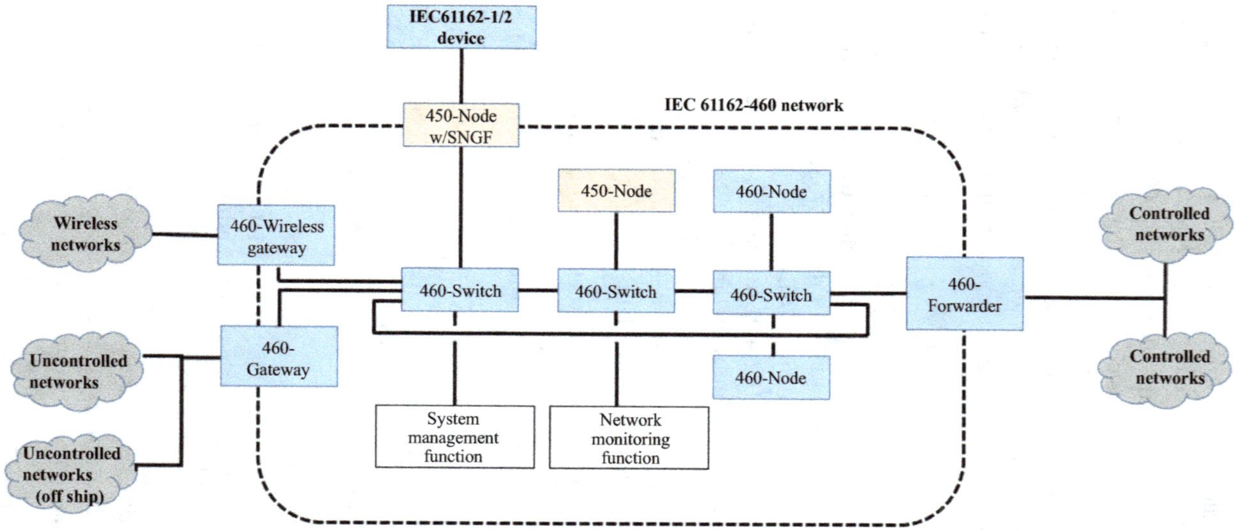

Figure 5.16 Network structure of IEC 61162-460

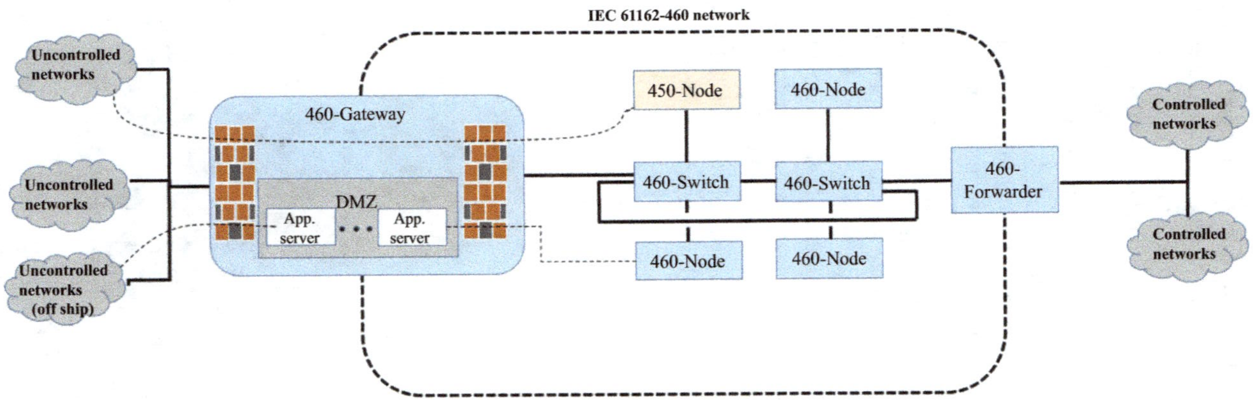

Figure 5.17 Structure of a 460 gateway

the 35th TC 8 Plenary Meeting was standardization in the field of computer applications (big data and processing, IoT, sensor, etc.) in shipbuilding and the operation of ships and marine technology. Dr. Xiaofeng, Wu reviewed the development of smart technologies and illustrated potential international standard demands and proposed that the realization of the smarter and greener future of shipping relies on a league of stakeholders and thus requires effective cooperation and collaboration. Figure 5.18 shows what Dr. Xiaofeng, Wu suggested. He categorized the domains of smart shipping as ship domain, ICT Infrastructure domain, ship operation domain, support services domain, and conceptual domain, which included all scopes of TC 8 technologies.

It may be mentioned that smart ships are safe, eco- and environment-friendly ships that support fault diagnosis, preventive maintenance, and aids in decision-making by utilizing ICT technology. This includes big data analysis and AI, enabling safe navigation by autonomous technology such as route planning and monitoring, route tracking and control, and collision avoidance. Collection and management of data and information during ship operation is necessary by close connection to the shore with marine and shore telecommunication technology, facilitating maritime commerce. To maximize benefits of smart shipping, it is important not to focusing only on smart ships and smart ports but also smart logistics. Recently, Maersk shipping company and IBM developed a smart logistic platform called TradeLens that utilizes a cloud computing platform and Kubernetes that was developed by Google. TradeLens

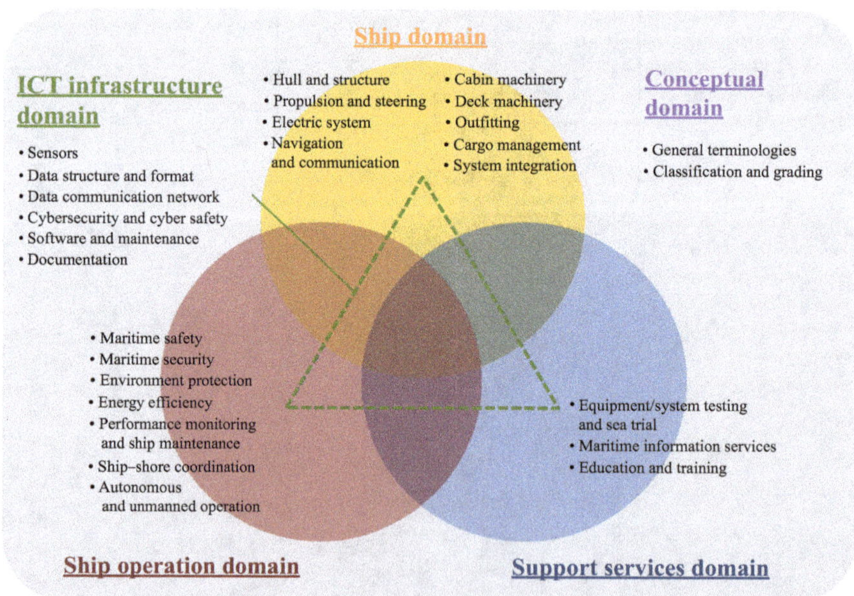

Figure 5.18 Domains and key fields that build up the technical framework of smart shipping [27]

provides services temporarily using IPv4 and IPv6, as IPv4 services are projected to deprecate due to M2M and IoT, as well as a lack of address space on IPv4.

Larger address spaces are necessary for smart ships, because they need the perceptions of navigational situation awareness, fluent data flow, cybersecurity, lots of information for cargo and machineries, as well as navigational information, and ultimately a fast and large transmission of information.

The IMEA/NMEA are preparing new ship networks using IPv6 to cover not only navigational data but also engine machineries and cargo data from OneNet. The IMEA/NMEA had plans to release OneNet at early of 2020. OneNet has the functionality to operate secure networks by authentication, data integrity, and encryption using TLS 1.2 and DTLS 1.2 of RFC that is the international standard legislated by the IETF.

ISO TC 8 WG 10 registered this new ship network as PWI 23816 to make a new international standard for ship networks in smart shipping with reference of IMEA/NMEA OneNet.

5.8 Cybersecurity guidelines of the IMO and BIMCO

In recent times, cyber security has claimed its place in the maritime industry's hot seat, as shipping companies reported significant losses due to cyberattacks; so naturally, the potential of a ship's system being hacked becomes a looming concern. The Baltic and International Maritime Council (BIMCO), the International Association of Lighthouse Authority, and ship classification societies such as DNV-GL, ABS (American Bureau of Ship classification society), LR of Shipping, KR (Korean Register of Ship Classification Society), and Class NK (Nippon Kaiji Kyokai) are looking to the IMO to enact guidelines for cybersecurity. The following section will talk about different technologies, standards, and guidelines as it pertains to cybersecurity.

5.8.1 IMO countermeasures for cybersecurity

The IMO announced the ISM Code standing for International Safety Management is the Resolution A.741(18) as amended by MSC.104(73), MSC.179(79), MSC.195 (80) adopted on November 4, 1993. The purpose of this code is to provide an international standard for the safe management and operation of ships and for pollution prevention. The code is expressed in broad terms so that it can have a widespread application. Different levels of management will require varying levels of knowledge and awareness of the items outlined.

The International Ship and Port Facility Security (ISPS) Code entered into force under SOLAS chapter XI-2 as on July 1, 2004. The ISPS Code has since formed the basis for a comprehensive mandatory security regime for international shipping. The code is divided into two sections, Part A and Part B. Mandatory Part A outlines detailed maritime and port-security-related requirements that SOLAS contracting governments, port authorities, and shipping companies must adhere to be in compliance with the code. Part B of the code provides a series of recommendatory guidelines on how to meet the requirements and obligations set out within the provisions of Part A [28].

Since the publication of the IMO's 2012 edition of the Guide to Maritime Security and the ISPS Code (the Guide), which was developed to assist SOLAS Contracting Governments, port facility personnel, and the shipping wider shipping industry, the IMO, through a Global Maritime Security program that is part of the organization's Integrated Technical Co-operation Programme, developed and implemented comprehensive technical cooperation projects and activities worldwide, with the Guide as a basis. The focus was primarily on assisting the states in the implementation, verification, compliance with, and enforcement of, the provisions of the IMO maritime security measures, including the ISPS Code and SOLAS chapter XI-2, counter-piracy initiatives, the SUA Convention (Convention for the Suppression of Unlawful Acts—IMO) that is about suppression of unlawful acts of violence against the safety of maritime navigation and long-range identification and tracking.

5.8.2 BIMCO countermeasures for cybersecurity

BIMCO organization was founded under the title of The Baltic and White Sea Conference in Copenhagen in 1905 and was the first organization to see the benefit in joining forces with other countries to secure better deals and standard agreements in shipping. As the organization grew and became more international, it was renamed as The Baltic and International Maritime Council and simply BIMCO.

BIMCO announced their third version of the Guidelines on Cyber Security Onboard Ships on December 7, 2018. Key updates in version 3.0 include [29]

- the requirement to incorporate cyber risks in the ship's safety management system;
- more detailed information related to the risk assessments of operational technology;
- increased guidance for dealing with the risks in the ship's supply chain; and
- case studies of verified cyber incidents on-board ships to highlight and illustrate potential problems.

BIMCO and the International Chamber of Shipping published a new cybersecurity workbook on November 5, 2019, meant to serve as a guide for the masters and officers on-board ships and thereby help them prepare for a potential cyber incident.

5.9 Security technologies standard of the IETF

5.9.1 Security Socket Layer (SSL) and Transport Layer Security (TLS)

The Security Socket Layer (SSL) was developed by Netscape in 1994 and was included in the Netscape Navigator web browser. It was designed to secure the data created in the application layer using a compression service. The browser was later updated with version 3.00 in 1995.

The TLS was developed by IETF based on SSL 3.0 and released TLS v1.0 as RFC 2246 in 1999 which is the same as SSL 3.1. The latest version of TLS is TLS 1.3 (RFC 8446) that was released in August 2018, but DTLS 1.3 that is a DTLS is still under development. Many libraries do not have DTLS 1.3 support yet. TLS is used in transmission control protocol (TCP) connections and DTLS is used in UDP connections.

5.9.2 Structure of TLS 1.2

Figure 5.19 shows the structure of TLS 1.2. The TLS protocol stack is laid between transport and application layer and composed with a handshake protocol, change cipher spec protocol, alert protocol, HTTP, and heartbeat protocol.

The heartbeat protocol is defined separately in RFC 6250 and HTTP provides transfer service of web client/server interaction.

The handshake protocol uses HTTP v1.1 to exchange a key. The handshake protocol is composed of four phases; the so-called client hello, server hello, client certificate with finished message, and server change cipher spec with finished message are shown in Figure 5.20. This protocol handles the cryptographic configuration that will be used during and after the handshake. The handshake protocol has the goals of having a client and server agree on a protocol versions, select cipher suites, optionally authenticate each other, and reliably negotiate a shared key as stated in section 7.3 of RFC 5246 [30].

1. The "client hello" message: The client initiates the handshake by sending a "hello" message to the server. The message will include which TLS version the client supports, the cipher suites supported, and a string of random bytes known as the "client random."
2. The "server hello" message: In reply to the client hello message, the server sends a message containing the server's TLS certificate, the server's chosen cipher suite, and the "server random," another random string of bytes that is generated by the server.
3. Authentication: The client verifies the server's TLS certificate with the certificate authority that issued it. This confirms that the server is who it says it is, and that the client is interacting with the actual owner of the domain.

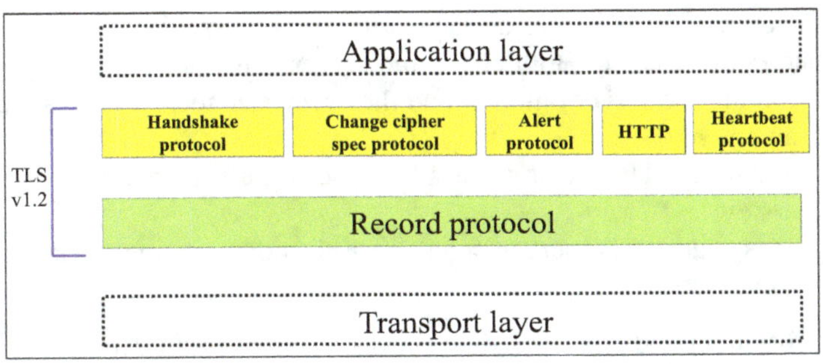

Figure 5.19 Structure of TLS 1.2 protocol

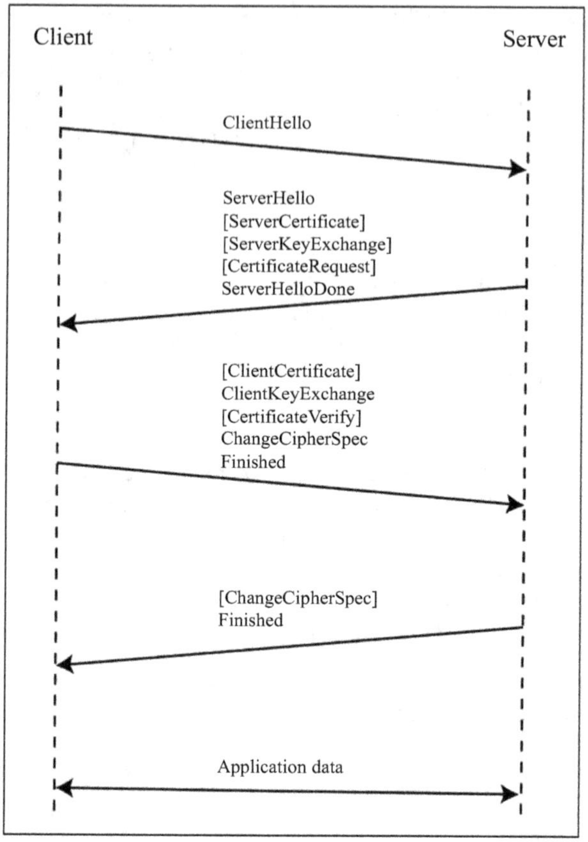

Figure 5.20 Message flow in the handshake protocol

The premaster secret: The client sends one more random string of bytes, the "premaster secret." The premaster secret is encrypted with the public key and can only be decrypted with the private key by the server. (The client gets the public key from the server's TLS certificate).

4. Private key used: The server decrypts the premaster secret.
   Session keys created: Both client and server generate session keys from the client random, the server random, and the premaster secret. They should arrive at the same results.
   Client is ready: The client sends a "finished" message that is encrypted with a session key.
   Server is ready: The server sends a "finished" message encrypted with a session key.

If both of client and server receive finished message and decrypt successfully the encrypted finished message, then secure symmetric encryption achieved. The handshake is completed, and communication continues using the session keys in secured communication.

5.9.3 Improved TLS

The IETF released TLS 1.3 in August 2018. In this section, the difference between TLS 1.2 and TLS 1.3 are described briefly in addition to identifying problems in TLS 1.2 [31].

Weakness of cybersecurity in TLS 1.2

Many weaknesses were found in TLS 1.2, described briefly as follows:

- FREAK (Factoring RSA Export Keys): cryptographic weakness using RSA Key. RSA Keys may be broken easily by anyone with access to relatively modest computing resources using the well-known Number Field Sieve algorithm.
- SLOTH (Security Losses from Obsolete and Truncated Transcript Hashes): This weakness was disclosed in 2015. It is a new transcript collision attack caused by insecure hash functions such as MD5 and SHA-1. It incurs authentication failures in the TLS and SSH and leads to impersonation attacks.
- DROWN (Decrypting RSA with Obsolete and Weakened Encryption): This weakness was disclosed in March 2016. It allows attackers to break confidentiality by revealing sensitive credentials such as passwords and credit card numbers.
- POODLE (Padding Oracle On Downgraded Legacy Encryption): This weakness was discovered by the Google security team October 14, 2014. The server chose the highest version of TLS but supported lower version for backward compatibility. Eavesdroppers intentionally negotiate each other to use vulnerable version and attack.
- LogJam and WeakDH: Diffie–Hellman key exchange is a popular cryptographic algorithm that allows Internet protocols to agree on a shared key and negotiate a secure connection. It is fundamental to many protocols, including HTTPS, SSH, IPsec, SMTPS, and protocols that rely on TLS. The Logjam attack allows a man-in-the-middle attacker to downgrade vulnerable TLS connections to 512-bit export-grade cryptography. This allows the attacker to read and modify any data passed over the connection. The attack is reminiscent of the FREAK attack but is due to a flaw in the TLS protocol rather than an implementation vulnerability and attacks a Diffie–Hellman key exchange rather than an RSA key exchange. The attack affects any server that supports DHE_EXPORT ciphers and affects all modern web browsers. Precisely 8.4% of the top 1 million domains were initially vulnerable. Attacks in 2015 showed that many TLS servers could be tricked into using small numbers for Diffie–Hellman, allowing an attacker to break the security of the protocol and decrypt.

5.9.4 Advances in TLS 1.3 [32,33]

TLS 1.3 was released in August 2018. The key goals of TLS v1.3 were as follows:

- clean up: remove unsafe or unused features;
- security: improve security with modern techniques;
- privacy: encrypt more of the protocol;
- performance: 1 RTT (return to trip) or 0 RTT; and
- continuity: backwards compatibility.

Because of weakness described previously, the following items are removed:

- key exchange: RSA;
- encryption algorithms: RC4, 3DES, Camellia;
- cryptographic hash algorithm: MD5, SHA-1;
- cypher modes: AES-CBC;
- other features:
  - TLS compression and session renegotiation;
  - DSA signatures (ECDSA $\geq$ 224 bit);
  - ChangeCipherSpec message type and "Export" strength ciphers; and
  - Arbitrary/Custom (EC)DHE groups and curves.

The following are some new and changed items:

- Cipher suites
- Handshake
- Hashed-key derivation function
- Key schedule
- Sessions

In TLS 1.2, there were many choices because cypher suites are combined with each item, such as the key exchange method, authentication, AEAD cipher mode, and PRF hash algorithm as displayed in Figure 5.21.

TLS 1.2 only has 37 cipher suites, but if the previous version is added, then it becomes 319 cipher suites. The structure of cipher suites were changed in TLS 1.3 as depicted in Figure 5.22.

TLS 1.3 supports the following five cipher suites:

- TLS_AES_128_GCM_SHA256
- TLS_AES_256_GCM_SHA384
- TLS_CHACHA20_POLY1305_SHA256

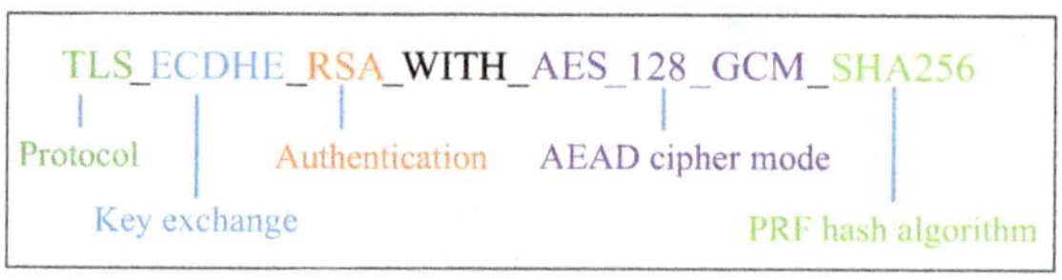

Figure 5.21 Structure of TLS 1.2 Cipher Suites combined with each item

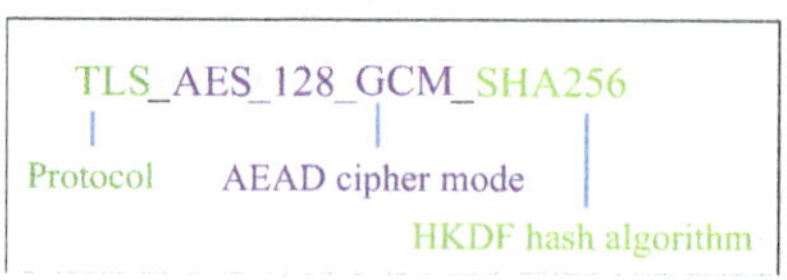

Figure 5.22 Structure of TLS 1.3 Cipher Suites combined with each item

- TLS_AES_128_CCM_SHA256
- TLS_AES_128_CCM_8_SHA256

The handshake has the following three goals:

- agree on a cipher suite
- agree to a master secret
- establish trust between client and server

TLS 1.3 made the handshake process brief and simple, reducing 2 RTT to just 1 RTT on the first time, and 1 RTT to 0 RTT upon resumption.

5.10 Conclusion

Many countries have now invested and are pursuing further developments in e-navigation services. The increase in the stringency of IMO environmental regulatory requirements makes these initiatives not only very relevant but also increasingly urgent. Both e-navigation and cybersecurity are huge focal points for shipping and ship building companies who depend on compliance with IMO requirements.

In direct response to IMO requirement for emission from the ship, the application of ICT technologies is continuing to grow in order to comply with obligations as it pertains to the emission of CO_2 and NO_x, as well as minimizing the spread of bio-objects. Shipping companies are also developing countermeasures to achieve lower fuel consumption and a more economic deployment of fleet through the application of advanced technologies and smart ship solutions.

Ship building companies and equipment manufacturers have refined their own solution platforms for smart shipping such as SVESSEL and KOGNIFAI through the use of IoT, M2M, AR (augmented reality)/VR (virtual reality), and MR (mixed reality) technologies. In order to preserve the strides made in the realm of technology, it just is important to look toward ways of keeping and creating more secure networks that are faster and can provide larger IP address space.

In smart shipping, shore and ship are intricately connected with the concentration affixed on the reliable exchange of comprehensive information going in both directions, to provide a safer, more economic, and greener means of navigation. As technologies continue to be developed and advancements are made, it is imperative that the application of new technologies is promptly adapted to for the success of the maritime domain.

List of abbreviations

VDES	VHF (Very High Frequency) Data Exchange System
SAT	Satellite
LTE	Long Term Evolution

DMX, DMA, DMB and DMC	grade of special light distillate fuel oil in ISO 8217 fuel standard
RMA30, RMB30, RMD80, RME180, RMF180, RMG380, RMH380, RMK380, RMH700 and RMK700	kinds of marine residual fuels in ISO 8217 fuel standard
HP SCR	High Pressure SCR (Selective Catalytic Reduction)
LP SCR	Low Pressure SCR
PM	Particulate Matter
BWM	Ballast Water Management
EGR	Exhaust Gas Recirculation
E/R	Engine Room
G/E	Generator Engine
FOC	Fuel Oil Consumption
DF	Dual Fuel
X-DF	name of two-stroke dual fuel engine
ME-GI	M-type Electronically Controlled Gas Injection engine
FGSS	Fuel Gas Supply System
TCP	Transmission Control Protocol
CAN	Control area Network
MS	Maritime Service

References

[1] Korean Ministry of Trade, Industry and Energy website, "4th industrial revolution on the sea, self-driving ship development project started" available online at http://www.motie.go.kr/motie/ne/presse/press2/bbs/bbsView.do?bbs_seq_n=163046&bbs_cd_n=81 (in Korean language) (accessed 22.01.2021).

[2] IMO website, "IMO takes first steps to address autonomous ships" available online at www.imo.org/en/MediaCentre/PressBriefings/Pages/08-MSC-99-MASS-scoping.aspx (accessed 22.01.2021).

[3] IMO website, "Marine Environment Protection Committee (MEPC), 74th session", 2019, available online at www.imo.org/en/MediaCentre/MeetingSummaries/Pages/MEPC-74th-session.aspx (accessed 22.01.2021).

[4] Hong-Ryeul, Ryu, 'Application of Eco-Friendly Ship Solution', Presented at 2018 Korea Maritime Week, Seoul, Korea, June 2018.

[5] WinGD website, "X-DF dual-fuel design" available online at www.wingd.com/en/technology-innovation/engine-technology/x-df-dual-fuel-design/ (accessed 22.01.2021).

[6] WinGD website, "Technology & innovation" available online at https://www.wingd.com/en/technology-innovation/ (accessed 22.01.2021).

[7] ChosunBiz website, page available at https://biz.chosun.com/site/data/html_dir/2020/05/22/2020052200874.html (in Korean language) (accessed 22.01.2021).

[8] Yung H., Yu, 'What Is Necessary for Smart Shipping and Smart Ship', Presented at ISO TC8 WG10 Smart Shipping, 2nd meeting of WG10, Tokyo, Japan, 19 April, 2018.

[9] IITP. "Analysis report-ship building". Seoul: IITP (Institute of Information & Communications Technology Planning & Evaluation) HQ Korea, 2017.

[10] Maritime Unmanned Navigation through Intelligence in Networks (MUNIN) website, available at www.unmanned-ship.org/munin/ (accessed 24.01.2021).

[11] YARA, Kongsberg, VARD and Norwegian government, "Autonomous ship 'Yara Birkeland' – How far has the industry reached?". Internet Press, 2018, available online at www.itfglobal.org/sites/default/files/resources-files/autonomous-ship-yara-birkelnad-how-far-has-the-industry-reached.pdf pp 1–6, (accessed 24.01.2021).

[12] DNV GL, "The ReVolt – A new inspirational ship concept" available online at www.dnvgl.com/technology-innovation/revolt/index.html (accessed 24.01.2021).

[13] DNV GL, "SIMAROS – An assurance framework for AI-decision support", available online at www.dnvgl.com/research/review-2020/featured-projects/simaros-safe-autonomous-shipping.html (accessed 24.01.2021).

[14] DNV GL, "ROMAS – Remote operations of machinery and automation systems" available online at www.dnvgl.com/research/review2018/featured-projects/romas-remote-operations-machinery-automation.html (accessed 24.01.2021).

[15] Esa, Jokioinen *et al.*, 'Remote and Autonomous Ships – The Next Steps', Advanced Autonomous Waterbone Applications (AAWA) Position Paper, Rolls-Royce plc., available online at www.rolls-royce.com/~/media/Files/R/Rolls-Royce/documents/customers/marine/ship-intel/aawa-whitepaper-210616.pdf pp 1–88 (accessed 24.01.2021).

[16] Norwegian University of Science and Technology (NTNU) website, "Autosea" page, available online at www.ntnu.edu/autosea (accessed 24.01.2021).

[17] Jukka, Merenluoto, "One Sea – Autonomous maritime ecosystem" available online at https://business.esa.int/sites/default/files/spacemoves2017/One%20Sea%20-%20Space%20Moves%21%202017-09-19.pdf pp 1–14, (accessed 24.01.2021).

[18] Takachika, Bunya, 'Activities of Smart Ship Application Platform 2 Project (SSAP2)', Symposium on Maritime Safety and Security (MSS) Applications, 13 June 2017, available online at www.jsmea.or.jp/ssap/assets/pdf/8c_SSAP_in_NMMT_Bremen(JSMEA20170613).pdf pp 1–25, (accessed 24.01.2021).

[19] Offshore energy website, "DNV GL, SDARI Launch Green Dolphin Bulk Carrier Concept", 2013, webpage available online at www.lngworldnews.com/dnv-gl-sdari-launch-green-dolphin-bulk-carrier-concept/ (accessed 24.01.2021).

[20] IEC, “IEC Standard IEC61162-1”. London: IEC (International Electrotechnical Commission), 2016.
[21] IEC, “IEC Standard IEC61162-2”. London: IEC (International Electrotechnical Commission), 1998.
[22] IEC, “IEC Standard IEC61162-3”. London: IEC (International Electrotechnical Commission), 2008.
[23] IEC, “IEC Standard IEC61162-450”. London: IEC (International Electrotechnical Commission), 2018.
[24] NMEA (National Marie Electronics Association), “NMEA 2000 standard”, 2018.
[25] Yung H., Yu, Ship Network Seminar of TTA (Telecommunication Technology Association of Korea), 2016.11.04.
[26] IEC, “IEC Standard IEC61162-460”. London: IEC (International Electrotechnical Commission), 2018.
[27] Xiaofeng, Wu, ‘Standardization Roadmap for Smart Shipping’, Presented at ISO TC8 WG10 2nd meeting, Tokyo, Japan, 20 April 2018.
[28] IMO ISM Code, “ISM Code and guidelines on implementation of the ISM Code 2010”. London, 2011.
[29] BIMCO, “The guidelines on cyber security onboard ship version 3”, 2018.
[30] Cloudflare website, “What happens in a TLS handshake? SSL handshake” available online at www.cloudflare.com/learning/ssl/what-happens-in-a-tls-handshake/ (accessed 24.01.2021).
[31] Yung H., Yu, ‘Security and Interoperability of Ship Network for Smart Shipping’, Presented at ISO TC8 Plenary, Copenhagen, Denmark, 19 September 2018.
[32] Andy, Brodie, “Overview of TLS v1.3”. London, UK: OWASP (Open Web Application Security Project), 2018.
[33] E. Rescorla RTFM, Inc., “The Transport Layer Security (TLS) Protocol Version 1.3 – Draft-ietf-tls-tls13-28”, Network Working Group, 20 Mar 2018, available online at https://tools.ietf.org/html/draft-ietf-tls-tls13-28 (accessed 24.01.2021).

Chapter 6
Maritime communications

Jillian Carson-Jackson[1] and Peter Pokorny[1]

6.1 Introduction

Our daily lives now include the expectation of 'connectiveness' – the use of digital communications to support our work environment, provide personal connection with friends and family, and support our entertainment through games, on-demand online television, streaming video, and more. Technology continues to develop at almost a breakneck speed, sometimes without us even understanding what the technology can do, or why. Where a faxed still image of a weather map was provided, there may be interactive weather maps with a need for bandwidth to support text, images, even video.

This chapter will look at the developments around maritime communications in a digital age; the existing and evolving roles and responsibilities for maritime communications; what is changing in the industry, and why; an overview of existing and evolving technologies for maritime use; and a glimpse into the future.

6.1.1 Addressing user requirements

The challenge to the maritime industry is to ensure technology solutions that address user requirements, which are changing almost as fast as the technologies. This includes the introduction of 'just-in-time' and 'any time' communications, within a networked environment of an Internet of Things (IoT). There is also the concern that those using current technology may not imagine what future technologies could provide, thus inhibiting innovation. While existing communications systems may be adequate to serve current needs, one must question if they will be able to address evolving requirements driven by increasing digitalisation of the maritime industry and autonomation of systems. This was first identified through the International Maritime Organization (IMO) work on e-navigation* [1] and is

[1]JCJ Consulting Pty Ltd, Bruce, ACT, Australia

*e-Navigation is defined as 'the harmonized collection, integration, exchange, presentation and analysis of marine information on board and ashore by electronic means to enhance berth to berth navigation and related services for safety and security at sea and protection of the marine environment'.

continuing through the development of Maritime Autonomous Surface Ships (MASS)† [2].

The challenge to provide systems which are available 24 h/day, in all weathers, in all parts of the globe continues, as technologies change and evolve.

Through its work on the development of the VHF data exchange system (VDES), the International Association of Marine Aids to Navigation and Lighthouse Authorities (IALA) has developed a set of VDES user requirements, which is cross-referenced against the IMO Maritime Services. These are presented in an IALA Guideline [3] which includes communications to support the exchange of information related to vessel traffic services (VTS); provision of Maritime Safety Information (MSI); telemedical assistance; chart and publications services; meteorological services; and real-time hydrographic and environmental data.

Studies of user requirements, as well as developing technologies, provide a basis for the analysis of candidate technologies and have led to the development of guidance on their implementation. IALA has developed a guideline for the reporting of e-navigation test beds [4] and maintains a database of reports. While these studies provide a basis for future work, the performance requirements, including the data throughput requirements and communications requirements to support e-navigation, continue to evolve.

6.1.2 Supporting safe, efficient, and pollution-free shipping

To support these activities, there is a need for robust, reliable, and redundant digital data exchange. Robust in that all the overall approach provides for a strong, well-built network. Reliability to ensure consistent communications can be trusted. Redundant in that there are multiple systems or options for transferring the information – even if there is a failure in one aspect, the communication can continue. In an ideal implementation, these work autonomously, choosing the best communications option for use, through a least-cost, always available model.

Of course, when we talk about ships, we need to be careful, as there are many specific ship types which require special radio/navigation and other electronic equipment. In addition, there are significant user groups of maritime communications that are not always covered by the term 'ship' – including many fishing vessels and pleasure craft.

International Convention for the Safety of Life at Sea (SOLAS) ships are often discussed as the primary responsibility at IMO, but in reality, SOLAS vessels are in the minority of vessels that exist,‡ as reported by UNCTAD [5,6], when one considers the

†Autonomous ships: For the purpose of the (IMO) regulatory scoping exercise, 'Maritime Autonomous Surface Ship (MASS)' is defined as a ship which, to a varying degree, can operate independently of human interaction.

‡World total propelled seagoing merchant vessels of 1,000 GT and above=52,684 at 1 January 2019 (Table 2.6, p. 47, *Review of maritime transport*, 2019, United Nations Conference on Trade and Development (UNCTAD)) and for vessels 100 GT and above; world total propelled seagoing merchant vessels of 100 GT and above=90,917 as at 1 January 2016 (Table 2.6, p. 45, *Review of maritime transport*, 2016, UNCTAD).

fishing fleets of the world. Those are the Food and Agriculture Organization (FAO) of the United Nations Global Record of Fishing Vessels, Refrigerated Transport Vessels and Supply Vessels records in excess of 5.2 million vessels as of July 2020 [7]. To this we need to add smaller fishing vessels and recreational vessels that are sometimes described as 'non-convention' vessels. Ensuring interoperability and compatibility between SOLAS and non-SOLAS vessels for Global Maritime Distress and Safety System (GMDSS) purposes is therefore a major challenge.

Within the 'SOLAS' fleet, there are differing communications requirements, including the monitoring of containers on a container vessel; emissions monitoring; reporting of weather observations through the World Meteorological Organization's Volunteer Observing Ships Scheme [8]; remote machinery monitoring; and requirements for high-capacity communications by passenger ships to name just a few.

Looking beyond the SOLAS fleet, fishing vessels and pleasure craft are other craft that have, in some cases, unique communications requirements. On some fishing vessels, there are more electronics on the bridge than on a regular SOLAS cargo ship, and some pleasure craft have much the same fit-out. This category also includes oceanographic and other research/special purpose vessels, non-SOLAS passenger vessels, and specialist vessels serving the offshore oil and gas industry, cable-laying ships, icebreakers, pilot vessels, tugs, defence vessels, and law enforcement vessels such as fisheries patrol and coastguard vessels. The suppliers to these industries cover more than one ship type or market segment.

The challenge that is facing the maritime radio communications sector is how to transfer more information, more often, to meet changing user requirements. With an emphasis on the effective and efficient use of the radiofrequency spectrum (spectrum), the maritime industry is continuing to challenge the current reliance on analogue communications (e.g. VHF voice) and digital data exchange, for example the automatic identification system (AIS). Spectrum is a finite source, and there is an increasing demand on the existing spectrum by other industries. However, while the spectrum is finite, there are options to better use the spectrum.

While we recognise that the arrangements for maritime communications for defence vessels are out of scope for this chapter, there is some overlap with non-military applications due to the dual-use nature of radio communications. This recognises the need for interoperability with non-defence vessels in specific instances – for example for search and rescue (SAR), operational deployments, and the ability of defence vessels to work with vessels of opportunity.

6.2 Digital communications – existing and developing

At the basic level, digitisation is the conversion of text, pictures, or sound into a digital form that can be processed by a computer. The concept of digitisation of the maritime industry is therefore transferring information – images, video, and text – in a digital format. The next opportunity through digitisation is then the exchange of that information through digital systems.

In the maritime world, there have been aspects of digital data exchange for some time, including digital selective calling (DSC), as part of the GMDSS and AIS. Using existing, and developing, technologies to support digital data exchange, there is opportunity to address effective, efficient, and consistent transfer.

6.2.1 Digital data communications – some considerations

With a global area of operation, and the need for multiple communication channels (redundancy), there are several digital data transfer technologies that can support shipping. These include both terrestrial and satellite options. In addition to providing communications and data exchange within identified areas of operation, there is a need to consider the level of automation for data exchange within each system and the interoperability of different systems. Some systems are autonomous and continuous, such as AIS, while others require human intervention to enter the data and then decide on the technology to use, or a requirement to confirm the data and transport means prior to sending.

Considerations for digital data communications include [9]

- latency,
- bandwidth and channel capacity, and
- life cycle costs.

Within these three specifics basic considerations of the technology – for example, can it both send and receive data; how does the system address priority of service; what is the area of operation; and what is the requirement for data to address commercial requirements?

6.2.1.1 Latency

Latency is defined, in a general sense, as the time interval between the stimulation and response, or a time delay between the cause and the effect. For digital technology, this is the delay between source and destination caused by the (mobile) network as described by Kurian [10]. Within the context of maritime communications, service latency is defined as time taken for the delivery of the total service from the initiation of the delivery of the service component to the completion of the delivery of that service component [9]. This may include elements outside the remit of the technology used for the communications. Latency may include, for example, the time required to refresh a system (i.e. a low Earth orbit (LEO) satellite), which can be called 'revisit time'.

There are many different requirements, existing and evolving, for digital data exchange within the maritime environment. Acceptable latency will depend on the specific requirement. For example, updates of nautical charts may be acceptable with longer delays, even deferred until within wireless network coverage, while safety information on the status of aids to navigation (AtoN) or SAR will require much shorter latency.

For example, the International Civil Aviation Organization (ICAO) *Required Communications Performance (RCP) Manual* [11] defines four parameters that can be used to characterise the RCP of any communications system (i.e. wired or wireless, terrestrial LOS, or satellite). These four parameters are defined as follows:

- *Availability*: The probability that an operational communication transaction can be initiated when needed.
- *Continuity*: The probability that an operational communication transaction can be completed within the transaction time.
- *Integrity*: The probability of one or more undetected errors in a completed communication transaction.
- *Transaction time*: The maximum time for the completion of the operational communication transaction after which the initiator should revert to an alternative procedure.

RCP specifications are identified by a designator (e.g. RCP 240), which indicates the RCP transaction time readily apparent. RCP 240 indicates a transaction time of 240 s, and RCP 400 indicates a transaction time of 400 s. Associated with each of these are continuity (probability), availability (probability), and integrity (acceptable rate/flight hour) [11]. As MASS will be reliant on robust, reliable, and redundant communications links, these concepts may be useful for specifying the performance requirements for systems to support MASS.

6.2.1.2 Bandwidth and channel capacity

In radiofrequency spectrum terms, bandwidth relates to any portion of the electromagnetic spectrum used for communications by radio waves or Hertzian waves [12]. There is a relationship between the amount of data that can be conveyed via a portion of the radiofrequency spectrum, known as the frequency 'band', and the width of that band, called the 'bandwidth'. Frequency here refers to cycles per second for a periodic waveform (SI unit 'hertz', SI symbol 'Hz') [13].

The more data that are needed to be conveyed, the greater the bandwidth is required. The relationship is non-linear and varies with such factors as the modulation scheme(s) used, data coding schemes, error correction, duty cycle, number of subcarriers, antenna size and type, diversity, access time and method, signal-to-noise ratio, and signal-to-interference ratio of the communications channel, to name just a few factors.

For simplicity, we will be using 'bandwidth' as it relates to the amount of data that can be sent from one point to another, within a context of a period of time, as an indication of channel capacity or 'throughput'. This is expressed as the rate, in bits[§] per second (bit/s or bps) or multiples of bits per second (kbit/s or kbps as well as Mbit/s or Mbps).

The requirement for bandwidth in maritime communications depends on the type, and amount, of data to be sent. For example, text-based data may be sent effectively with relatively low bandwidth requirements, while images or video require a much higher bandwidth.

The maritime environment has traditionally had significant limits on the available bandwidth once a vessel goes beyond a terrestrial-based network system.

[§] A 'bit' is shorthand for a 'binary digit', and in our context it refers to the rate at which information is passed.

The evolution of satellite-based technologies, including geostationary earth orbit (GEO), geostationary satellite orbit (GSO) or geosynchronous orbit (GSO), and LEO satellites, is providing increased digital data exchange capabilities for emergency and general ship operations. Developments in medium Earth orbit (MEO) satellites have, to date, mainly been focussed on emergency support, for example, MEO SAR and Global Navigation Satellite System (GNSS), but MEO-SATs are now providing additional digital data exchange capability and are being used in combination with GEOs ('multi-orbit' solution). LEOs are also being used in the Polar regions, including via store and forward techniques in combination with GEOs, for time-tolerant data transfers [14].

6.2.1.3 Life cycle costs

When considering digital data capabilities for the maritime environment, there are a number of cost elements to consider. These include

- equipment purchases,
- licence charges for ship stations (ship earth stations for satellite),
- subscription (user) charges for service, and
- maintenance agreement costs.

Depending on the area of operation, these costs could increase for vessels that are beyond terrestrial communications links. The expectation for communications includes the automated approach for least-cost routing supporting seamless communications [15].

Some of the rules managing seamless data communications could be programmed into the system, for example those concerned with safety of navigation or priority communications under GMDSS. Other rules might be set by the ship's operator in accordance with company procedures or the operator's commercial model.

6.3 Maritime communications in a digital age

As technology is developing, there is an increasing demand for digital data exchange. This was realised in the development of the GMDSS and has continued through not only communications to/from shipboard personnel but also shipboard equipment (equipment monitoring) and cargo monitoring. The developments of the IoT are providing the expectation for the interconnectivity of objects, sensors, and processes to facilitate efficient, safer, and secure ship transits to always be available. The developments in 'smart ports' include the link between ship, shore, and the hinterland to encompass the entire supply chain.

6.3.1 Introduction

Digital communications permeate our daily lives – both for work and for recreation. The introduction of 'any-time, any-where' access to information through the IoT; text-based or image-based communications; geospatial locating; and more is driving demand for faster, more robust, and more integrated communication solutions.

Within the e-navigation framework, solutions will depend on robust, reliable and redundant communications for data transfer. These communications will be required for ship-to-ship, ship-to-shore and shore-to-ship. The requirements for data capacity reflect current needs (known) as well as developing needs (unknown) and, as we have seen in other industries, are expected to change over time. Several e-navigation test beds have been carried out to identify data transfer requirements. These provide a basis for initial work on changing digital data communications links and lay the groundwork for further developments within a rapidly changing maritime environment.

It is expected that the user of the Digital Communications Technologies will benefit from a system that identifies the most appropriate transfer mechanism, within a least-cost routing process. In this manner, the users will not be required to choose the technology as the system will do this for them, within a set of parameters that could include notification of, for example, instances where transfer of data may be delayed or an additional cost may be incurred.

In all aspects of society, we are seeing the demand for communications expanding exponentially and use cases changing over time as data throughput becomes available. The maritime community is diverse, ranging from small pleasure boat users, domestic vessels, fishing vessels, coastal traffic, and large vessels plying the open ocean. Beyond the requirements for data by vessels, data exchange supports ocean data acquisition buoys and other hydrographic and oceanographic sensors.

There is a wide range of access to maritime communications technologies, with some smaller vessels having no, or limited, communications, while other vessels, such as super yachts, having access to significantly advanced communications. Vessels that come under the SOLAS Convention focus on communications related to carriage requirements or those that have proven economic benefits. Vessels that are required to address the obligations of the SOLAS Convention are required to have interoperable systems, meeting stringent cybersecurity requirements, and may not benefit from the cutting-edge technologies available to non-SOLAS vessels.

6.3.2 Coastal communication networks

With the advent of the GMDSS, and greater access to satellite communications, the extensive shore infrastructure that characterised the pre-GMDSS era has dramatically decreased. The ability of higher data rate communications via satellite has all but eliminated the use of narrowband direct printing (NBDP) or radio telex. Medium frequency (MF) NBDP [16] is widely used for GMDSS Navigational Telex (NAVTEX) for SOLAS/GMDSS sea area A2, but there are acknowledged limitations on the amount of data which can be transmitted in a 10-min time slot. MF navigational data (NAVDAT) is a likely candidate to augment/replace NAVTEX due its greater data capacity [17] among other benefits.

Morse code radio communications by SOLAS vessels have not been a requirement for over 20 years, after having been used for nearly a century (however, some countries, at the time of writing, such as China, Philippines, and the Russian

Federation, continue using Morse code radio to some extent within their fleets for non-GMDSS communications) [18]. Dedicated maritime channels for Morse radio communications have been removed from the International Telecommunication Union (ITU) Radio Regulations (RR) at World Radiocommunication Conference (WRC)-12 (relevant provision effective from 1 January 2017) [12], and where Morse is used, are not afforded protection under the RR, from digitally modulated emissions¶ [19,20].

The number of high frequency (HF) maritime coast stations has significantly diminished. Major MF and HF maritime coast stations which provided commercial services for radiotelephony, NBDP, and Morse code had lost their economic model and have been largely dismantled.

Nevertheless, HF NBDP is still a GMDSS requirement for certain ships depending on their area of operation [21]. HF NBDP is a follow-on mode form of distress and safety communications following the initial DSC alert or safety announcement, but it is rarely or if ever used, and possibly will be removed as a requirement as part of the GMDSS modernisation. Broadcasting of MSI via HF NBDP remains in use in some parts of the world, but HF NAVDAT would allow increased MSI data to be broadcasted to ships [22]. The results of a successful field trial of HF NAVDAT via Digital Radio Mondiale in Alaska by the United States Coast Guard were published in 2018 [23]. China and France have also conducted field trials at MF and HF, respectively [24,25].

Significant VHF radiotelephony coast station networks are in use globally. Many channels are used to support VTS, local port services, and ship-to-ship communications. VHF maritime mobile frequencies are also used on aircraft for maritime activities, including pilot transfer and SAR.

However, a significant new infrastructure has sprung up, namely coastal VHF AIS base stations and repeaters. These AIS stations are generally operated (though not exclusively) by the competent maritime authority. These stations, being connected to a central node(s) via communications links of various types, have the potential of being upgraded with VDES equipment, thereby enabling a VHF e-navigation communications layer along coasts [26,27].

To support e-navigation, the terrestrial component of VDES (VDE-TER) now is an internationally agreed set of channels for 'international VDES' which is globally harmonised, plus a significant number of additional channels for regional use [12]. Additionally, a satellite component for VDES (VDE-SAT) was agreed at WRC-19, which is also globally harmonised [28].

6.3.3 Mobile broadband

All mobile broadband systems (third generation (3G), fourth generation (4G), and fifth generation (5G™)) are based on the ITU's IMT (International Mobile Telecommunications) standards. There is no agreed definition of 'broadband' as this

¶Digital modulation is the modulation of a sinusoidal carrier by a digital signal, as against an analogue signal. Simple digital modulation is digital modulation in which the radiofrequency signal can assume four or fewer values of frequency, phase, or amplitude. Many modern modulation schemes are not simple but are spectrally more efficient.

is subjective – depending on user requirements. Broadband combines connection capacity and speed. As requirements change, the definition of broadband changes.

As a basic reference, anything faster than 1.5–2 Mbps transmission capacity is considered as being broadband [29].

The development of technology continues to follow Moore's Law where the number of transistors in a dense-integrated circuit doubles every 2 years, which was updated to indicate a doubling every 18 months. With the demand to transfer more information, faster, over longer distances, there are the limitations of physics. Innovative processes are addressing the ongoing demand for better, faster, and stronger (robust) transfer opportunities.

6.3.4 The Internet of Things

The IoT has entered common usage through the growth of global interconnectivity. When discussing IoT, however, it is important to consider the following definitions in use:

- IoT is a concept encompassing various platforms, application, and technologies implemented under a number of radio communication services [30].
- A global infrastructure for the information society, enabling advanced services by interconnecting (physical and virtual) things on the basis of existing and evolving interoperable information and communication technologies [30].

In essence, the goal of IoT is to digitise the world – connecting people, processes, data, and things.

As an IoT, it is also necessary to understand what these 'things' are. There are two types of 'things':

- Physical things that exist in the physical world and are capable of being sensed, actuated, and connected (e.g. industrial robots, goods, and electrical equipment).
- Virtual things that exist in the information world and are capable of being stored, processed, and accessed (e.g. multimedia content, application software) [30].

The IoT relies on intelligent connections that facilitate the automated exchange of data elements. Examples of intelligent connections include

- connectivity (digitise access to information) – e-mail, web browser, and search;
- networked economy (digitise business process) – e-commerce, digital supply chain, and collaboration; and
- immersive experiences (social, mobility, cloud, video).

The digital data elements are transferred through a transport layer. Therefore, to address the IoT from a radio propagation point of view, the transport of data can be divided into fixed, short range, and long range.

Fixed and short-range technologies include

- radiofrequency identification, Bluetooth®, Zigbee™, and Wi-Fi™

and long-range technologies include

- non-Third-Generation Partnership Project (3GPP) Standards – low-power wide-area network (LPWAN)
- 3GPP[||] Standards

3GPP has introduced a 'maritime vertical' and has started a study item for maritime to move vertical domains towards the 3GPP under SA Working Group 1 [31] and further studies are underway. 3GPP has noted the work of the IMO on e-navigation, modernisation of the GMDSS, and the Republic of Korea's SMART Navigation Project [32].

The Republic of Korea's SMART Navigation Project arose out of a decision by its Ministry of Ocean and Fisheries, on the selection of Long-Term Evolution (LTE™) technology to implement e-navigation for non-SOLAS ships and to reduce the information gap at sea by providing mobile broadband services over LTE [32].

The Smart Navigation Project includes LTE in the 700-MHz frequency band, and the choice was taken with the backdrop of GMDSS modernisation, e-navigation for SOLAS ships, ITU's IMT-2020,[**] and 5G standardisation to satisfy IMT-2020 over the same system (or technologies under the 3GPP framework). The strategy is that coverage within 100 km of the coast will be provided by LTE, and from 100 to 300 km will be covered by 'Digital GMDSS'. The VDES will be used (including terrestrial and by satellite) alongside LTE as explained by An [33].

The working processes within 3GPP are significantly different to that traditionally used in the maritime environment. This includes the speed of development of standards and the reliance on economies of scale. The maritime industry cannot offer the same economies of scale as the major players in the 3GPP space, but the important work has started.

6.4 Roles and responsibilities in maritime communications

The standards world can be bewildering in its complexity, and standardisation participants are representatives from the industry, government regulators, academia, and volunteer experts, many of which serve multiple standards organisations

[||] The 3GPP unites seven Telecommunications Standards Development Organizations (The Association of Radio Industries and Businesses (Japan), ATIS, China Communications Standards Association, ETSI, TSDSI, Telecommunications Technology Association (Korea), Telecommunication Technology Committee (Japan)). The technologies include technologies such as 2G, 2.5G, Universal Mobile Telecommunications Service, 3G, High Speed Packet Access, LTE, 4G, LTE Advanced, LTE Advanced Pro, Next generation, related 5G standards, and evolved Internet protocol (IP) Multimedia Subsystem.

[**] Resolution ITU-R 55-1 (REV.2012), *Naming for International Mobile Telecommunications*, resolves (1) that the term 'IMT-2000' should encompass also its enhancements and future developments, (2) that the term 'IMT-Advanced' should be applied to those systems, system components, and related aspects that include new radio interface(s) that support the new capabilities of systems beyond IMT-2000; and (3) that the term 'IMT' should be the root name that encompasses both IMT-2000 and IMT-Advanced collectively.

directly or in liaison roles. Maritime equipment is only a small subset of the total number of standards these organisations create and maintain.

6.4.1 Standards (regional and national/international bodies)

The standards regime which has evolved for maritime radio communication/navigation and other shipboard electronic devices and installations is long-standing and involves the close cooperation between IMO, ITU, IALA, the International Hydrographic Organization (IHO), the International Electrotechnical Commission (IEC), the International Special Committee on Radio Interference, the International Organization for Standardization (ISO), the European Telecommunications Standards Institute (ETSI), National Marine Electronics Association (NMEA), Radio Technical Commission for Maritime Services, and the International Association for Marine Electronics Companies (CIRM), and this is not necessarily the exhaustive list. The CIRM represents many of the world's maritime electronic manufacturers, system integrators, software companies, service providers, and regulators. The International Marine Electronics Alliance is a non-profit alliance which serves to be the NMEA technical interface with international standards bodies and organisations [34]. IALA is also progressing its decision to become an intergovernmental organization (The IGO Project) [35].

The IMO and ITU are UN specialist agencies, and regional standards bodies, such as ETSI, play a major role in the development of maritime radio communications and navigation equipment standards which put into practise and interpret the decisions made in other forums like the IMO and ITU, so that equipment can be manufactured. Some regional standards bodies are de facto international standards bodies for some standards. The Institute of Electrical and Electronics Engineers (IEEE) is also responsible for many standards. The complexity of IoT standardisation, for example, is shown in Figure 6.1.

Convergence in the market for communications in the maritime shipping industry is unavoidable and may occur faster if the standards regime were able to respond faster. This may frustrate manufacturers who are constrained to produce equipment which must be first type approved, with the associated costs and delays. Manufacturers and suppliers always need to prepare for the next revision of the technical specification, as government regulations may make their current stock of equipment obsolete, and many participate in the standards-making process to ensure that a testing standard is implementable. Although newly type-approved equipment may not need to be fitted to existing ships, ships newly constructed or undergoing major refits should thereby receive the latest equipment.

Many 'standards' are not called standards at all, and there are many different types. For example, IMO MSC.1/1609, *Guidelines for the standardization of user interface design for navigation equipment,* refers to 5 IMO Performance Standards, 6 IMO Guidelines, 1 IMO procedure, 18 ISO/IEC standards, and 1 IHO standard [37]. Standards developed by the ITU are called Recommendations (with a capital 'R'). ISO and IEC are international standards development organisations, with

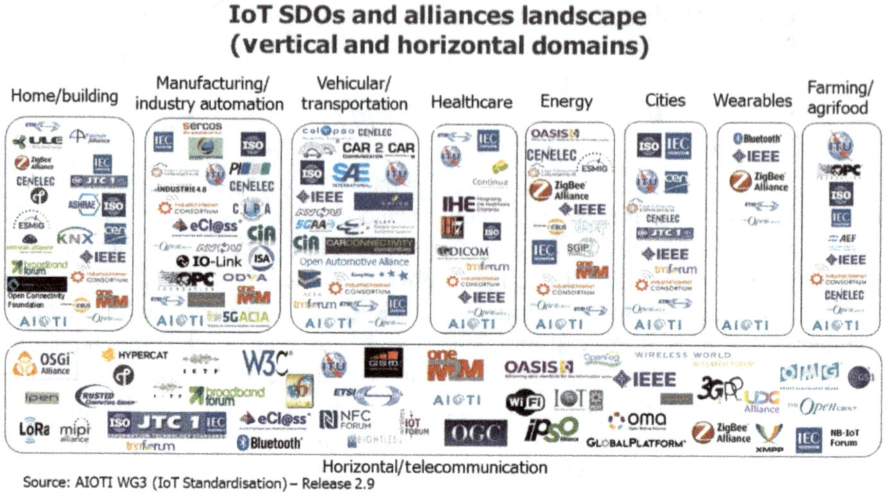

Figure 6.1 IoT SDOs and alliances landscape (vertical and horizontal domains). Source: Alliance for Internet of Things Innovation WG03, IoT Relation and Impact on 5G, Release 3.0 [36]

representation via national committee structures composed of government, industry specialists, testing houses, and other experts. IEC Technical Committee 80 (TC 80) develops testing standards on the basis of IMO Performance Standards, and other instruments, such as IMO Resolutions of its Assembly or its Maritime Safety Committee. IEC TC 80 also maintains liaison with other relevant standards committees, industry organisations, IMO, and ITU [38]. There are also joint technical committees between the two organisations.

The ITU is organised into three sectors: The Standardization Sector (ITU-T), Radiocommunication Sector (ITU-R), and Development Sector (ITU-D). In former times, the ITU-T was the CCITT,[††] and the ITU-R was the CCIR.[‡‡] References to CCITT and CCIR standards are still encountered today.

The ITU-T and ITU-R both develop and maintain a range of standards used in all telecommunications and radio communications equipment used in the maritime industry. ISO and IEC develop detailed testing and other standards for not only communications equipment but also for electrical devices, machinery, cabling, lighting, switchboards, electromagnetic compatibility, bridge layout, paint, medical equipment, fire-fighting equipment, cargo equipment, lifting appliances, and containers, to name but a few. In addition, a vast array of standards for information technology (IT) finds application in the maritime industry. From this viewpoint, the radio communications standards comprise a relatively small subset.

[††]Consultative Committee on International Telephony and Telegraphy (CCITT).
[‡‡]International Radiocommunication Consultative Committee (CCIR).

A useful history and discussion on the navigation equipment standards for merchant ships is by Beattie [39], which discusses the evolution in the 1980s from national to international standards for such equipment, but Beattie also mentions the then Future GMDSS. An agreement between ISO and IEC allocates the responsibility of each organisation for certain navigation equipment [39].

6.4.2 Cybersecurity

The Internet in all its complexity is a triumph of standardisation and enhances global trade, but with it has come the need for cybersecurity. The IMO has developed guidelines on cyber risk management referencing the ISO/IEC 27001 standard on security management systems. The IMO has adopted Resolution MSC.428(98) on 'Maritime Cyber Risk Management in Safety Management Systems' [40], and MSC-FAL.1/Circ.3 on 'Guidelines on maritime cyber risk management approved by the Facilitation Committee' [41].

IEC TC 80 has published several standards relating to the security of equipment interfaces using the Ethernet protocol, used on the command bridges of ships (IEC 61162-450[§§] and IEC 61162-460[¶¶]), and another on general requirements (IEC 63154[‖‖]) are in progress at the time of writing. Maritime industry trade literature indicates a growing market for cybersecurity companies who can provide an ongoing risk management in cybersecurity to ship operators.

6.4.3 Pressure on the radiofrequency spectrum (spectrum)

The maritime community acknowledges that the spectrum for radio communications in all its forms is a finite resource. Even where spectrum is available, it is always difficult to achieve international harmonisation, something which is essential for a global industry like shipping. Once harmonisation is achieved, there are still competing services wishing to share the spectrum used by incumbent services, and if maritime spectrum has been allocated and is not being used, it may become a candidate for reallocation to other services.

For these reasons, all the communications requirements for the IMO's GMDSS modernisation and e-navigation are focussed on using existing maritime allocations more efficiently. Maritime and aviation spectrum mostly is internationally harmonised for operational and safety reasons. However, this has not prevented governments and economists attempting to put a value on spectrum and implement pricing regimes of various types on existing usage, seeking to maximise economic benefit to governments [42]. In distress and safety situations, the

[§§]IEC 61162-450, Maritime navigation and radiocommunication equipment and systems – Digital interfaces – Part 450: Multiple talkers and multiple listeners – Ethernet interconnection.

[¶¶]IEC 61162-460, Maritime navigation and radiocommunication equipment and systems – Digital interfaces – Part 460: Multiple talkers and multiple listeners – Ethernet interconnection – Safety and security.

[‖‖]IEC 63154, IEC Maritime navigation and radiocommunication equipment and systems – Cybersecurity – General requirements, methods of testing and required test results.

'laws' of economics no longer apply and efforts to quantity economic benefit seem counter-productive.

The maritime mobile service spectrum is fully shared between all users (ship and coast stations). In this respect, sharing in this context is not the sharing between services in Article 5 of the ITU RR, but operationally shared in the real world, operating on what economists call a 'private commons basis' [42].

As there is little or no opportunity for additional 'maritime' spectrum (or at least where *maritime mobile service* is the primary service) and since there are constraints to implementing technological change due to the many stakeholders in the maritime industry, two options present themselves:

- Implement more efficient use of existing maritime spectrum, which means that legacy systems should be retired, forcing more spectrum efficiency equipment to be adopted.
- Any additional data communications requirements must be found outside the 'maritime' spectrum, which necessitates the maritime user becoming another user of the global broadband user spectrum.

The IMT community has been successful in recent years in obtaining additional spectrum (displacing or sharing with other services, such as satellite services), so the maritime community should become more aware of the potential for these technologies for operational use.

6.4.4 Innovation in maritime communications

Vessels not directly subject to the provisions of SOLAS ('non-SOLAS' or 'non-Convention') are in some cases able to take advantage of technological change at a faster rate than SOLAS vessels. Many useful features of non-type-approved equipment may not be available to SOLAS vessels, but not all such features necessarily receive support at fora like the IMO or ITU, or they may receive support at lower levels, only to be rejected at higher levels.

Examples of such features include the ability to directly open a communications channel unambiguously with another vessel via a chart plotter/display by clicking on its AIS target icon which, in turn, sends a VHF DSC message to the other vessel. Non-SOLAS vessel equipment sometimes combines equipment types which tend not to be the case for SOLAS vessels, for example, an integrated AIS transceiver combined with a VHF marine radio and chart plotter. An example of this is provided in Figure 6.2 [43].

There have been cases of innovative suggestions at IMO, which do not find immediate favour due to the wariness of shipowners and operators, and others, since a new requirement at some future date may have economic and other implications. Some features not accepted can be quite subtle but important, and keywords become critical ('must' or 'shall' replaced by 'may'). Practitioners in maritime safety and SAR are also justifiably conservative in their approach to implement change, as the whole world must be brought along to achieve harmonisation under the *International Convention on Maritime Search and Rescue 1979.*

Figure 6.2 Integrated AIS/VHF/monitor 'Cortex'. Reproduced, with permission, from Reference [43]

However, innovation can be thought of as the finding of solutions to problems and not simply change for changes' sake.

6.4.4.1 Constraints

SOLAS Chapter IV/regulation 14 requires all (GMDSS) equipment in the chapter to be type approved by the administration. To avoid possible misunderstanding by Port State Control Officers, this has been clarified in 2006 by the IMO in circular MSC.1/1221, Validity of type approval certification for marine products [44]. This circular is a key document which influences shipowners, shipbuilders, manufacturers, and the associated supply chain for the equipment installed on a ship subject to the SOLAS Convention. Essentially this means that once a type-approved product is fitted to a ship, it remains type approved irrespective of the expiry of any type approval certificate.

In practical terms, this means that equipment generally will not be replaced or upgraded unless the equipment fails or can no longer be repaired, where the ship is newly constructed, or an existing ship undergoes a major refit. Replacement of radio equipment with new features or updated standards may take 30 years to occur and is closely linked to the operational lifetime of a ship. This contrasts with the ITU RR where certain maritime communication ITU-R Recommendations which are incorporated by reference are required to be the 'most recent version' or a near-recent version incorporated by reference in those regulations. A submission from the IEC to the IMO, the IMO Sub-Committee on Radiocommunications, and SAR (COMSAR) in 2012 (COMSAR 16/9) stated that the average age at which ships are withdrawn from service is 32 years. If a change is made to the design of, for example, a GMDSS radio distress alert button, it may take some 30 years to implement this change globally in the SOLAS fleet [45].

An exception to this occurs when a service provider ceases to provide a service, and the required equipment is no longer required, or where new equipment is required. The service provider provides due notice of this cessation, and in some cases, has extended the operational life of the service where there has been insufficient take-up of the replacement/upgraded service and equipment.

Examples of cessation of service include Inmarsat L-band emergency position-indicating radio beacon (EPIRB), Inmarsat-A, Inmarsat-B, Inmarsat-D, D+, and Inmarsat-M/mini-M services, the first three being approved GMDSS services (Inmarsat Fleet 33, Fleet 55, and Fleet 77 are planned to be retired towards the end of 2020 at the time of writing) [46]. Another example is the cessation of satellite monitoring of 121.5 MHz distress beacons by the COSPAS-SARSAT satellite system [47]. Newer and even more capable services replace these services, but it shows how dynamic is the communications world, constantly adapting to accommodate the need for increased data rates. A submission from Australia to COMSAR 16 lists many of the technological changes from 1992 to 2012 that have occurred in radio communications and radio navigation related to the maritime industry [48].

Service providers also have their own constraints where the equipment needed to provide the service also has a limited service life and costs of software and hardware maintenance are to be considered. There are also competing pressures in the satellite segment which may drive choices made by a satellite service provider.

Changes to the GMDSS satellite segment (recognised mobile satellite service) are overseen by the International Mobile Satellite Organization (IMSO), an intergovernmental organisation which oversees the performance of all components and the availability of the mobile satellite communications systems recognised by IMO and also is responsible for ensuring the performance of the long-range identification and tracking (LRIT) system as required under SOLAS Chapter V/regulation 19-1 [49].

In the satellite industry, where specific providers have unique equipment, they may become ‘certified’ via arrangements with the industry, and type approval and certificates of compliance can be obtained through accredited ‘testing houses’. Examples of such equipment include satellite terminals, and their antenna equipment. IEC TC 80 publishes several standards to facilitate such certification for SOLAS purposes.

Due to the processes of the ITU and standards bodies, it may take 8–10 years for any changes in a standard to be reflected in a government’s updated regulatory instrument.

6.5 What is changing and why

The modern expectation of being connected everywhere, all the time, is becoming a ‘human right’ [50]. Political discussions identify the ‘digital divide’ where access to the Internet, which requires communications capabilities, is seen as critical to 'combating situations inequality . . . '. This access to broadband, and the Internet, is challenging in many maritime environments.

There are a number of requirements for maritime communications which tend to address one of three key areas: safety, operational, or commercial. The key requirements for each area are identified in Table 6.1, as adapted from the IALA Maritime Radio Communications Plan (MRCP) [9]. There are also requirements for automated, machine-to-machine communications, as well as communications with human intervention. In addition, as indicated in the IMO e-navigation strategic implementation plan (SIP) [51], there are six main areas of operation:

1. port areas and their approaches;
2. coastal waters, confined and restricted areas;
3. open sea and open areas;
4. areas with offshore and/or infrastructure developments;
5. polar areas; and
6. other remote areas.

There are many different communications options available; however, the technologies are often developed to address a specific application. In addition, different communications options are effective in different areas of operation. This means that both ship and shore stations need to implement a range of technologies to address the different requirements.

Table 6.1 Requirements for maritime communications

	Safety	**Operational**	**Commercial**
Position reporting	X	X	
Aids to navigation	X	X	
Maritime Safety Information	X	X	
Weather reporting/provision	X	X	
GNSS/DGNSS	X		
VTS reporting	X	X	
Distress/urgency/safety calls	X		
Ship reporting	X	X	X
Notice of Arrival (NOA)	X	X	X
Notice of Readiness (NOR)		X	X
Chart updates	X	X	
Equipment monitoring		X	X
Emissions monitoring		X	X
Voyage orders			X
Commercial port services		X	X
Crew/passenger manifest	X	X	X
Crew training	X	X	
Crew personal communications		X	
Access to publications and port handbooks	X	X	
Access to vessel equipment manuals		X	X
Long-range identification and tracking (LRIT)	X	X	

To address the changing and increasing, requirements for maritime communications, the international maritime community is focusing on three key areas: spectrum, digitalisation, and standardisation.

6.5.1 Radiofrequency spectrum (spectrum)

Spectrum is limited by the constraints of physics. This means that spectrum is a finite resource. Within the maritime industry, there is a need to share more information, more often, in all areas of operation. This puts pressure on the radio spectrum available, which is a finite resource. In addition, the maritime industry is competing with other global sectors to access the spectrum.

To address the increasing demand for digital communications, the maritime environment has reviewed the existing use of radio spectrum through reassigning spectrum for specific uses (e.g. AIS), grouping of spectrum to form larger bands (e.g. VDE), and reviewing the manner in which the spectrum is accessed.

The review of spectrum requirements has been developing over time, as presented in the IALA MRCP [9]. Activities that specifically affect spectrum requirements include the GMDSS modernisation and the development of MASS.

6.5.1.1 GMDSS modernisation

This summary is included to show the deliberate (and lengthy) processes used in international organisations which use consensus building to achieve global solutions.

The IMO adopted the requirements of the GMDSS under SOLAS Chapter IV in 1988; after the IMO Assembly, at its 11th session in 1979, it decided that a new GMDSS should be established.[†††] The GMDSS implementation period occurred from 1992 to 1999, whereupon it was considered fully implemented.

Once fully implemented, the IMO sought to keep monitoring the operation of the GMDSS and other maritime radio communication matters [52]. The IMO's List of Outputs for the 2020–21 Biennium includes consideration of developments in GMDSS satellite services, the GMDSS Master Plan, and revision of SOLAS Chapters III and IV for the purposes of the modernisation of the GMDSS with a completion date of 2021 [53].

In 2008, in a submission from the United Kingdom, the 'Modernisation of the GMDSS' as a term appeared at IMO, and wholesale changes to the GMDSS were proposed [54]. There had been prior calls for changes to some components of the GMDSS, such as the replacement of NBDP (radio telex) by the CIRM in 2004, the use of the IP technology in the GMDSS by Japan in 2004 [55], using GMDSS as a data communications network for e-navigation by the Republic of Korea [56], and

[†††]Resolution 6 of the *International Conference on Maritime Search and Rescue 1979* invited the Inter-Governmental Maritime Consultative Organization (IMCO, the previous entity name for IMO prior to 1982) to develop a GMDSS that includes telecommunications provisions for the effective operation of the search and rescue plan prescribed by the Annex to the *International Convention on Maritime Search and Rescue (Final Acts).* The Conference was convened because of IMCO Resolution A.406(X) of 17 November 1977 to consider the adoption of a convention on maritime search and rescue.

in 2006, MSC 81 included in COMSAR's work program, a low-priority item on replacements for NBDP (radio telex) in the MF/HF bands with a target completion date of 2008. In 2008, the United Kingdom raised issues regarding DSC, a key component of the GMDSS [57]. Also in 2008, the United States also raised issues regarding DSC [58], and Norway used the term 'VHF data exchange and electronic mail' which described a coastal VHF system covering the Norwegian coast and offshore installations [59].

In 2008, the United States proposed to the IMO a review of the GMDSS in its paper titled 'GMDSS Modernisation', noting that the planning for the GMDSS (known as the FGMDSS or Future GMDSS in its preparatory phase) had started more than 25 years previously [60].

In 2009, in response to a submission from Chile, France, Norway, United Kingdom, and the United States [61], MSC 86 agreed to include under the GMDSS work item, a 'scoping exercise to establish the need for a review of the elements and procedures of the GMDSS' with two sessions to complete the work, including the recommendation to use the resources of the Joint IMO/ITU Experts Group in the cause of this exercise [62].

In May 2010, MSC 87 [63] agreed to include the 'scoping exercise to establish the need for a review of the elements and procedures of the GMDSS' as an urgent matter at MSC 89 in 2011. The scoping study was to be completed at COMSAR 16. The paper set out six general principles, the last of which was that the review was to be divided into two phases. The first phase would last 3–5 years to address near-term issues, and a long-term strategic plan for the second phase should be in development for evolving the GMDSS for the subsequent 10–15 years [64].

The 40 or so items which had already been identified were to be reduced to no more than 15 and aggregated into no more than 6 categories for completion during the first 5 years of the review. Action on many items of the 40 could be deferred to Phase 2. Six categories were proposed in Phase 1.

In 2012, MSC 90 noted that the scoping study had been finalised by the COMSAR on how any review of the elements and procedures of the GMDSS might be implemented. It identified the compelling need[‡‡‡] [39] for a review and modernisation of the GMDSS [65]. The wide-ranging Draft Work Plan was set out by COMSAR 16 [66] and included the justification for the project and alignment with IMO's objectives, which was subsequently approved by MSC 90 in May 2012 [65].

The Terms of Reference for the Correspondence Group on the Review of the GMDSS had a target completion year of 2017. The Sub-Committee on Standards of Training and Watchkeeping (now HTW) was to consider issues relating to the Human Element. In the Work Plan, the mandatory checklist for consideration of human element issues by IMO bodies was duly completed. The output was to be described in SMART terms (specific, measurable, achievable, realistic, and

[‡‡‡]The term 'compelling need' is often found in the records of IMO meetings. The history of how this term was to be applied to the objectives of the IMO can be traced back at least to IMO Resolution A.500 (12), 20 November 1981, as recalled by Beattie [39], who gives further background.

time-bound), and a summary of the external organisations which needed to be consulted was included, including standards-making bodies [67].

In 2013 at Sub-Committee on Navigation, Communications and SAR (NCSR) 1, the interim report of the Correspondence Group had been split into two parts, one being a high-level review and the other the detailed review [68]. The IMO/ITU Experts Group and the Correspondence Group agreed that the functional requirements were to be the basis for discussion of issues and a preliminary outline of the detailed review of the GMDSS was developed [68].

It is worthwhile to note that already at COMSAR 11, it was not only GMDSS communications functions to be reviewed but also the harmonisation of SAR requirements because of technological change [69].

The COMSAR had a standing SAR Working Group, which has been retained in the new NCSR[§§§] formed by the merging of the COMSAR and the Sub-Committee on Safety of Navigation.

The terms 'Phase 1' and 'Phase 2' were not used for the grouping of tasks, so one can perhaps group what has been agreed as at NCSR 4 (2016) at the completion of the scoping study, high-level review, detailed review, and draft modernisation plan, as Phase 1, and what has not been agreed for one reason or another, by NCSR 4 and beyond, as Phase 2.

The coordinated plan of work for the modernisation of the GMDSS at NCSR 4 showed that there are several unresolved issues, and NCSR plans the work to be finished at NCSR 8 in 2021 [70].

The Secretary-General of the IMO has noted that IMO regulations should not require ships to carry obsolete equipment, and the modernised GMDSS should enable the use of modern communications equipment [71].

At the time of writing, the IMO 2020–21 Biennium Agenda [53] shows the revision of SOLAS Chapters III and IV by the NCSR for the modernisation of the GMDSS running until 2021, some 13 years after the term 'modernisation' first appeared in 2008.

It is planned that this work will be completed by 2022 (which will also allow it to be considered by the ITU at WRC-23) and so appear in the SOLAS amendments planned for 2024, an outcome some 16 years in the making, or 20 years, if one includes the early calls for examination of the replacement for MF/HF NBDP (radio telex).

6.5.1.2 Maritime Autonomous Surface Ships

MASS is a term agreed at IMO (for the purpose of the scoping exercise) to describe the development of autonomous technologies to support maritime transport.

[§§§] In 2013, the IMO Maritime Safety Committee (MSC) agreed to restructure its Sub-Committee structure by reducing the number of subcommittees from nine to seven. Apart from the potential benefit of interrelated issues between the form Sub-Committee on Safety of Navigation (NAV) and COMSAR Sub-Committees, one reason was achieving savings by reducing the number of meetings, engagement of translators, professional staff, and cost reductions for Member States, among others (MSC 92/26 para. 22.6.2, and MSC 92/INF.2). Previously, the subcommittees were reduced from 11 to 9 in the 1996–97 biennium.

Autonomous does not necessarily mean 'unmanned' but does indicate the increase in automated systems. The IMO is considering the levels of autonomy that would be relevant in the commercial shipping environment.

Whether 'autonomous' ships became a commercial reality outside of a small number of examples or in limited areas is unknown. Irrespective of whether autonomous ships become an engineering fancy or become commercially viable, or acceptably safe to entrust cargo to, one thing is certain, there will be no widespread implementation of autonomous (or near-autonomous ships) without reliable radio communications, not only to/from the autonomous ship, but also to/from other vessels and shore stations in their vicinity.

At ITU (ITU-R Working Party 5B), in the study period 2012–15, some preliminary work [72,73] was started on what were then known as unmanned maritime systems (UMSs). This utilised some of the extensive work at ITU-R for WRC-15, on unmanned aircraft systems (UASs), for example, in [74] and [75]. However, the UMS study was not progressed at the time, but considering the work underway on MASS at IMO and other places, the ITU will be following developments. The ITU's work on UAS could usefully inform the work on MASS at IMO and elsewhere.

Although there is already much written about autonomous ships at this early stage, and some field trials have demonstrated success with more trials on selected smaller vessels underway, more research on the required resilient communications links is needed, and how they are categorised and defined. Whether such communications links would be considered a *safety service* as defined by the ITU would need to be determined and is worthy of an agenda item for a future World Radiocommunications Conference (currently being held every 4 years), as would the spectrum requirements and the associated regulatory provisions.[¶¶¶] The need for these issues to be examined has also recently been raised by IMSO at IMO [76].

A survey of communications and networks for autonomous marine systems by Zolich *et al.* acknowledges the lack of high data rate links as a major constraint [77]. Höyhtyä *et al.* [78] provide some options for communications for MASS vessels to consider in order to meet their needs in various sea areas, and in different phases of a ship's voyage, and recognise the challenges of ship-hijacking, cybersecurity, and resilience in various failure modes.

It is estimated that a maximum data transmission rate of 6 Mbps is required for a MASS vessel, depending on the type of sensor data used (most data-intensive). Options include a whole range of techniques from HF, VHF, high-altitude platform stations (HAPS),[‖‖‖] HAPS as IMT base stations,[****] satellite (LEO, MEO, GSO or GEO, and mega satellite constellations), 4G, 4G LTE, and 5G even. Rødseth and

[¶¶¶] *Safety service*: Any radiocommunications service used permanently or temporarily for the safeguarding of human life and property. No. **1.59,** ITU *Radio Regulations 2016.*

[‖‖‖] (HAPS) *High altitude platform station* is defined as a station located on an object at an altitude of 20–50 km and at a specified, nominal, fixed point relative to the Earth. No. **1.66A,** ITU *Radio Regulations 2016.*

[****] HIBS is being studied at the ITU under WRC-23 Agenda item 1.14.

Kvamstad [79] estimate that the minimum data rate requirement for an autonomous ship is 4 Mbps and note that the larger portion of the data requirement is from ship-to-shore, i.e. from ship to the shore control centre.

A word is needed about 'cognitive radio' (CR). CR networks allow a secondary use to utilise vacant spectrum left by a primary user without interfering with their transmission. To know when to occupy or leave the spectrum for the primary user, the secondary user must be able to predict in advance the spectrum availability of the primary user. Various methods have been studied, including Bayesian online learning as explained by Mikaeil [80]. This technology has been studied, a great deal, but its commercial reality appears not yet to have been realised, although it potentially can be used widely, which is examined by Pelechrinis *et al.* [81]. One potential use for maritime CR has been proposed by Tang *et al.* where a preliminary study concluded that a CR-AIS VHF network where each node shared its local decisions could perform spectrum sensing of AIS spectrum without interfering with AIS services, and thereby solve spectrum overloading [82].

Höyhtyä *et al.* [78] use the term 'connectivity manager' to describe a communications controller which can determine the most appropriate choices for the end-to-end communications channel and resolving the best utilisation of resources. In the paper, the GMDSS and the IMO's e-navigation initiative are both mentioned. The 'Provision of a system for automatic source and channel management on-board for the selection of most appropriate communication means (equipment) according to criteria such as bandwidth, content, integrity, costs is listed as Task 15 in the IMO e-navigation SIP [51]. Other researchers, via the EfficienSea2 project, such as Bronk *et al.* [15] describe a 'hybrid communication network concept' to address the coastal, on-board segment, and satellite segment, which includes cellular networks (2G, 3G, LTE, WiMax^{TM}, Wi-Fi, VDE, and VDE-SAT) and very small aperture terminals. To enable seamless roaming, a mechanism or algorithm is proposed for a control unit to handle the choice of communications link, which includes weightings for transmission cost, duration of link availability, data rate, and radio signal power level in various scenarios.

Table 6.3 shows indicative ranges for some technologies over land, e.g. LTE, 5G; however, performance over a sea path is a different story, at least away from ports, since the high-throughput modes of modem techniques often use multiple-in–multiple-out technology to multiply the capacity of radio links using more than one antenna at the transmitter and receiver to maximise the benefits of multipath propagation modes. In the open sea, this phenomenon is absent, except near coasts or in straits, as noted by Røste *et al.* [83]. However, the shore-based user constraints on antenna/terminal size are often not present aboard many ships, so by using a variety of different antenna types and/or larger sizes, and a variety of access technologies, it is possible to extend broadband coverage well offshore, using LTE operating in the frequency bands of 800 MHz, 1,800 MHz, and 2.6 GHz. It is estimated that by using larger and different antennas aboard ship, ranges at 2,600 MHz between 82 and 103 km and 800 MHz between 87 and 113 km can be obtained [83].

Field trials of a maritime mesh network using WiMax (IEEE 802.16) technology by Hao-Min *et al.* [84] demonstrate that data communication is effective across the

Strait of Singapore, with a data rate of 5.5 Mbps in less than ideal conditions. WiMax as an e-navigation communications solution has been proposed by Choi *et al.* [84,85].

IMO's 2018 e-navigation 'Update 1' on the e-navigation SIP [86] includes, under Task 15, a deliverable (in part) on 'Guidelines on seamless integration of all currently available communications infrastructure and how they can be used and what future systems are being developed along with the revised GMDSS'. In that 2018 update, the status of Task 15 is listed as 'Under consideration'.

Under MASS, irrespective of category, AIS will continue to be important. This has implications for regulators, ship operators, and shipmasters to ensure that the AIS information is always correct [87].

It is not clear whether the current arrangements used in the maritime industry will be adequate for this new world, considering the existing issues that arise such as when ships are reflagged and/or renamed, creating some confusion. In addition, many smaller vessels do not have IMO numbers and therefore may not have a unique identifier, although the FAO is progressing implementation of a unique vessel identifier for fishing vessels, of which an IMO number can be used [88]. Additional identification schemes other than those currently broadcasted by AIS may need to be prescribed for MASS vessels via alternative communication channels, not necessarily visible to such ships, but only remote operators.

The technical availability targets for the quality of service for these MASS links are stated to be 'always available', but this is impossible to achieve in practise. IMO Resolution A.953(23) on the World-Wide Radio Navigation System uses three levels of requirements depending on area, volume of traffic, and/or significant risk, as summarised in Table 6.2.

Table 6.2 shows a one-way level of performance. Where ships are to be controlled remotely, or autonomously, the previous factors need to be carefully considered and specified considering that two-way performance requirements will be required together with parameters such as maximum unplanned and planned outages. For some definitions of link performance, there are differences between ITU and ICAO definitions, as noted by ICAO [89]. However, none of these can be calculated when the linking system and technologies are unknown.

In a world where full access to the World Wide Web and 24 h/day e-mail is not yet currently provided to crews of many ships, let alone the fishing fleets of the world, the challenges for new robust data communications links for MASS vessel operations seem large.

It appears, therefore, that an overall requirements document is required for the communications links for MASS, followed by a call for proposals to meet these requirements, which could then be evaluated and informed by field trials. The regulatory provisions at IMO (carriage requirements and Performance Standards) and ITU needed to bring these into existence need to be studied.

The maritime industry has made use of non-GMDSS satellite communications 'to bring the office on-board the ship' for many decades now; however, the potential applications such as MASS, greater use of mobile stations which the ITU defines as 'maritime Earth stations in motion' [73], and 'Earth stations on vessels' need to be included in the formalisation of e-navigation and communications for MASS.

Table 6.2 IMO World-Wide Radio Navigation System performance requirements specified in IMO Res. A.953(23)

Parameter	**Navigation in those harbour entrances, harbour approaches, and coastal waters with a high volume of traffic and/or significant degree of risk**	**Navigation in those harbour entrances, harbour approaches, and coastal waters with a low volume of traffic and/or a less significant degree of risk**	**Navigation in ocean waters**
Position accuracy	$\leq$ 10 m (95% probability)	$\leq$ 10 m (95% probability)	$\leq$ 100 m (95% probability)
Update rate of computed and displayed position data	$\geq$ Once every 10 s $\geq$ Once every 2 s if used for AIS, graphical display, or direct control of the ship	$\geq$ Once every 10 s $\geq$ Once every 2 s if used for AIS, graphical display, or direct control of the ship	$\geq$ Once every 10 s $\geq$ Once every 2 s if used for AIS, graphical display, or direct control of the ship
Coverage	Local	Local	Global
Availability	$\geq$ 99.8% (2 years)	$\geq$ 99.5% (2 years)	$\geq$ 99.8% (30 days)
Continuity	$\geq$ 99.97% (3 h)	$\geq$ 99.85% (3 h)	NA
Time-to-alarm	$\leq$ 10 s	$\leq$ 10 s	As soon as practical by MSI

This is where 'goal-oriented' requirements may be preferable to prescriptive standards. The role of classification societies and protection and indemnity clubs will be crucial as not only will they be key players on all aspects of MASS vessel design, construction, certification, and operation but can be expected to approve shore infrastructure and operation for MASS, subject to any host country requirements for the establishment of such facilities. It may well be that provision of shore facilities for MASS may need to go beyond guidelines.

6.5.2 Digitalisation

We are living in a connected world, with increasing access to digital data. There are two key terms that we are becoming familiar with: digitisation and digitalisation. While similar, there are specific differences: digitisation is the representation of an object, image, sound, documents, or signal (usually an analogue signal) in a digital manner (converting analogue source to a numerical, or digital, format). Digitalisation, as defined by the Gartner Glossary, is the use of digital technologies to change a business model and provide new revenue and value-producing opportunities; it is the process of moving to a digital business [90]. In the maritime environment, we have seen digitisation for many years and are now seeing an increasing approach to digitalisation.

Initial examples of digitisation of analogue systems began with the development of the GMDSS, making the use of digital data over an existing analogue VHF, HF, and MF channels to support DSC, but using analogue techniques. The development of AIS to assist in positive identification of vessels within a VTS area, and incorporated into SOLAS to support safety of navigation, is a further example of the existing work on digitisation, as well as digitalisation.

Potential new techniques to replace/augment legacy GMDSS NAVTEX at MF is MF NAVDAT and at HF is HF NAVDAT. Field trials have occurred for these technologies, and nearly all the regulatory provisions in the RR are complete. These potential candidates are being considered by IMO as part of the modernisation of the GMDSS project.

The development of the IoT, big data analytics, machine learning, and even artificial intelligence (AI) is driving a need to collect more data, identify value-added analysis, and then transfer this data for use to address safety, operational, and commercial requirements.

Examples of data exchange to support maritime operations include

- automated position reporting (AIS, LRIT, other);
- monitoring of vessel traffic in ports, remote, and sensitive areas;
- provision of electronic certificates;
- developments in maritime 'single window' to support safety, efficiency, and commercial aspects;
- shared data for optimisation of port activities, for example Port Collaborative Decision Making;
- cargo chain logistics;
- engine efficiency monitoring;
- and more.

The developments in digitalisation in maritime are benefiting from other industries, including retail, travel, and banking. The challenges in the maritime sector relate to the existing legacy systems, the extreme operational environment, large number of standards bodies within the sector, a variance in adoption of digital technologies, and a need to clearly define use cases.

Digitalisation is seen as a key to increasing safety and security while supporting operational efficiency. To assess the existing activity towards digitalisation, and identify where to focus in the future, organisations use a digital maturity model [91]. There are several models available, including the Seer four-stage model of nascent, emerging, connected, and multi-moment. In the maritime environment, the Port of Rotterdam [92] has identified a similar four-step model, where the digital developments fit within the following levels:

1. digitisation of individual parties in the port;
2. integrated systems in a port community;
3. logistics chain integrated with hinterland; and
4. connected ports in the global logistics chain.

Through the developments in digital data exchange to support digitalisation of the maritime industry, we see an increasing requirement for reliable, robust, and redundant digital communications capabilities.

6.5.3 Standardisation

Standardisation is a critical element in interoperability of systems. Within the maritime environment, there are a number of standards development organisations, including international, regional, and national bodies. These are addressed in Section 6.4.1.

6.6 Overview of communication services

There are multiple communications systems that can be used to support the needs of the maritime industry. These range from low bandwidth (<10 kbps) to broadband systems (>1.5/2 Mbps). The requirement for digital data exchange, not only in ports and coastal areas but also in Polar and other remote areas, necessitates a wide range of technology solutions. The user requirement for digital communications includes audio, visual, and data, with an expectation for seamless and continuous access through the most appropriate system, using least-cost routing within an intuitive operational environment as explained by Bronk *et al.* [15].

Potential solutions include maritime specific solutions as well as cross-industry solutions. While some communications options are well developed with implementation either within the maritime industry or in other industries, there are a number of developing solutions. Digital data technologies can be roughly grouped by bandwidth/data rate: 0–10 kbps; 10–500 kbps; 510–1,000 kbps; 1–1,000 Mbps; and more than 1,000 Mbps.

Some technologies span different categories, for example, GEO or GSO satellite technologies can include Inmarsat-C (600 bps) store-and-forward as used in the GMDSS; to high-speed, high-capacity Inmarsat broadband.

Standards for 5G, LTE, and Global System for Mobile Communications (GSM™) are all contained within the work of 3GPP. LEO satellites include 'baby LEOs' and 'big LEOs' with different capacities. In addition, some terrestrial technologies such as the VDES contain both terrestrial (VDE-TER) and satellite (VDE-SAT) elements, with around 300 kbps potentially possible.

An overview of these is included in Table 6.3.

As noted in Section 6.5, the IMO e-navigation SIP identified six areas for operation, from port and approaches, coastal operations to polar areas and other remote areas. A graphic representation of different technologies potentially suitable for use in the IMO areas is provided in Figure 6.3.

An indicative overview of different technologies and data rate referenced against the sea areas identified by the IMO is provided in Table 6.4. With multiple systems available in each area of operation, the potential exists to address the requirement for redundant, reliable, and resilient digital communications.

6.6.1 Satellite technologies

Satellite communications provide a mature set of digital data exchange options and are provided commercially. These services may have global, regional, or spot beam coverage. The satellite links can support broadband digital communications supporting voice, e-mail, crew calling, remote monitoring, telemedicine, and chart and weather updates.

Satellites operate in three different levels of orbit, or altitudes above the Earth's surface [96]. These include

- HEO (high Earth orbit), altitude 40,000 km in apogee;
- GEO or GSO, altitude 35,786 km;
- MEO, altitude approx. 8,000–20,000 km;
- LEO, altitude approx. below 400–2,000 km.

Satellite communications in the VHF band are commercially available and can provide services for short message service, weather, and tracking. In the maritime, VHF band satellite communications capabilities are included as a part of VDES, known as VDE-SAT.

It is not possible in this chapter to do justice to the many and varied services available via satellite providers to the maritime community.

6.6.2 Terrestrial technologies

The maritime industry makes the use of several existing, established terrestrial communications technologies, including

- UHF – analogue and digital (including radar, EPIRB homing, and on-board voice communications);

Table 6.3 Comparison of Digital Communications Technologies for maritime use

Communication technology	Data rate (typ.)	Infrastructure	Coverage estimation	Maritime/non-maritime spectrum
Very high capacity broadband (>1 Gbps)				
5G New Radio (NR)	10 Gbps (peak)	Base stations	200 m	Non-maritime
Gigabit LTE	>1 Gbps	Access points	200 m	
Wi-Fi (IEEE 802.11ac)	1.3 Gbps (local)	Mobile devices	In-building/house	
High capacity broadband (1-1,000 Mbps)				
MEO-SAT*	Up to 1,000 Mbps	Satellites Satellite Earth stations	Regional Spot beams	Non-maritime
LTE (1,800 MHz)	1 Gbps (peak)	LTE base stations Mobiles	15 km approx. 87–113 km NOTE 1 (47–61 NM)	Non-maritime
LTE (800 MHz) (64QAM)	100 Mbps (peak)		5.5 km (urban) 8 km (suburban) 18 km (rural)	NOTE 1 Using larger and different antennas for ship terminals
WiMax (IEEE 802.16)	75 Mbps	Routers, access points	5 km (2.7 NM)	
GEO-SAT (broadband)†	50 Mbps (up)/5 Mbps (down)	Satellites Satellite Earth stations	Hemispheric Regional Spot beams	
Medium capacity (501-1,000 kbps)				
GEO-SAT (medium capacity)†	Up to 1,000 kbps	Satellites Satellite Earth stations	Hemispheric Regional, spot beams	Maritime/non-maritime
MEO-SAT*	Up to 1,000 kbps	Satellites Satellite Earth stations	Regional, spot beams	Non-maritime
4G	600 Mbps	4G base stations Mobiles	30 km (16 NM)	Non-maritime
IoT LPWAN technologies‡	Up to 1 Mbps (variable symmetry)	Cloud services, cellular base stations, radio access points via ISM bands – dependent on technology	Variable to 15 km (urban) 50 km (rural)	

Low capacity (10-500 kbps)				
GEO-SAT (low capacity)[†]	150–444 kbps	Satellites Satellite Earth stations	Global (<82° lat.) Regional, spot beams	Non-maritime
IoT LPWAN technologies (see footnote[‡])	Various – up to 1 Mbps	Can include Cloud services, cellular base stations, access points, radio via ISM bands – dependent on technology	5 km max (urban) 15 km max (suburban)	
NAVDAT (HF)	20 kbps (approx.)	Similar to NAVTEX	Global (subject to skywave)	Maritime RR Appendix 17 channels
NAVDAT (MF)	12–18 kbps	Similar to NAVTEX	460–560 km (250–300 NM)	Maritime
VDE-TER	38.4–76.8 kbps (25 kHz) 76.8–153.6 kbps (50 kHz) 153.6–307.2 kbps (100 kHz)	Base stations Mobile radios VHF data link	Line of sight, approx. (15–65 NM)	Maritime RR Appendix 18 channels
VDE-SAT	307 kbps	Mobile radios VHF data link	Line of sight, approx. 28–120 km (15–65 NM) Satellite component provide LEO coverage	Maritime RR Appendix 18 channels
LEO (Big LEO)[§]	Up to 134 kbps	Satellites Satellite Earth stations	Global coverage, via spot beams, regional	Non-maritime
Low capacity (10–500 kbps)				
GEO-SAT (very low capacity)[‖]	1,200 bps–10 kbps	Satellites Satellite Earth stations	Global (<82° lat.) Regional, spot beams	Non-maritime/maritime
GSM(Note: some networks have been discontinued)	9.6–40 kbps	GSM base stations Mobiles	Line of sight, approx. 37 km (20 NM)	Non-maritime
AIS-TER	9.6 kbps	AIS/VDES radio VHF data link RR Appendix 18 channels	Line of sight, approx. 28–120 km (15–65 NM)	Maritime

(Continues)

Table 6.3 (*Continued*)

Communication technology	Data rate (typ.)	Infrastructure	Coverage estimation	Maritime/non-maritime spectrum
ASM-TER	9.6 kbps	AIS/VDES radios VHF data link RR Appendix 18 channels	Line of sight approx. 28–120 km (15–65 NM)	Maritime
Digital UHF	9.6–19.2 kbps	Base stations Mobile radios	Line of sight approx. 28–120 km (15–65 NM)	Maritime/non-maritime
Digital VHF	9.6–19.2 kbps	Base stations Mobile radios VHF data link RR Appendix 18 channels	Line of sight, approx. 28–120 km (15–65 NM)	Maritime
NAVTEX (MF)	100 bps	Coast stations Shipboard receivers	Coastal up to 830 km (450 NM)	Maritime

Source: Adapted from the IALA MRCP [9], with additional sources [12,15,22,27,28,32,83–85,93–95]).

*For example, SES O3b.

†For example, Intelsat, Inmarsat, Thuraya.

‡LPWAN technologies include the following technologies: Dash7, eMTC, EC-GSM-IoT, LoRa, LoRaWAN, LTE-M™, NB-IoT, Sigfox™, Ingenu, Telensa, Weightless, Zigbee (IEEE 802.15.4 g/k). The performance of each of these vary, but the upper bound of data transfer rate is about 1 Mbps, with variable symmetry (i.e. different uplink and downlink capacity). The infrastructure includes cloud services, cellular base stations, radio access points via ISM bands – dependent on the technology. Some of these technologies are denoted as 'long range' which might extend to 15 km (urban) or 50 km (rural), and some are focussed on machine-to-machine communication using very low power, and low data rate, with the objective of longevity using small battery-powered devices.

§For example, Globalstar, Iridium.

‖For example, Inmarsat-C.

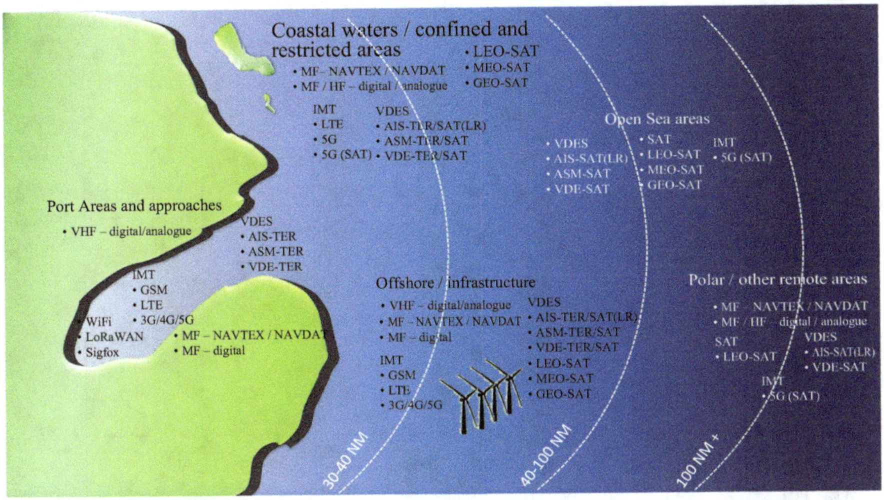

Figure 6.3 Graphic of maritime communications options for IMO areas

- VHF – analogue and digital (including AIS, DSC, and voice);
- HF – digital and analogue (including DSC, NBDP, and voice); and
- MF – analogue and digital (including NAVTEX, DSC, and voice).

Different technologies that have been developed for other industries have been implemented or are being reviewed for use in the maritime environment. These include some higher bandwidth options that may be suited to port environments.

- Digital private mobile radio (PMR) (to replace analogue PMR for on-board communications): terrestrial trunked radio (formerly Trans-European Trunked Radio), Project 25 (trunked radio system), digital mobile radio;
- 4G mobile: LTE;
- LPWAN; and
- 5G.

In addition, some other technologies which can support digital data exchange for maritime operations at very short ranges (i.e. normally less than 100 m) include Wi-Fi, Bluetooth, Zigbee Alliance, and ultra wideband (UWB), for example, aboard a ship's bridge, control room, or machinery space. While the geographic range of these technologies is very limited, they use globally allocated frequencies, with no specific need for consideration if they are used in the maritime environment [9].

3GPP (see Section 3.4) is now rolling out the 5G technology, which includes both terrestrial and satellite capability, with (proposed) ubiquitous, global high speed, and large bandwidth capabilities.

Table 6.4 Comparison of maritime communications options for IMO areas – ship/shore

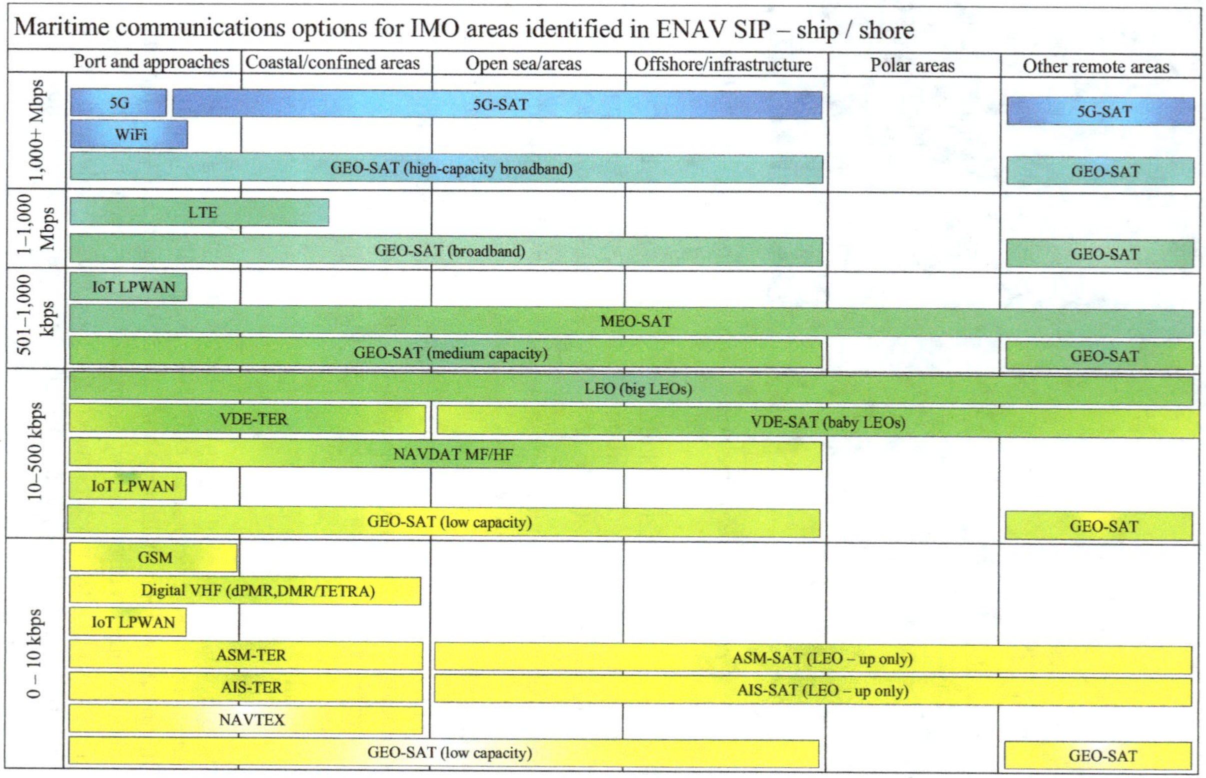

6.6.3 Communications technologies to address requirements

As the requirements for communications increase, the need to enable multiple, reliable, and robust options for communications increases. As presented in Figure 6.3, there are a number of technologies that can provide services in the different areas. With data transmission requirements ranging from small text-based data to large image-based files, and telemedicine, there is an opportunity to benefit from different technologies in each of the geographic areas.

In addition to the ship–shore/shore–ship element, there is a need for ship-to-ship communications. These include the relevant terrestrial options of VHF radio and AIS, as well as satellite options using suitable, and compatible, systems.

When reviewing the communications options for digital data exchange, several factors should be considered. These include

- area of operation,
- bandwidth available versus bandwidth required,
- availability of the technology (where used in the world), and
- European Union (EU) technology readiness level.

6.7 A glimpse into the future

We live in an age which requires more data to be transferred at ever-increasing rates. This demand is supported by advances in technology.

While the requirement is evident, and the demand is for more options, more speed, and more data, the current structure for the development of technology within the maritime environment may not necessarily support quick innovation and adoption of technologies.

How do we move from the current structure, with independent, purpose-built technologies developed through a highly structured process to a future which supports innovation?

6.7.1 Getting from here to there

In the context of GMDSS modernisation, e-navigation, and even communications for MASS operations, one major challenge is defining what the target outcome should be, due to the uncertainty of several issues outstanding.

Other challenges are that significant projects are running in parallel, within the IMO and externally, such as e-navigation, VDES (IALA and ITU), and new satellite provider(s) (a second GMDSS satellite service provider has recently been approved, but another is being examined by IMSO at the time of writing). Further, the technology in high-tech industries such as the maritime oil–gas exploration/exploitation industries continues to increase, and there are continual incremental changes in the technology used in maritime SAR, such as second-generation COSPAS-SARSAT distress beacons.

From the literature, there are references to the term 'enterprise system modernisation'. The enterprise system is often the IT system of a large enterprise, and a significant body of work exists on which approach is appropriate when modernising legacy IT systems while keeping enterprise operations essentially unchanged throughout the process. The 'enterprise' could also apply to the communications systems for the whole shipping industry (and all others at sea), such as the GMDSS.

There are similarities to the challenges of the 'enterprise' modernisation to the modernisation challenges for the GMDSS, communications for e-navigation and MASS, some of which might include the following:

- Some functions of the existing 'legacy' system need to continue unchanged.
- The existing legacy systems for existing ships and non-shipborne infrastructure will continue through a lengthy transition period.
- The user experience must not degrade but improve in the target system.
- New technologies will allow more features, but care must be taken not to complicate operation of the system, for both users, technical and administrative purposes.
- Interconnections between the existing and external systems are increasing, and the original separation of the original system from other systems is becoming blurred – making the task of modernisation more challenging.
- The target system is desired to be cheaper to operate and maintain.
- The new system should be scalable.

In Figure 6.4, Zhao [97] illustrates the three major components for enterprise system modernisation.

The three major components are:

- *processes* – provide guidance on activities and steps involved for enterprise system modernisation,
- *approaches and methodologies* – provide technical references to support the processes or 'how to do it', and
- *portfolio, programme, and project management* – for operational management, life cycle governance, organisational support, roadmap create, and execution planning.

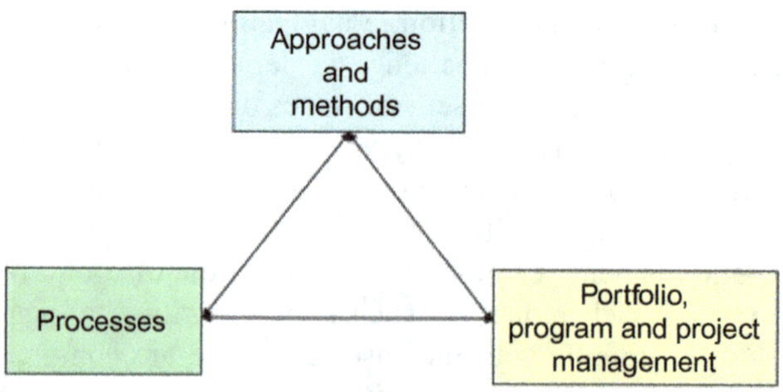

Figure 6.4 The three major components for enterprise system modernisation, from Zhao [97]

In Figure 6.5, the enterprise system modernisation concept model adapted for GMDSS Modernisation, e-navigation, and MASS communications projects, adapted from Zhao [97].

6.7.2 Rethinking the approach to standards

The drawbacks of the current technologies used by the shipping industry are well described by Valčić *et al.* [93]. The paper hints that there needs to be a more holistic approach to achieving the complete communications solution in the face of relentless change of technology, such as described by Weintrit [98], which is scalable, secure, standardised, and suitable.

For the shipowner/ship operator, faced with the task of safely delivering cargo (human and otherwise) from Point A to Point B today and tomorrow with utmost despatch, a bewildering choice of not quite suitable solutions apparently is available now (at a cost), but it seems that there is always a new system or solution 'on the horizon'. The complexity of the challenge is readily seen in an EfficienSea2 report by Valette [99]. WRC-19 has since approved the satellite component for the VDES, complementing the terrestrial component agreed at WRC-15 [12]. The WRC-19 agreed for the regulatory provisions for GMDSS satellite services (recognised satellite service providers) to occur in the frequency band 1,621.35–1,626.5 MHz (transmit and receive), allowing an additional recognised satellite service provider [28].

There are indications that another approach may be useful such as to list all requirements and call for complete proposals from interested parties, which can then be evaluated by an expert panel. This approach has worked well for the mobile broadband wireless access community.

Parallels can be drawn with the process used by IMSO to evaluate the Iridium application to become a recognised satellite service provider in the GMDSS. IMO amended its previous requirements document(s), to provide for non-geostationary satellite providers to be included in the GMDSS. After further discussions, consensus was achieved that provided for IMSO to create an expert panel to evaluate the Iridium proposal. The difficulty for e-navigation is that no communications specification exists. In contrast, the communications requirements for GMDSS are specified, and are being updated now at IMO via proposed revisions to SOLAS Chapters III and IV. When communications for MASS is considered, this would appear to be too difficult a problem to include in the mix, and accordingly, could be kept separated, but cannot be totally so, due to the essential interoperability required for non-autonomous (or not-so autonomous) shipping.

A similar parallel can be drawn to the review of the STCW 78 Convention[††††] of which was carried out by an expert group of seven government representatives (seven countries), three international non-government organisations, and four scientific institutes, all drawn from across the world, as noted by Dirks [100].

[††††] *International Convention on Standards of Training, Certification and Watchkeeping for Seafarers 1978.*

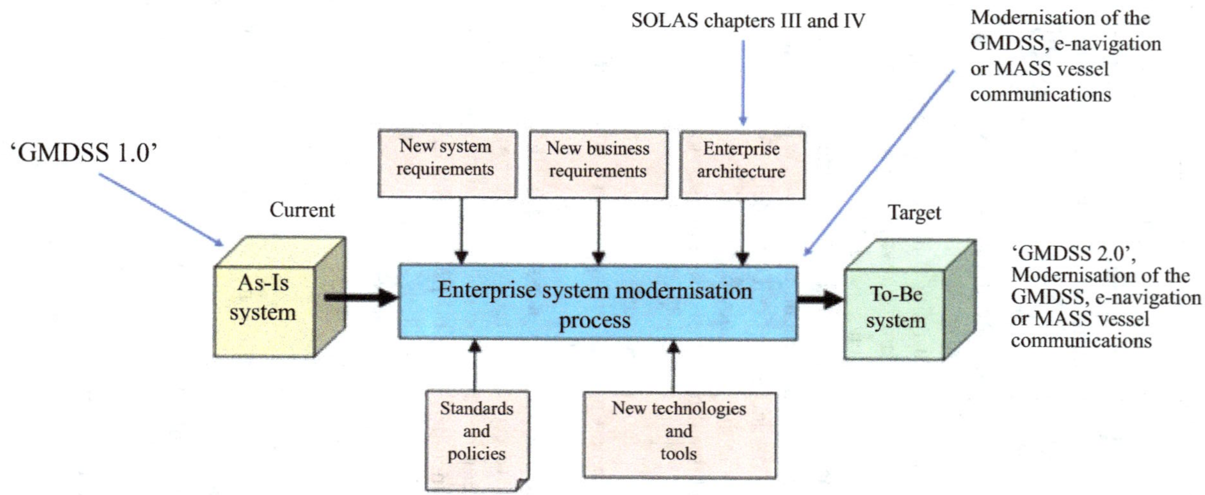

Figure 6.5 Enterprise system modernisation annotated for use as a concept model for GMDSS Modernisation, e-navigation, and MASS communications projects, from Zhao [97]

Another parallel can be drawn by the EU's EfficienSea projects. The projects had a number of deliverables, including a report on e-navigation maritime communications infrastructure, assessed independently from IMO, by Valette [99] and later by Jensen and Pedersen [101].

If we examine the standardisation process used for IMT-2020, the following is noted:

1. Development of the standard(s) itself is essentially performed by one organisation, the ITU.
2. Development of the standard essentially is done by one Working Party (WP 5D) – within Study Group 5.
3. WP 5D traditionally meets in different locations around the world (twice per year).
4. Initial workshop is held.
5. Approximately 18-month proposal submission window.
6. Outside resources are used for evaluations of technologies, and external entity (3GPP) involved in all stages feeding documents into ITU process.
7. ITU produces the report(s).
8. Radio interface specifications are then developed outside ITU.
9. Radio interface specifications are brought into the ITU who develops and publishes the ITU Recommendation(s).

The above process seeks to achieve global harmonisation, and the final Recommendation(s) are to be drafted taking into account the Common Patent Policy of the ITU-T/ITU-R/ISO/IEC on intellectual property rights [102].

The IMT process is different to that used by IMO for projects like the Review and Modernisation of the GMDSS. It is noteworthy that private industry plays a key role in the evaluation process. One evaluation group, called the ATIS[‡‡‡‡] Wireless Technologies and Systems Committee IMO-2020 Evaluation Group, had some 49 voting members, mostly from industry, with over 150 members [103]. 3GPP at the time of writing has over 700 member companies.

The pace of technological change is such that the current standards approach used by the maritime industry should ideally accommodate industry-developed technologies and focus on solutions (goal-based) rather than a bottom-up approach. Figure 6.6 tries to capture this approach, where both maritime-specific technology standards and multi-industry standards find their place.

Another interesting activity related to successful computer programming and interoperability are 'hackathons' (or hackfests/codefests) and 'plugfests' [104], which are being used successfully for those involved in software development and related communications (such as 5G) to ensure interoperability. These are often held by universities, but sometimes within companies or other entities, to promote innovation and involvement, and often have a competitive element and a specific focus.

[‡‡‡‡]Alliance for Telecommunications Industry Solutions (ATIS).

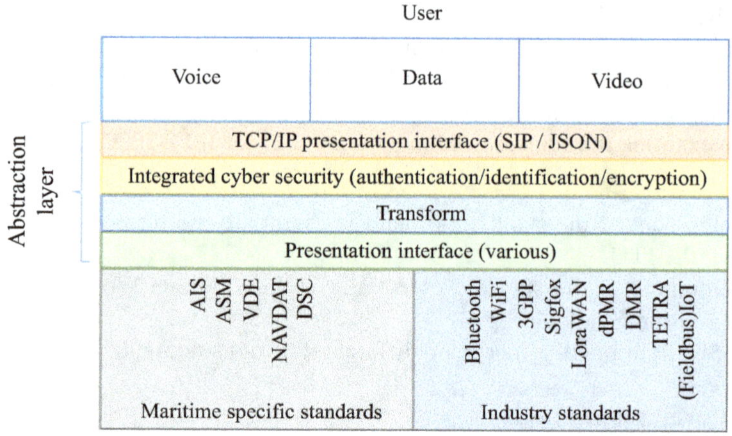

Figure 6.6 A working model to support digital communications in the maritime industry

6.7.3 A model to support digital communications in the maritime industry

Within the maritime environment, the opportunity to make best use of multiple technologies can provide an opportunity to focus on the expected outcome of the communications exchange, rather than the technology. A possible model for future maritime communications standards is presented in Figure 6.6. This approach identifies both maritime-specific technologies and broader industry-developed technologies, and the standardisation approach.

The opportunity to move from a technology-specific presentation interface through a transformation layer into a generic presentation interface enables the user to interact with information that may be audible, data, or visual without the need to know the specific technology used, as in Figure 6.6. The outcome-based or goal-based approach supports innovation while addressing the operational requirements within the industry.

6.7.4 Blockchain and maritime communications

The potential use of blockchain technology (blockchain) is generating much research activity, and the maritime industry is no exception, due to its role in supply chain management. Blockchain is the technology which underlies the crypto-currency Bitcoin (and other, later cryptocurrencies), but it is being investigated to see whether it can be applied to improve business practises, including contracts, approval processes, and trade settlements.

Blockchain's distributed 'ledger' has been likened to a distributed database, as against a centralised database. The advantage of blockchain is that it does not depend on a central trusted authority (or database) but depends on a distributed database which allows peer-to-peer transactions without relying on a third party.

This is called a 'trustless environment'. Each participant in the network votes and decides (consensus) which transaction or block of transactions is valid [105].

Whether blockchain (outside of cryptocurrencies) becomes a practical solution to these issues is it still unknown, and significant doubts as to its usefulness have been expressed by Bunduchi, who also says that even before blockchain implementation in the maritime industry, research has praised the technology beyond its technical limitations [105].

In discussing the possible benefits in a practical maritime context, Meland *et al.* [106] compare the public key infrastructure (PKI) with blockchain for different scenarios. For two scenarios, MSI and port state control (PSC), Meland *et al.* [106] concluded that PKI is the preferred solution. For MSI, this relates to the expectation that the broadcasts to ships will, in future, need to be electronically signed to prove the identity of the originator. In the PSC scenario, the increasing automation of information submitted will require authentication, proof of transmission, and verification. The IMO already use PKI in its LRIT system to validate data received from LRIT data centres at the LRIT International Data Exchange [107].

However, in the third scenario, *Ship Certification* in the *Approval Letter* process for a new-build ship, the current process of sequential exchange of documents, drawings, and designs of ships is communication rich and can easily go awry. It was concluded that traceability and transparency to processes could be provided more appropriately by blockchain than PKI [106].

It is said that blockchain has a property unique to itself, which is the possibility to include 'smart contract' functionality. This means that scripts can be executed 'by themselves' to conduct trade and move value without 'middle-men', downtime, censorship, or fraud. In this way, international trade transactions and negotiations can be digitalised, such as Bills of Lading, Letters of Credit, Certificates of Origin, commercial invoices, and packing lists. However, Bunduchi says that smart contract functionality can also be achieved using *stored procedures* in relational databases [105] citing Coffin (2011) and Mensah (2006).

The Meland *et al.* analysis recognised that ships may not have continuous access to the Internet, and where access is available, it may be slow, unstable, and expensive [106]. It recognises also that ships operate in a low-cost business environment where expensive technical solutions may not be viable. The advantage of PKI over blockchain is that PKI can store certificates offline, whereas with blockchains, data validation depends on online access [106].

Others believe that blockchain technology, although potentially useful, has some unanswered questions, not the least of which are significant legal questions, as discussed by Herian [108].

However, despite the significant spectrum available for 5G, there will be spectrum scarcity. Even here, blockchain can be used for spectrum management to more efficiently use spectrum management models by using blockchain 'miners' which can authenticate user access to spectrum with reduced latency, using blockchain-enabled spectrum sensing platforms. CRs can also be used to share certain blocks of data [109].

There are different types of blockchain, such as public blockchain, permissioned (or private) blockchain, blockchain where types of voting tools used, and many other variations. While there may be technical benefits to using blockchain for spectrum management, and business benefits to support trade, there are limitations that need to be considered, such as governance, and whether the benefits outweigh traditional database methods [110]. Nevertheless, in a 5G (and beyond) communications world, new requirements arise, beyond that needed with 4G. In 5G, security and privacy will be more complex, and there appears to be a consensus that blockchain will find wide application due to asymmetric cryptography and its use of public and private key infrastructure [109].

6.7.5 The future looks 'smart'

The rise of digital connectivity supports the development of the 'smart port' where all elements within the maritime and port ecosystem interact in a seamless manner, where actors are both providers and consumers, of data. There are many benefits to this smart maritime ecosystem, as described by Lind *et al.* [111], which include

- shared situational awareness;
- use of dynamic slot allocation for just-in-time arrivals, departures, and operations;
- sensor data on AtoN, weather, tides, and current meters;
- exchange of electronic certificates; and
- lowered administrative burden for reporting during port visits.

As experience is gained and technology continues to evolve, it is anticipated that the demand for 'smart' services will continue to grow.

The increasing reliance on digital data is leading to a review of the competencies required by maritime professionals. Skills related to technology networks, data analysis, and AI are becoming core skills to support a connected maritime industry. The Maritime University of Szczecin, Poland, has recently developed a relevant IT degree program in Marine Informatics [112].

Maritime Informatics forms an intersection of professional maritime operations and digitalisation. Maritime Informatics supports collaboration of all those involved in the maritime and port ecosystem.

This chapter has looked at the expectations for connectivity throughout the maritime environment. There are many existing and developing approaches to digital data exchange that support a maritime IoT environment. The coverage area varies; however, there are opportunities to enable full, global connectivity using multiple systems.

Abbreviations used and glossary of terms

2G	second generation
3G	third generation
4G	fourth generation

5G	fifth generation
3GPP	Third-Generation Partnership Project
AI	artificial intelligence
AIS	automatic identification system
AIS-TER	AIS (via terrestrial)
ASM	application-specific message/ing
ASM-TER	ASM (via terrestrial)
Bluetooth®	Bluetooth® technology
CIRM	International Association for Marine Electronics Companies
COMSAR	Sub-Committee on Radiocommunications and Search and Rescue (now part of NCSR)
COSPAS-SARSAT	International Satellite System for Search and Rescue
DSC	digital selective calling
EPIRB	Emergency Position Indicating Radio Beacon
ETSI	European Telecommunications Standards Institute
EU	European Union
FAO	Food and Agriculture Organization of the United Nations
GEO	geostationary earth orbit
Globalstar	Globalstar, Inc.
GSO	geostationary satellite orbit or geosynchronous orbit
GHz	gigahertz
GMDSS	Global Maritime Distress and Safety System
GNSS	Global Navigation Satellite System
GSM™	Global System for Mobile Communications
HEO	high Earth orbit
Hz	hertz
HTW	Sub-Committee on Human Element, Training and Watchkeeping (formerly STW)
IALA	International Association of Marine Aids to Navigation and Lighthouse Authorities
ICAO	International Civil Aviation Organization
IEC	International Electrotechnical Commission
IEEE	Institute of Electrical and Electronics Engineers
IHO	International Hydrographic Organization
IMO	International Maritime Organization
IMSO	International Mobile Satellite Organization
IMT	International Mobile Telecommunications
Inmarsat	Inmarsat Global Limited

Intelsat	Intelsat Corporation
IP	Internet protocol
Iridium	Iridium Communications Inc.
ISM	Industrial, Scientific and Medical applications (of radio-frequency energy)
ISO	International Organization for Standardization
IT	information technology
ITU	International Telecommunication Union
kHz	kilohertz
km	kilometre
LEO	low Earth orbit
LPWAN	low-power wide-area network
LRIT	long-range identification and tracking
LTE™	Long-Term Evolution
MEO	medium Earth orbit
MHz	megahertz
MRCP	Maritime Radio Communications Plan (IALA)
NAVTEX	Navigational Telex
NAVDAT	navigational data
NBDP	narrowband direct printing (radio telex)
NCSR	Sub-Committee on Navigation, Communications and Search and Rescue
NM	nautical mile (1,852 m)
NMEA	National Marine Electronics Association
PMR	private mobile radio
PSC	port state control
RCP	required communications performance
RR	Radio Regulations (ITU)
SAR	search and rescue
SES O3b	SES S.A.
SI	Système International (The International System of Units)
SOLAS	International Convention for the Safety of Life at Sea
Thuraya	Thuraya Telecommunications Company
TSDSI	Telecommunications Standards Development Society, India
UN	United Nations
UNCTAD	United Nations Conference on Trade and Development
VDE	VHF data exchange
VDE-TER	VHF data exchange – terrestrial
VDE-SAT	VHF data exchange – satellite

VDES	VHF data exchange system
VTS	vessel traffic services
Wi-Fi™	wireless network protocols based on IEEE 802.11 standards
WiMax™	WiMax™ Technology
WRC	World Radiocommunication Conference

References

[1] International Maritime Organization, *E-navigation*. [Cited July 2020]; Available from: http://www.imo.org/en/OurWork/safety/navigation/pages/enavigation.aspx. Note: IMO meeting documents of past sessions are available on IMO's website (www.imo.org) using the public user access and the document reference given under the relevant Committee or Sub-Committee.

[2] IMO, *IMO takes first steps to address autonomous ships*. 2018 [Cited August 2020]; Available from: http://www.imo.org/en/MediaCentre/PressBriefings/Pages/08-MSC-99-MASS-scoping.aspx.

[3] International Association of Marine Aids to Navigation and Lighthouse Authorities, *VHF data exchange system (VDES) overview G1117*. 2017, Saint-Germain-en-Laye, France. Available from: https://www.iala-aism.org/product/vhd-data-exchange-system-vdes-overview-1117/.

[4] IALA, *Planning and reporting of e-navigation testbeds 1107*. 2016, Saint-Germain-en-Laye, France. [Cited June 2020]; Available from: https://www.iala-aism.org/product/reporting-of-results-of-e-navigation-testbeds-1107/.

[5] United Nations Conference on Trade and Development, *Review of maritime transport 2016*. 2016, New York: UNCTAD.

[6] UNCTAD, *Review of maritime transport 2019*. 2019, New York: United Nations Conference on Trade and Development.

[7] Food and Agriculture Organization of the United Nations, *Global record of fishing vessels, refrigerated transport vessels and supply vessels*. 2020 [Cited 12 July 2020]; Available from: http://www.fao.org/global-record/information-system/en/.

[8] World Meteorological Organization, *The Voluntary Observing Ship (VOS) Scheme*. 2016; Available from: https://www.wmo.int/pages/prog/amp/mmop/JCOMM/OPA/SOT/vos.html.

[9] IALA, *Maritime Radio Communications Plan, MRCP*. 2017, Saint-Germain-en-Laye, France. Available from: https://www.iala-aism.org/product/maritime-radio-communications-plan/.

[10] Kurian, A., *Latency analysis and reduction in a 4G network*. [Master's thesis on the Internet]. 2018, Delft, The Netherlands: Delft University of Technology. [Cited April 2020]; Available from: https://repository.tudelft.nl/islandora/object/uuid:e1badd8d-a384-49a1-b958-a0c1e499c539?collection=education.

[11] International Civil Aviation Organization, *Doc 9869, performance-based communication and surveillance (PBCS) manual*. 2nd edn. 2017, Montréal, Canada: ICAO.

[12] International Telecommunication Union, *Radio regulations 2016*. 2016, Geneva: ITU.

[13] Bureau International des Poids et Mesures, *SI brochure: the international system of units (SI)*. 2019, BIPM.

[14] Kepler Communications, *Polarconnect, bringing superior connectivity to the Earth's polar regions*. 2020 [Cited July 2020]; Available from: https://www.keplercommunications.com/solutions/polarconnect.

[15] Bronk, K., Lipka, A., Niski, R., Wereszko, B., and Wereszko, K., *Hybrid communication network for the purpose of maritime applications*. The International Journal of Maritime Engineering, 2017. Vol. 159: p. 115–122. DOI: 10.3940/rina.ijme.2017.a1.396tn. Available from: https://www.researchgate.net/publication/318593735_Hybrid_Communication_Network_for_the_Purpose_of_Maritime_Applications.

[16] ITU, *Operational and technical characteristics for an automated direct-printing telegraph system for promulgation of navigational and meteorological warnings and urgent information to ships*. Recommendation ITU-R M.540-2, 1990, Geneva: ITU. Available from: https://www.itu.int/rec/R-REC-M.540/en.

[17] ITU, *Characteristics of a digital system, named Navigational Data for broadcasting maritime safety and security related information from shore-to-ship in the 500 kHz band*. Recommendation ITU-R M.2010-1, 2019, Geneva: ITU. Available from: https://www.itu.int/rec/R-REC-M.2010-1-201901-I/en.

[18] United Kingdom Hydrographic Office, *Admiralty list of radio signals, maritime radio stations, 2017/2018*. Vol. NP281. 2017, Taunton, Somerset, United Kingdom: UKHO.

[19] International Electrotechnical Commission, *Electropedia*. 1998 [Cited August 2020]; Available from: http://www.electropedia.org/.

[20] ITU, *Characteristics of digital fixed wireless systems below about 17 GHz*. Recommendation ITU-R F.1101-0, 1994, Geneva: ITU. Available from: https://www.itu.int/rec/R-REC-F.1101-0-199409-I/en.

[21] ITU, *Technical characteristics for a high frequency direct-printing telegraph system for promulgation of high seas and NAVTEX-type Maritime Safety Information*. 1990, Geneva. Available from: https://www.itu.int/rec/R-REC-M.688-0-199006-I/en.

[22] ITU, *Characteristics of a digital system, named Navigational Data for broadcasting maritime safety and security related information from shore-to-ship in the maritime HF frequency band*. Recommendation ITU-R M.2058-0, 2014, Geneva: ITU. Available from: https://www.itu.int/rec/R-REC-M.2058-0-201402-I/en.

[23] Gonin, I., Dahlen, A. H. S., Johnson, G., and Grayson, C., *HF Digital Radio Mondiale (DRM) broadcast summary report for long range dissemination of maritime information*. R&DC UDI #1732, 2018, New London, CT, United States: United States Coast Guard. Available from: http://www2.unb.ca/gge/test/SWL/2018-0568a%20DRM%20REPORT%20final%20PDF%20UDI%201732.pdf.

[24] IMO, *Response to matters related to the Radiocommunication ITU-R Study Group and ITU World Radiocommunication Conference, test measurements of NAVDAT system under real conditions*. China, France, NCSR 7/INF.13, 2019, London: IMO.

[25] IMO, *Response to matters related to the Radiocommunication ITU-R Study Group and ITU World Radiocommunication Conference*. China, France, NCSR 7/12/6, 2019, London: IMO.

[26] IALA, *VHF data exchange system (VDES)*. 2020 [Cited June 2020]; Available from: https://www.iala-aism.org/technical/connectivity/vdes-vhf-data-exchange-system/.

[27] ITU, *Technical characteristics for a VHF data exchange system in the VHF maritime mobile band*. Recommendation ITU-R M.2092-0, 2015, Geneva: ITU. Available from: https://www.itu.int/rec/R-REC-M.2092/en.

[28] ITU, *World Radiocommunication Conference 2019 (WRC-19) final acts*. 2019, Geneva: ITU.

[29] ITU, *The birth of broadband*. 2003 [Cited June 2020]; Available from: https://www.itu.int/osg/spu/publications/birthofbroadband/faq.html.

[30] ITU, *Overview of the Internet of Things*. Recommendation ITU-R Y.2060, 2012, Geneva: ITU. Available from: https://www.itu.int/itu-t/recommendations/rec.aspx?rec=11559.

[31] 3GPP, *3GPP TS 22.819 V16.2.0 (2018-12). 3rd Generation Partnership Project; Technical Specification Group Services and System Aspects; Feasibility Study on Maritime Communication Services over 3GPP system; Stage 1 (Release 16)*. 2018, Sophia Antipolis, Valbonne, France. Available from: https://www.3gpp.org/ftp/Specs/archive/22_series/22.819/.

[32] Koo, H., *Maritime communication services over 3GPP systems*. 2017, Sophia Antipolis, France: ETSI Future Evolution of Marine Communication. Available from: https://docbox.etsi.org/Workshop/2017/20171107_FUTURE_EVOL_MARINE_COM/MartimieCOMover3GPPServices_KOO.pdf.

[33] An, K., *E-navigation services for non-SOLAS ships*. International Journal of e-Navigation and Maritime Economy, 2016. Vol. 4: p. 13–22. DOI: 10.1016/j.enavi.2016.06.002.

[34] International Marine Electronics Alliance, *Promoting maritime safety worldwide*. 2018 [Cited June 2020]; Available from: https://www.imea-marine.org/.

[35] IALA, *Submission by the International Association of Marine AIds to Navigation and Lighthouse Authorities (IALA), External Relations (d) Relations with non-governmental organizations*. A.29/INF.8, 2015, London: IMO.

[36] Alliance for Internet of Things (AIOTI WG03). *IoT relation and impact on 5G, release 3.0*. 2020, AIOTI. [Cited September 2020]; Available from: https://aioti.eu/wp-content/uploads/2019/10/AIOTI-WG3-SDOs-Alliance-Landscape-IoT-LSP-standrad-framework-R2.9-Published.pdf.

[37] IMO, *Guidelines for the standardization of user interface design for navigation equipment*. MSC.1/Circ. 1609, 2019, London: IMO.

[38] IEC, *TC 80 maritime navigation and radiocommunication equipment and systems*. 2020 [Cited June 2020]; Available from: https://www.iec.ch/dyn/www/f?p=103:7:0::::FSP_ORG_ID:1271.

[39] Beattie, J., *Navigation equipment standards for merchant ships*. Journal of Navigation, 1985. Vol. 38(3): p. 384–404. (Racal-Decca Ltd). DOI: 10.1017/S0373463300032756. Available from: https://www.cambridge.org/core/article/navigation-equipment-standards-for-merchant-ships/2A1188E6607AB1B616CA815885E88837.

[40] IMO, *Maritime cyber risk management in safety management systems*. IMO Resolution MSC.428(98), 2017, London: IMO.

[41] IMO, *Guidelines on maritime cyber risk management*. MSC-FAL.1/Circ.3, 2017, London: IMO.

[42] Ofcom, *Applying spectrum pricing to the maritime sector, and new arrangements for the management of spectrum used for radar and aeronautical navigation aids – a statement*. 2010 [Cited April 2020]; Available from: https://www.ofcom.org.uk/__data/assets/pdf_file/0020/45353/statement.pdf.

[43] Vesper Marine, *Integrated AIS/VHF/monitor 'cortex'*. [Cited September 2020]; Available from: https://www2.vespermarine.com/cortex.

[44] IMO, *Validity of type approval certification for marine products*. MSC.1/1221, 2006, London: IMO.

[45] IEC, *Specifications of distress alert buttons*. COMSAR 16/9, 2012, London: IMO.

[46] Inmarsat Global Limited, *Fleet 33, 55, 77 and MPDS, closure dates announcement*. 2017 [Cited June 2020]; Available from: https://www.inmarsat.com/fleet-33-55-77-mpds/.

[47] COSPAS-SARSAT, *International satellite system for search and rescue*. 2020 [Cited June 2020]; Available from: https://cospas-sarsat.int.

[48] IMO, *Global Maritime Distress and Safety System. Scoping exercise to establish the need for a review of the elements and procedures of the GMDSS, comments on the report of the Correspondence Group – technological change the GMDSS*. Australia, COMSAR 16/3/8, 2012, London: IMO.

[49] Moin, A., *IMSO Director General's message*. 2020 [Cited June 2020]; Available from: https://imso.org/imso-director-general/#more-206.

[50] Australian Human Rights Commission, *A right to access the Internet*. 2020 [Cited June 2020]; Available from: https://humanrights.gov.au/our-work/8-right-access-internet.

[51] IMO, *Draft e-navigation strategic implementation plan (in report to the Maritime Safety Committee)*. NCSR 1/28, Annex 7, 2014, London: IMO. [Cited June 2020]; Available from: http://www.imo.org/en/OurWork/Safety/Navigation/Documents/enavigation/SIP.pdf.

[52] IMO, *Long-term work plan of the organization (up to 2006)*. IMO Resolution A.879(21), para. III.9, 2000, London: IMO.

[53] IMO, *List of outputs for the 2020-2021 biennium*. IMO Resolution A.1131 (31), Annex, p. 7, 2019.

[54] IMO, *Developments in maritime radiocommunication systems and technology, modernization of the GMDSS*. United Kingdom, COMSAR 12/7/2, 2008, London: IMO.
[55] IMO, *Developments in maritime radio communication systems and technology, use of IP technology for the GMDSS*. Japan, COMSAR 9/8/1, 2004, London: IMO.
[56] IMO, *GMDSS as a data communication network for e-navigation*. Republic of Korea, COMSAR 11/14, 2006, London: IMO.
[57] IMO, *ITU maritime radiocommunication matters, digital selective calling (DSC) operational capabilities*. United Kingdom, COMSAR 12/4/7, 2008, London: IMO.
[58] IMO, *Radiocommunications ITU-R Study Group 8 matters, digital selective calling (DSC) operational capabilities*. United States, COMSAR 12/4/8, 2008, London: IMO.
[59] IMO, *ITU maritime radiocommunication matters, radiocommunication ITU-R Study Group 8B matters, new ITU-R recommendation for VHF data exchange and electronic mail using narrow-band techniques*. Norway, COMSAR 12/4/5, 2008, London: IMO.
[60] IMO, *Developments in maritime radiocommunication systems and technology, GMDSS modernization*. United States, COMSAR 13/7/2, 2008, London: IMO.
[61] IMO, *Request for scoping exercise to establish the need for a review of the elements and procedures of the GMDSS*. Chile, France, Norway, United Kingdom, United States, MSC 86/23/11, 2009, London: IMO.
[62] IMO, *Report of the Maritime Safety Committee on its eighty-sixth session*. MSC 86/26, para. 23.20, 2009, London: IMO.
[63] IMO, *Report of the Maritime Safety Committee on its eighty-seventh session*. MSC 87/26, para. 24.27, 2010, London: IMO.
[64] IMO, *Global Maritime Distress and Safety System (GMDSS), GMDSS modernization*. United States, COMSAR 15/3/10, 2010, London: IMO.
[65] IMO, *Report of the Maritime Safety Committee on its ninetieth session*. MSC 90/28, para. 8.10, 2018, London: IMO.
[66] IMO, *COMSAR 16 report to the Maritime Safety Committee*. COMSAR 16/17, Annex 2, 2012, London: IMO.
[67] IMO, *Introduction to the guidelines on the organization and method of work of the MSC and MEPC and their subsidiary bodies*. 2013 [Cited June 2020]; Available from: http://www.imo.org/en/About/strategy/Documents/IMO%20Guidelines%20MSC-MEPC%20(final).pdf.
[68] IMO, *Consideration of ITU World Radiocommunication Conference matters, report of the ninth meeting of the Joint IMO/ITU Experts Group on maritime radiocommunications matters, including information on review and modernization of the GMDSS (agenda item 13), and consideration of radiocommunication ITU-R Study Group matters (agenda item 16)*. NCSR 1/17, para. 16, 2013, London: IMO.

[69] IMO, *COMSAR 11 report to the Maritime Safety Committee*. COMSAR 11/18, para. 6.20, 2007, London: IMO.

[70] IMO, *NCSR 4 report to the Maritime Safety Committee*. NCSR 4/29, Annex 11, Appendix 1, 2017, London: IMO.

[71] Kitack, L., *Address of the Secretary-General at the opening of the seventh session of the Sub-Committee on Navigation, Communications and Search and Rescue (NCSR)*. 2020 [Cited June 2020]; Available from: http://www.imo.org/en/MediaCentre/SecretaryGeneral/Secretary-GeneralsSpeechesToMeetings/Pages/NCSR-7-opening.aspx.

[72] ITU, *Characteristics of unmanned maritime systems (unpublished), working document towards a preliminary draft new report ITU-R M.[MAR-UMS]*. Annex 10 to Document 5B/883, 2015, Geneva: ITU.

[73] ITU, *Satellite issues: Earth stations in motion (ESIMs)*. 2019; Available from: https://www.itu.int/en/mediacentre/backgrounders/Pages/Earth-stations-in-motion-satellite-issues.aspx.

[74] ITU, *Characteristics of unmanned aircraft systems and spectrum requirements to support their safe operation in non-segregated airspace*. Report ITU-R M.2171, 2009, Geneva: ITU.

[75] ITU, *Characteristics and spectrum considerations for sense and avoid systems use on unmanned aircraft systems*. Report ITU-R M.2204, 2010, Geneva: ITU.

[76] IMSO, *Regulatory scoping exercise for the use of Maritime Autonomous Surface Ships (MASS), comments on document MSC. 102/5/1 – potential gaps and themes regarding connectivity, cybersecurity and the implication of MASS on search and rescue*. MSC 102/5/28, 2020, London: IMO.

[77] Zolich, A., Palma, D., Kansanen, K., *et al.*, *Survey on communication and networks for autonomous marine systems*. Journal of Intelligent & Robotic Systems, 2018. Vol. 95: p. 789–813. DOI: 10.1007/s10846-018-0833-5. Available from: https://link.springer.com/content/pdf/10.1007/s10846-018-0833-5.pdf.

[78] Höyhtyä, M., Huusko, J., Kiviranta, M., Solberg, K., and Rokka, J., Connectivity for autonomous ships: architecture, use cases, and research challenges. *International conference on information and communication technology convergence (ICTC)*. 2017, Jeju, South Korea: Institute of Electrical and Electronic Engineers. p. 345–350. DOI: 10.1109/ICTC.2017.8191000.

[79] Rødseth, Ø., Kvamstad, B., Porathe, T., and Burmeister, H. Communication architecture for an unmanned merchant ship. In *2013 MTS/IEEE OCEANS – Bergen*. 2013. DOI: 10.1109/OCEANS-Bergen.2013.6608075.

[80] Mikaeil, A., *Bayesian online learning-based spectrum occupancy prediction in cognitive radio networks*. ITU Journal: ICT Discoveries, 2017. (Special Issue No. 1). Available from: https://www.itu.int/en/journal/001/Documents/itu2018-11.pdf.

[81] Pelechrinis, K., Krishnamurthy, P., Weiss, M., and Znati, T., *Cognitive radio networks: realistic or not?* ACM SIGCOMM Computer Communication Review, 2012. Vol. 43. DOI: 10.1145/2479957.2479965. Available from: https://

www.researchgate.net/publication/235323988_Cognitive_Radio_Networks_Realistic_or_Not.

[82] Tang, C., Kandeepan, S., Hourani, A., Munari, A., and Berioli, M., Spectrum sensing for cognitive maritime VHF networks. *OCEANS 2014 – TAIPEI*. 2014. DOI: 10.1109/OCEANS-TAIPEI.2014.6964419.

[83] Røste, T., Yang, K., and Bekkadal, F., Coastal coverage for maritime broadband communications. *OCEANS – BERGEN*. 2013. DOI: 10.1109/OCEANS-Bergen.2013.6608109. Available from: https://www.researchgate.net/publication/261490359_Coastal_coverage_for_maritime_broadband_communications.

[84] Lin, H.-M., Ge, Y., Pang, A.-C., and Pathmasuntharam, J.S., Performance study on delay tolerant networks in maritime communication environments. *OCEANS'10 IEEE SYDNEY*. 2010, Sydney: IEEE. DOI: 10.1109/OCEANSSYD.2010.5603627.

[85] Choi, M., Park, S., Lee, Y., and Lee, S., *Ship to ship maritime communication for e-navigation using WiMAX*. International Journal of Multimedia and Ubiquitous Engineering, 2014. Vol. 9: p. 171–178. DOI: dx.doi.org/10.14257/ijmue.2014.9.4.18. Available from: http://citeseerx.ist.psu.edu/viewdoc/download?doi=10.1.1.640.3810&rep=rep1&type=pdf.

[86] IMO, *E-navigation strategy implementation plan – update 1*. MSC.1/Circ. 1595, 2018, London: IMO.

[87] IMO, *Revised guidelines for the onboard operational use of shipborne automatic identification systems (AIS)*. IMO Resolution A.1106(29), 2015, London: IMO.

[88] FAO, *Unique vessel identifier (UVI)*. 2020 [Cited July 2020]; Available from: http://www.fao.org/global-record/background/unique-vessel-identifier/en/.

[89] ICAO, *ITU inputs regarding FSS performance*. ACP WG-F/29 WP-08, 2013, Montreal, Canada. [Cited June 2020]; Available from: https://www.icao.int/safety/acp/ACPWGF/ACP-WG-F-29/ACP-WGF29-WP08_paper%20on%20RCP%20for%20UAS%20v6a.doc.

[90] Gartner Inc., *Gartner Glossary*. 2020 [Cited June 2020]; Available from: https://www.gartner.com/en/information-technology/glossary/digitalization.

[91] Coor, J., *An introduction to the digital maturity model*. Seer Interactive. [Cited June 2020]; Available from: https://www.seerinteractive.com/blog/introduction-to-digital-maturity/.

[92] Gardeitchik, J., Buck, W., and van der Deijl, A., *Move forward: step by step towards a digital port*. [White Paper] 2019 [Cited June 2020]; Available from: https://www.portofrotterdam.com/en/port-forward/step-by-step-towards-a-digital-port.

[93] Valčić, S., Mrak, Z., and Gulić, M., *Analysis of advantages and disadvantages of existing maritime communication systems for data exchange*. Scientific Journal of Maritime Research, 2016. Vol. 30: p. 28–37. DOI: 10.31217/p.30.1.4. Available from: https://hrcak.srce.hr/file/236732.

[94] Adelantado, F., Vilajosana, X., Tuset-Peiro, P., Martinez, B., Melià-Seguí, J., and Watteyne, T., *Understanding the limits of LoRaWAN*. IEEE Communications Magazine, 2017. Vol. 55. DOI: 10.1109/MCOM.2017.1600613.

[95] Foubert, B. and Mitton, N., *Long-range wireless radio technologies: a survey*. Future Internet, 2020. Vol. 12: p. 13. DOI: 10.3390/fi12010013.

[96] Matas, A., Orbit/spectrum, ITU International Regulatory Framework. *UN/Nepal workshop on GNSS applications*. Presented by Mitome, T. Kathmandu, Nepal, 2016, Geneva: ITU. Available from: http://www.unoosa.org/pdf/icg/2016/nepal-workshop/IDM/5_13_1415_a.pdf.

[97] Zhao, Y., *Solution for enterprise system modernization*. 2013, Digital Enterprise Architecture Forum. [Cited June 2020]; Available from: https://architechllc.blogspot.com/2013/03/solution-for-enterprise-system_11.html.

[98] Weintrit, A., *e-Nav, is it enough?* TransNav, International Journal on Marine Navigation and Safety of Sea Transportation, 2016. Vol. 10: p. 567–574. DOI: 10.12716/1001.10.04.04. Available from: http://www.transnav.eu/pdf/0679.pdf.

[99] Valette, J., *Deliverable 2.4: Analysis report on available and emerging communications technologies*. [Cited June 2020]; Available from: https://efficiensea2.org/wp-content/uploads/2018/04/Deliverable-2.4_Analysis-report-on-emerging-communication-technologies.pdf.

[100] Dirks, J., *Decision making in the International Maritime Organization – the case of the STCW 95 convention*. 2001, United Kingdom: European Consortium for Political Research (ECPR 2001), University of Kent. Available from: https://ecpr.eu/filestore/paperproposal/512c0b96-d29d-418d-8568-f8dc51f539aa.pdf.

[101] Jensen, J. and Pedersen, P., *Deliverable 3.1: Analysis report, maritime communication and infrastructure*. EfficienSea 2, 2020. Available from: https://efficiensea2.org/wp-content/uploads/2018/04/Deliverable-3.1_Analysis-report-on-Maritime-Cloud-Communication-and-infrastructure.pdf.

[102] Honda, Y., Submission and standardization process (including IPR treatment), consensus building and GCS. *Workshop on IMT-2020 terrestrial radio interfaces Munich, Germany, 4 October 2017, Ericsson*. 2017. Available from: https://www.itu.int/en/ITU-R/study-groups/rsg5/rwp5d/imt-2020/Documents/S01-4_Standardization%20Process%20for%20IMT-2020_Rev.pdf.

[103] Khatibi, F., ATIS WTC IMT-2020 Evaluation Group (WTSC-USA). *Workshop on IMT-2020 terrestrial radio interfaces*. Presented by Pica, F. 2017, Munich, Germany: Alliance for Telecommunications Industry Solutions. Available from: https://www.itu.int/en/ITU-R/study-groups/rsg5/rwp5d/imt-2020/Documents/S03-2_ATIS%20WTSC%20USA.pdf.

[104] 3GPP, *Small cell forum completes 1st LTE Plugfest*. 2013 [Cited July 2020]; Available from: https://www.3gpp.org/news-events/partners-news/1507-small-cell-forum-completes-1st-lte.

[105] Bunduchi, D., *Emperor's new blockchain: an overview of the technology adoption within the maritime industry*. 2019, Gothenburg, Sweden: Chalmers University of Technology. Available from: https://hdl.handle.net/20.500.12380/300162.

[106] Meland, P., Frøystad, C., Drugan, O., and Rødseth, Ø., *PKI vs. blockchain when securing maritime operations*. 2019. Available from: https://www.researchgate.

net/publication/332717931_PKI_vs_Blockchain_when_Securing_Maritime_Operations.

[107] IMO, *Long-range identification and tracking system – technical documentation (Part 1)*. MSC.1/Circ. 1259/Rev.8, 2020, London: IMO.

[108] Herian, R., Legal recognition of blockchain registries and smart contracts. *Blockchains & smart contracts legal and regulatory framework*. Paris, France, 2018, Milton Keynes, United Kingdom: The Open University. DOI: 10.13140/RG.2.2.12449.86886/1. Available from: http://oro.open.ac.uk/59481/.

[109] Nguyen, D., Pathirana, P., Ding, M., and Seneviratne, A., *Blockchain for 5G and beyond networks; a state of the art survey*. IEEE Communications Surveys & Tutorials, 2019. arXiv:1912.05062v1. DOI: arXiv:1912.05062v1. Available from: https://arxiv.org/pdf/1912.05062.pdf.

[110] Weiss, M., Werbach, K., Sicker, D., and Bastidas, C., *On the application of blockchains to spectrum management*. IEEE Transactions on Cognitive Communications and Networking, 2019. Vol. 5(2): p. 193–205. Available from: https://ieeexplore.ieee.org/abstract/document/8703084.

[111] Lind, M., Gardeitchik, J., Carson-Jackson, J., Haraldson, S., and Zuesongdham, P., *'Get Smart', developing smart maritime ecosystems, Seaways*. 2020, London: The Nautical Institute. p. 2.

[112] IMO, *Introduction to the Polish approach to include IT competencies in maritime education and training, HTW 7/INF.9*. Poland, 2020, London: IMO.

Chapter 7

Emerging radio communication technologies in the maritime domain – an overview with a critical evaluation of technologies' promises

Jan-Hendrik Oltmann[1]

7.1 '5G' has it all for society – has it?

Now, '5G' has finally arrived – or the IMT for 2020 and beyond as the fifth generation of mobile telecommunication technology is called officially by the International Telecommunication Union (ITU) – International Mobile Telecommunication, to be precise [1]. The promises made even in official IMT-2020 standards – which one would expect to be conservative in their approach – are high.[*]

The promises most relevant to shipping are summarised here in 2-fold angles of perspective as follows:

- In terms of three fundamental usage scenarios ([1], Section 4, pp. 11f):
  - *'Enhanced mobile broadband':* This usage scenario would allow for 'human-centric' applications that require features such as multimedia (including 3D video and ultra high definition) or massive data transfer (i.e. in the order of gigabytes) in a very short time (i.e. seconds). These applications would be supported to be seamlessly available either at hot-spots with high user density and very high required traffic capacity as well as in wide area mobility situations.
  - *'Ultra-reliable and low-latency communications':* This usage scenario allows for applications that require extremely high reliability and extremely low latency of any throughput. Applications like these are regularly

[1]Federal Waterways and Shipping Administration, Germany.

Disclaimer: This study reflects the professional opinion of the author alone and does not necessarily represent the position of the Administration.

[*]The term 'IMT-2020' is used instead of '5G' (except for within quotes), because '5G' is a designator for a certain development stage of telecommunication technology and also tends to be used as a marketing term. IMT-2020 is used as an abbreviation of 'IMT for 2020 and beyond' throughout. It is an ITU term correlated with a substantial specification documentation that sets the requirement threshold for any and all applicable '5G' technology candidates; i.e. industry-developed '5G' technology candidates need to meet the requirements of IMT-2020. Hence, IMT-2020 is the stronger term.

found in wireless control of processes which would not easily tolerate any continuity disruption, such as (automated or even autonomous) manoeuvring and remote steering of vehicles (including vessels).
  - *'Massive machine-type communications (MTCs)':* This usage scenario would not have the rigid requirements of the previous use case regarding reliability and latency but would allow for a huge number of small, self-contained, long-life but 'always connected' devices to transmit simultaneously in a small area. This usage scenario is often correlated with the Internet of Things (IoT).

- In terms of absolute key target capabilities and their anticipated absolute values ([1], Section 5, pp. 12ff):
  - *'Peak data rate':* This is the data rate that the transmission to a single device may expect to achieve under ideal conditions. It is expected to lie in the order of 20 Gbit/s.
  - *'User-experienced data rate':* This is the more practical value; a transmission to a device or user may achieve everywhere throughout the planned coverage area. It is expected to lie in the order of 100 Mbit/s.
  - *'Latency':* This is the time needed for a packet to be delivered from reception by the IMT-2020 system at the source device to its delivery by the IMT-2020 system at the destination device of the communication link. It is expected to lie as low as 1 ms.
  - *'Mobility':* This parameter defines the maximum speed at which a device may travel while still seamlessly communicating via the IMT-2020 system with a defined Quality of Service (QoS). This is expected to lie in the order of 500 km/h.
  - *'Connection density':* This parameter defines the 'total number of connected and/or accessible devices per unit area' and is expected to lie in the order of 1 million devices per square kilometre.
  - *'Area traffic capacity':* This parameter defines the 'total traffic throughput served per geographic area' and is expected to lie in the order of 10 Mbit/s and square metre.

Voices from the information and communication technology industry summarise this promise in the following superlative terms even as the 'intriguing characteristics of 5G': 'ultra-fast data rate', 'ultra-low latency', 'ultra-reliability', 'ultra-connectivity', 'ultra-responsiveness', and 'ultra-densification' [2].

For all of the earlier, it is important to note that IMT-2020 is supposed to demonstrate those capabilities for voice and non-voice applications alike, as required by the use case under consideration.

It also appears that some important prerequisites to fulfilment are already in place, namely:

- The relevant technical standards are in place, namely ITU, have created a hefty documentation regarding land networks and radio communication interfaces alike,[†] and also industry consortia such as 3GPP have produced their share [11].

- The required frequency arrangements have been made by ITU [10] and national authorities [12].

So, one might jump into a quick conclusion that IMT-2020 is the solution for everything in terms of telecommunications. And apparently rightly so: by default, the IMT-2020 addresses the whole of society and its communication needs – the 'networked society' is the buzz word and the driving persistent vision ([1], Scope), whatever this may mean in detail. And will not the IoT or the MTCs – as the IoT is also known by a preferable synonym – and which is supposed to be a substantial part of that 'networked society', also be implemented in the seas, waterways, and ports (compare [13], 78–88, for instance)?

But what do these promises of IMT-2020 actually mean to the maritime domain? This needs to be understood to the maximum extent possible at this point in order to assess the potential impact of IMT-2020 on telecommunications in the maritime domain – and to determine whether the big promise holds true. ***If so***, this, in turn, would prompt an urge to abandon any other telecommunications' development effort or project currently under way in the maritime domain.

7.2 The relevance of IMT-2020 for maritime applications and its advantages for shipping

It must not be forgotten that the maritime community is part of society at large – be it on a local, national, regional, or even global level. Hence, on one hand it is justifiable to consider what developments in society at large – here in the field of telecommunications – might mean for the maritime community, specifically regarding today's perceived or real maritime telecommunication needs. The more maritime users will experience the high-quality IMT-2020 features actually delivered in their daily personal lives, the more they will expect a similar quality of experience (QoE) at their maritime workplaces – the longer the more.

On the other hand, it is necessary to recognise that the maritime community has always been reluctant accepting new things, in particular when these new things have the potential of replacing existing things and even, if accepted, only gradually.

[†]ITU has published many documents dealing with many different facets of the various IMT-systems, including IMT-2020, and their interdependencies. From a strategic, operational, and overview perspective, a good entry point into that bulk of documents is provided by a few documents, though: [1] together with [3] provides the IMT-2020 use cases and the requirement base in an overview manner; [4] focuses on the audio–visual capabilities, while [5] focuses on the use of IMT-2020 for MTCs; [6] provides IMT-systems traffic estimates 2020–30; [7] introduces the radio communication technologies supporting IMT-2020 in an overview manner; [8] shows the technical architecture and topology – here it starts to get sophisticated; and [9] explores the feasibility of, in particular, broadband applications by using frequency bands above 6 GHz, i.e. those frequency bands for which '5G' has become iconic and for which the image in the public mind has already associated them with range and applicability 'limitations of 5G'. A comprehensive list of IMT-related documents at ITU alone is contained in [10], Attachment 3.

New things always needed and still need to demonstrate real advantages and therefore are required to answer – in the wording of the International Maritime Organisation (IMO) – a 'compelling need'. Therefore, IMT-2020 needs to demonstrate real benefits for the maritime domain *specifically*.

Therefore, the present author is about to scrutinise the earlier IMT-2020 promises regarding maritime applications, staying on the operational level to the possible maximum extent – specific technical issues will only be addressed as far as necessary.

Like many other telecommunication systems developed over the past decades, IMT-2020 has two major components: the terrestrial component and the satellite component. During the development process of any system that comprises these two components, it is verified that the two components do not interfere with each other – to say the least – but ideally seamlessly complement each other in terms of functionality – to say the utmost. For the terrestrial component of any telecommunication system, a terrestrial infrastructure is required which is situated on earth – or from a maritime perspective, on shore – to a large degree. This allows for relatively many infrastructure sites with relatively easy set-up and access and holds true – in principle – also for sites in remote, even marine places like lighthouses and offshore structures. This, in turn, allows for cost-efficient set-up of those shore infrastructure sites, thus providing low-cost communications for users; when public administrations are setting up the shore infrastructure, the safety-related communication was and is free of charge for the user as a principle. The cost situation is different, when airborne sites need to be used, e.g. on aircrafts or High-Altitude Platforms. Also, it is different when the satellite component is introduced – at least up to now (2020), the user is charged with substantial communication costs. This situation holds true at least up to present. There are under way moves in the satellite domain which may change the above situation in the future. Considering this, the terrestrial component of any telecommunication system is to be considered first. The satellite domain will be discussed after that.

Taking the fulfilment of the earlier promises of IMT-2020 for granted – which is justified considering the technical progress in the field of telecommunications over the past few decades – eventually *for where terrestrial IMT-2020 will be available*, *near-shore maritime applications* depending on IMT-2020 shore infrastructure for the terrestrial component of IMT-2020 *will work*:

- For *ship movements under shore and in ports*, both of which are considered critical for the safety of a vessel's voyage, IMT-2020 will fulfil the safety critical requirements due to the earlier promises.[‡] By the same token, Vessel Traffic Services (VTS), Ship Reporting Systems (SRSs) using electronic means, as well

[‡]It should be noted that there is an inherent trade-off between mobility requirements on one hand and achievable reliable coverage on the other: while vessels certainly will not even come close to exploiting the mobility parameter promise – consider the speeds of even high-speed crafts compared to those of high-speed trains for which the IMT-2020 mobility requirement was set up in the first place – the margin thus gained may be used for exploiting the 'seamless wide area coverage' promise; compare 'macro rural' deployment scenario ([14], Section 2.2).

as envisaged applications of IMO's e-navigation strategy may benefit substantially from IMT-2020 in coastal waters and should seriously consider employing the terrestrial component of IMT-2020 in the future *as their default.*

- Also, for *port operations* in terms of cargo turnover to maintain the supply chain and to fulfil logistic/commercial requirements, IMT-2020 may be a prime candidate due to – besides other features – its promise of 'ultra-densification', i.e. to be able to also handle high-density telecommunication situations when vessels are crowded with thousands of people (as in the case of cruise ships) or participating[§] cargo items (in the case of large container vessels) or participating machinery, all within a waterway potentially itself fully fitted with a substantial number of IoT sensors – all of this regularly within or adjacent to city environments which add their own high densification already, too.
- Finally, all applications that require the promise 'ultra-low-latency' *on-top* of the earlier, i.e. all maritime applications that would fit into the 'ultra-reliable and low-latency communications' usage scenario, will be satisfied by the IMT-2020 promise, although many of these applications are still in the research fields:
  - Augmented Reality applications supporting the bridge team;
  - remote control of ship handling of vessels even to the extent of using Virtual Reality applications to actually remotely steer vessels from ashore or from a different vehicle; and
  - highly automated or even autonomous ships – Maritime Autonomous Surface Ships (MASS) in IMO parlour.

Not even confined to near-shore, the maritime domain may get benefitted largely by the abstract concepts of the terrestrial component of the IMT-2020, in particular when it comes to the convergence of data processing (IT) and tele- or radio communications, a specific underlying progressive feature of IMT-2020.[¶] In particular, the maritime domain may get benefitted

- by employing the 'ultra-connectivity' promise and closely connected to this by employing the 'follow-the-user' principle: the demand for that connectivity in the maritime domain is already proven by the existence of vanguard developments like the Maritime Connectivity Platform [15] that may get benefitted by absorbing relevant existing parts of the IMT-2020 definition framework.
- by key service-oriented architectural concepts, but with their most fundamental distinction in Core Network and Radio Access Network (RAN) alone: the RAN would allow for several terrestrial Radio Access Technologies (RATs) to be combined to a heterogeneous network (HetNet) of RATs and would thus give the terrestrial component of IMT-2020 a system flexibility which is critical in particular for the needs of the maritime domain, as will be shown later.

[§]That is participating in the IMT-2020 telecommunications.

[¶]These abstract concepts have been described in considerable detail, for example in the ITU-T Y-Series of Recommendations.

- by employing the 'smartness' promise: machine learning and big data analytic techniques are employed extensively by IMT-2020 – by necessity, namely to achieve the performance goals.

From the earlier considerations, one could draw first *intermediate conclusions* as follows:

- ***IMT-2020 indeed has the potential to cover the ground floor of maritime ship–shore/ship–shore applications in particular when shore is close or when vessels are in harbours, by its terrestrial component alone.***
- The IMT-2020 mindset as sketched earlier has resulted in a powerful functionality set-up. This appears 'good enough' or even superior to whatever was there in the maritime domain before in terms of functionality of shore-based infrastructure. Hence, *that powerful functionality set-up should be adopted by the maritime community*, and the maritime community should endorse the powerful concepts *at their appropriate places within the overarching maritime communications architecture* when further developing their strategies regarding the modernisation of maritime tele-/radio communications (including the modernisation of the Global Maritime Distress and Safety System (GMDSS), e-navigation implementation, and the introduction of MASS).
- But also the 'unavoidability' of IMT-2020 when it comes to *user expectations*, i.e. when QoE-derived measures and concepts are applied to the maritime domain – as opposed to 'only' providing a high QoS – should be recognised and accepted: it has been a bitter 'running gag' at e-navigation conferences for years, pointing to bridge team members using their private smartphone – even though 'only' available in '3G' or '4G' technology so far – in lieu or only grudgingly in addition to the community blessed, but limited functionalities of the mandatory shipboard equipment; with all the associated risks for navigation of that behaviour. This is not to condone the usage of private smartphones for safety-critical purposes of navigation, but the point is that the 'QoE gap' between the daily telecommunication experience of shipboard users in their private lives on one hand and the professional equipment they encounter in their professional lives on the other will be widened with the advent of the IMT-2020. While a high QoE can and will be delivered by the IMT-2020 in an economical way to life of society at large, *the maritime community needs to proactively address the topic of introducing and maintaining an equally high QoE for mandated shipboard equipment, thus mitigating the risks created by an ever increasing 'QoE gap'.*

7.3 The availability challenge to IMT-2020 when applied to the maritime domain, and mitigation options

After gaining an understanding of the big promise of IMT-2020 in terms of applications and operational concepts as applied to the maritime domain, it is necessary to look at the *one* critical challenge for the ultimate delivery of that big promise to the maritime domain, and the options to mitigate that challenge. Again, considerations are confined to the terrestrial component of the IMT-2020 at this stage.

7.3.1 Weaknesses of the default terrestrial IMT-2020 RAN when applied to the maritime domain

When it comes to the radio communication part of it, the terrestrial RAN of IMT-2020 has been developed with certain RATs as the 'default' carriers of the IMT-2020 promise, i.e. those RATs are presently envisaged or about to be commercially available in large volumes.

These default RATs were developed having in mind radio scenarios on land or indoors, not specifically considering the rough environment and spatial extension of the maritime domain. These default RATs will operate in the frequency bands as given by [10] and Radio Regulations.[||] While these frequencies allow for a huge radio communication bandwidth and therefore for huge data rates as introduced earlier, the range of each individual station is limited by two main factors: line-of-sight and path loss, even when assuming no other factors detrimental to range.

While line-of-sight is mainly dependent on the respective antenna heights of the participating stations, path loss is a major limiting factor regarding range at these frequencies. Within an onshore IMT-2020 RAT, the limited range is compensated by setting up a grid of stations, each of which having a neighbouring station in several directions, thus ideally resulting in a cell structure for the terrestrial IMT-2020 RAT that allows also for shrinking of the cell sizes due to the expected density of participants – the more participants, the smaller the cell size.

When putting up a RAT for sea areas or stretches of inland waterways such as estuaries, rivers, and canals, often it is possible for only linear cell grid geometries to cover the sea area or waterway from shore sites, i.e. a pearl-string-like cell grid geometry along a coast or a waterway is the result. For inland waterways and certain sea bight situations, the alternate usage of the two shores might improve the coverage. Thus, the range of individual RAT shore stations may be sufficient for the necessary coverage of an inland waterway and a certain sea bight situation. However, the penetration of offshore sea areas by such shore stations may be very limited in terms of range by cell grid geometry alone.[**]

Hence, the big challenge for the terrestrial IMT-2020 RAN when applied to the maritime domain is the availability of the default terrestrial RATs of the IMT-2020 for substantial parts of the maritime domain, i.e. for all those sea areas beyond line-of-sight to the nearest shore infrastructure site.[††]

[||]Reference [16a] as amended by [16b].

[**]For the default terrestrial IMT-2020 RATs on top come additional limitations introduced by the adverse specifics of the marine path loss (as opposed to the land path losses for which the default terrestrial IMT-2020 RATs were designed).

[††]It should be noted that the *actual deployment* of the necessary number of shore stations according to the coverage planning for any terrestrial RAT is *not* an intrinsic feature of that RAT as a technology set-up as such. That is local availability deficiencies due to missing local deployment are not discussed here. Deployment issues will be discussed in a further section.

7.3.2 A composite scenario for the application of IMT-2020 to the maritime domain

The earlier consideration leads to a second intermediate conclusion, namely that the 'default' terrestrial RATs of the IMT-2020 RAN *alone* are 'not good enough' to cater for *all* of the needs of the maritime domain at large, simply because of missing availability in principle. This intermediate conclusion prompts the idea to investigate ***a composite scenario*** for the terrestrial component of the IMT-2020 RAN, namely ***supplementing the default terrestrial RATs by other radio communication technologies in those maritime areas, where default terrestrial RATs are not available.***

These supplementary terrestrial radio communication technologies are those that have been developed specifically for the maritime domain *or* that have been adapted from land mobile to maritime mobile radio communications, i.e. *terrestrial RATs for the maritime domain*. The author of this study would like to call these supplementary RATs as ***Maritime RATs***, for ease of usage, still having in mind their terrestrial nature in terms of radio communications. Also, the author would like to call the resulting terrestrial composite scenario ***HetNet-Maritime***, to indicate that this is a terrestrial HetNet tailored to the specific needs of the maritime domain. Figure 7.1 provides an overview of the resulting architecture.

Regarding Figure 7.1, it should be noted that it is a system engineering diagram showing a *functional* hierarchy of components. Hence, the high-level purposes and functionalities would be at the top of the figure, the interface to shipping (not drawn) would be at the lower end of the figure, and any and/or all *radio* communications with participating ships and other mobile entities would take place only via the two RATs parts, both subdivided by 'terrestrial' and 'satellite' (if any). In addition, shore-based parts of the shipping industry would be interfaced by shore networks thereby having direct access to higher levels of the overarching maritime (radio communication) architecture. It should be noted that the relative position of any entity in this diagram does not imply any value assessment for any entity shown or – in the case of shipping – not shown.

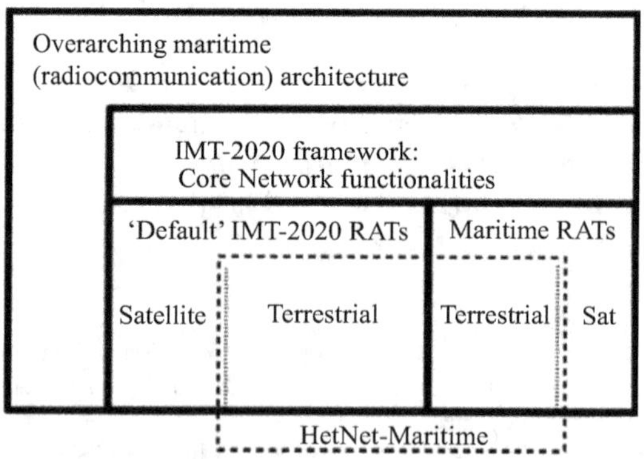

Figure 7.1 Proposed composite scenario for the application of IMT-2020 to the maritime domain. Source: Present author

7.4 Candidate Maritime RATs to supplement terrestrial component of IMT-2020

Hence, the question is what Maritime RATs may supplement the default terrestrial IMT-2020 RATs when applied to the maritime domain, i.e. what ***candidate Maritime RATs*** are there? This consideration is the subject of this section.

IMT-2020 was designed to cater for multimedia transmissions, including voice communication. Contrary to that, some of the candidate Maritime RATs to supplement the default terrestrial IMT-2020 RATs may be technologies designed for ***non-voice***[‡‡] applications only, while others would be supportive of ***voice***, too. This distinction is carried through the considerations of this section.

7.4.1 Overview on non-voice Maritime RATs

This section now presents in an overview manner some (known) candidate *non-voice* Maritime RATs with at least some recognisable/recognised potential for the maritime domain to be precise; at this stage, only the terrestrial components will be considered, even if any candidate Maritime RAT would comprise a satellite component, too.

Each candidate Maritime RAT is briefly introduced by its features relevant for the discussion at hand, while for more specific details references are given to the appropriate documentation of that candidate Maritime RAT. Their order of appearance reflects the present author's perceived maturity and thus applicability to the maritime domain.

7.4.1.1 The VHF data exchange system and the NAVDAT system

The present author introduces the VHF Data Exchange System (VDES) and the NAVDAT system here on a peer level, although they are quite different in regard to their respective technologies and applications. The reasons for that peer-level approach are as follows:

- Both candidate Maritime RATs share the same level of technological development and have acquired the necessary radio frequency allocations. They have arrived at the test bed stage based on internationally recognised technical documentations;[§§]
- Both candidate Maritime RATs share the same level of IMO recognition, i.e. of the competent international body for operational issues of safety of shipping.[¶¶]

[‡‡]A term borrowed from the mobile-satellite service domain.

[§§]For the VDES, the entry point for a system overview is [17,18], while [19] provides specific information on the satellite component of the VDES and [20] explains the frequency channel plan of the VDES; for frequency allocations for VDES refer to [16b], App. 18, footnotes 'm', 'mm', 'w', and 'z' in particular. Regarding NAVDAT, the operational and technical aspects are provided in [21–24] for two relevant maritime MF and HF frequency bands; for frequency allocations for NAVDAT refer to [16b], 5.79 and 5.82c as well as App. 17.

[¶¶]Both systems, NAVDAT and VDES, were introduced at IMO some years ago. As of 2020, there has been proposed to create IMO Performance Standards for NAVDAT and VDES equipment, respectively, which are regularly considered a recognition for the use of such equipment regarding safety of navigation. Also, it has been proposed that IMO allows one for the usage of VDES equipment as a fulfillment of the mandatory AIS carriage requirement by ships.

From this alone, one could draw the conclusion that both candidate Maritime RATs have standing and recognition in the maritime community which cannot – in practical terms – be undone, at least not in an intermediate or even short-term perspective. Hence, it is justified to assess that ***these two systems will have their firm place within any HetNet-Maritime.***

The envisaged maritime applications for and features of the two candidate Maritime RATs are different though, as follows:

VDES: From an operational point of view, the VDES has as its core the Automatic Identification System (AIS) and the Application Specific Messages (ASM).

The AIS as such was introduced originally for three purposes [25]: for collision avoidance (among vessels, i.e. ***ship–ship***), for VTS, i.e. in particular for safety-related interactions between shore centres of VTS authorities and vessels, i.e. ***ship–shore/shore–ship***, and for SRSs, i.e. for the fulfilment of the certain notification requirements for vessels passing through a coastal state's waters or approaching a coastal state's port, again ***ship–shore/shore–ship***.

The most important feature of the AIS – in particular in regard to the considerations at hand – is its capability to have *self-initiated, direct* and *purpose-tailored* non-voice radio communication between any *close* participating mobile entities anywhere on the globe, i.e. no shore infrastructure, satellite, or any other intermediate is required to fulfil the collision avoidance purpose of the AIS. As the radio communication protocol is engaged, certain latency is incurred for that radio communication, which lies in the order of seconds to a minute.

The ASM enhance the basic capabilities of the AIS by allowing for the transmissions of certain types of safety-related information by tailored messages, ship–ship, shore–ship, and ship–shore, some of which have been even standardised for that purpose by IMO [26].

AIS uses a line-of-sight radio communication link in the maritime VHF band ([16b], Appendix 18 refers), which renders the AIS essentially a relative short-distance radio communication means, mostly depending on the antenna heights of participating stations, i.e. ranging from literally sea level – in the case of buoy-mounted stations or Autonomous Maritime Radio Devices[||||] – to airborne or even satellite stations. Path loss over sea is regularly not an issue; range is fundamentally only limited by the range of sight. AIS can be construed as a 1G digital cellular radio communication system.

Over time, an AIS satellite component was introduced on the AIS frequencies, allowing for, among other applications, global tracking of AIS-equipped vessels by satellites as well as bidirectional ASM exchange via satellite. The terrestrial component of the AIS is not relying on or dependent on its satellite component.

The basic idea of the VDES is to provide much more capacity for the ASM mode of radio communication, support the AIS and ASM satellite interactions in a

[||||] Compare [27] which is incorporated by reference into the ITU-R Radio Regulations as of 2019; [16b], Appendix 18, Footnotes 'f' and 'r' refer in particular.

much better way (uplink and downlink in both cases), and ultimately protect the core AIS functionalities. The same maritime VHF band is used by the VDES terrestrial component. By default, the VDES inherits the fundamental features of the AIS, including the independence of the VDES terrestrial component from the VDES satellite component.

NAVDAT: The essential purpose of NAVDAT is to broadcast ***shore–ship*** the Maritime Safety Information in accordance with IMO stipulations over large sea areas, using non-voice digital radio communication protocols for a one-directional terrestrial transmission over two dedicated HF and MF frequency bands, respectively. Due to these frequency bands capable of achieving long-range terrestrial transmissions not limited by line-of-sight, the resulting range is in the order of several hundreds of kilometres. When different transmissions from prudently situated, distant shore stations are coordinated internationally by a time schedule, substantial ocean areas can thus be covered. The data rate of these transmissions is very small compared to the default terrestrial IMT-2020 RATs, however, but fully sufficient for the operational purpose. NAVDAT has no satellite component.

7.4.1.2 LTE-Maritime

LTE-Maritime is an application of existing and mature terrestrial IMT-Advanced (4G/LTE) technology to the maritime domain with the goal to provide a data rate to ***shore–ship/ship–shore*** applications in the order of 1 (Uplink)/3 (Downlink) MBit/s for non-voice applications within a coverage range of up to 100 km from ashore [28]. That data rate would be higher than what, e.g., VDES or NAVDAT might achieve. To achieve that LTE-Maritime is required to have IMT-Advanced technology to effectively and successfully deal with the specifics of the maritime domain, namely, among others, with the challenges of a coastal-line-dependent geometry for shore stations (as opposed to the cell structure for rural or city networks), and the specifics of the marine radio path losses over the envisaged distance (as opposed to radio path features over land for rural or city networks). LTE-Maritime has reached the test bed stage and has demonstrated its feasibility in principle. However, the envisaged coverage range of 100 km from ashore can only be reached when shore antenna heights are in the order of 500 m, i.e. when putting shore stations on mountains along the coast.[***] LTE-Maritime is therefore an option for those sea areas where this condition prevails. There may be many places globally, where this holds true; but there are also many sea areas adjacent to low countries, where LTE-Maritime would not exceed the capabilities of the envisaged default terrestrial IMT-2020 RATs. Also, it may be assumed that default terrestrial IMT-2020 RATs may reach similar radio coverage figures when similar antenna heights are applied. Hence, LTE-Maritime seems to be a presently commercially readily available solution for those sea areas with high sites along the coast, in particular when deployed to a closed user group, but LTE-Maritime may be surpassed in the medium- to long run by the default terrestrial IMT-2020 RATs as

[***]The radio coverage estimates of LTE-Maritime for different shore antenna heights, receiver heights, and different receiver types are given in detail in Figure 5 of [28].

soon as they will have reached the same degree of maturity as IMT-Advanced in similar configuration settings.

7.4.1.3 Terrestrial short-range RATs potentially applied to the maritime domain

Apparently entirely independent from the IMT-2020 developments, a group of terrestrial radio communication technologies has been developed over the past few decades which should be considered as potential non-voice Maritime RATs. The ITU generically calls this group of technologies Short-range Radiocommunication Devices (SRDs) on one hand [29] and Low-Power Wide-Area Networks (LPWANs) on the other [30].††† Due to their design purpose, these technologies fall into the broader category of MTC aka IoT, i.e. they intend to support similar or even identical use cases in the MTC domain as the default terrestrial IMT-2020 RATs introduced earlier, but with a different technological approach.‡‡‡

According to [29], SRDs comprise the following diverse technologies: Broadband Radio Local Area Networks (Broadband-RLANs), balise and loop transmission systems at railways, inductive loop systems, wireless audio applications and radio microphones, RF identification systems, medical implant communication systems, and radar level gauges. SRD technologies 'are permitted to operate on a non-interference and non-protected basis' ([29], Section 2; compare also Section 5 regarding specific frequency bands) and are generally not allowed on any frequency band that is allocated to 'safety of life services, including radio navigation' ([29], Section 5).

Obviously, only few of those systems may be relevant for the topic at hand here, namely those systems that can provide MTC between shore or close-to-shore devices (i.e. within the waterway already) and their counterpart devices carried onboard vessels. In particular, one such family of technologies may be Broadband-RLAN. Compared to the dimensions of even 'under shore' in the maritime domain, the range of these technologies is small, in particular when considering the adverse propagation characteristics over sea. Hence, the idea was investigated in two test beds in 2006/2007 and 2010 to use multi-hop communications, using one or more vessels as relay stations, thus creating a Maritime Mesh Network based on Broadband-RLAN technology [32]. It was claimed that it should be feasible to achieve working links of up to 44 km at antenna heights of 16 m above surface and an operating frequency of 5.8 GHz ([32], Section 5). Hence, the Broadband-RLAN technology employed by a Maritime Mesh Network may qualify as a candidate Maritime RAT for 'under shore'. Beyond ports and harbour approaches, the vessel density required for a Maritime Mesh Network may not be available any more.

LPWANs sound promising for extending MTC coverage, but it should be noted that 'wide area' is only meant to mean 'wider than the range of above SRDs', i.e. remaining limited to 'under shore' similarly to the Maritime Mesh Network approach. However, there may be margin to employ LPWAN technologies for

†††The present author will introduce these technologies by using the generic terms, only; not least to avoid any complications when using brand names however colloquial these may be already.

‡‡‡For an overview compare [5,31] together.

narrow bandwidth MTC applications in the maritime domain, and thus LPWAN may also qualify as a candidate Maritime RAT for 'under shore'.

Finally it can be concluded that several members of the SRD and LPWAN families of technologies may complement the terrestrial default RATs of IMT-2020 in very-close-proximity-to-shore scenarios, but only a few may qualify as a Maritime RAT in the Maritime-HetNet. Certainly, they cannot – by their very physical limitations – compete with VDES and/or NAVDAT in terms of coverage and, due to a presently missing regulatory allocation of a protected operating frequency bands, in terms of operational protection level; the latter of which – not being a law of physics – may be negotiable by appropriate concurrent regulatory efforts at IMO and ITU in due course, however. Also, the notion of the maritime domain tapping into MTC for on-route safety of navigation of vessels and vessel traffic management purposes needs to gain acceptance at IMO, first.

7.4.2 Overcoming the 'voice gap'

Only non-voice candidate Maritime RATs were discussed regarding their maritime application so far; but what about the voice communication in the maritime domain?

- First of all, it is the firm conviction of the present author that voice communication will stay for profound operational reasons and despite recurring discussions on topics like 'voiceless VTS'[§§§] – at least *as long as there are humans involved in the relevant VTS processes.*
- However, maritime voice communication, at present, is still using traditional and robust analogue technology on the radio link. Will the analogue technology stay in the maritime domain on the long run? *Certainly not*, even if 'only' for reasons of radio frequency management. Thus, ITU WRC-2019 started to put a limitation on analogue usages of certain frequency channels in the maritime mobile service.[¶¶¶]

It is a fact that digital voice transmission has not been addressed over the past decades as intensively by the maritime community as digital data transmission. There have been some initial considerations, though, based on the 'simple fact' that voice can always be resembled as 'data' and thus 'any' appropriate broadband radio transmission technology could be used 'in principle'; it would serve the purpose of digital voice transmission. Already by 2013, [35] applied the voice encoding concepts of RLAN to the maritime MF/HF frequency band within the GMDSS, which would have had also the benefit of being a protected frequency band ([16a] as amended by [16b], Appendix 17). However, [35] deliberately remained bound by the usage of existing MF/HF shipboard equipment and thus fails to consider the

[§§§]The notion of a 'voiceless VTS' has been around for decades, making its appearance in documents now and then (compare, e.g. [33], Annex 1, section 1), but only with the advent of the communication technologies under discussion here, it has gained fresh momentum (compare, e.g. [34], Annex 1, Nos. 1 and 2).

[¶¶¶]Reference [16b], Appendix 18, Footnote 'w', last paragraph, refers.

vocoder technology application to any other relevant maritime frequency band (beyond the scope of GMDSS) for example. To apply the required vocoder technology to an appropriate radio transmission technology, operating within the well-protected Maritime VHF frequency band ([16b], Appendix 18) would render a candidate Maritime RAT capable of digital voice communications. But there may be additional options.

The default terrestrial IMT-2020 RATs have in common that they are based on the digital cellular concept. Hence, this prompts the question whether there are technologies *not* using a cellular set-up that may render candidate Maritime RATs.

There is such a family of technologies which also stems – not surprisingly – from the land telecommunications sector, and which was designed of delivering voice and data channels *concurrently*. This is generically called Conventional Digital Land Mobile Radio (CDLMR) by ITU, and [36] provides an up to date generic system survey to a variety of different system approaches (together with references to more specific documents in many cases). Due to the fact that a single voice channel of the CDLMR technologies introduced there may require a minimum bandwidth of 6.25 kHz (FDMA) or 12.25 kHz (TDMA), only these technologies are compatible with the existing frequency channel spacing options of the Maritime VHF frequency band [37] and thus render candidate Maritime RATs when thus mapped to it – in fact, this notion was first introduced for one of those CDLMR as early as 1997 ([33], Section 2.2 +Annex 2).

Besides the CDLMR, some digital broadband members of the SDR family of technologies, as introduced earlier, may be a candidate for voice applications in the maritime domain, too, namely the Broadband-RLAN technologies.[‖‖‖] Applying them to carry voice – as originally intended by [35] – the digital equipment designed for the land mobile applications as well as their presently allocated frequency bands could be used for maritime purposes for very short range voice communications, besides their already existing capability to transmit data. Unfortunately, the very short range, among other factors, would prevent these Broadband-RLAN technologies from being candidates for on-route ship-to-ship voice communication. Also, questions would need to be answered as to the usage of unprotected frequency bands for potentially safety-related maritime voice communication.

Concluding: Default terrestrial IMT-2020 RATs certainly support the voice domain but are limited to 'under shore' applications as shown in previous sections. However, there are some candidate Maritime RATs capable of digital voice transmission, too, in principle: their potential sea penetration from ashore may be *much shorter* – as in the case of readily available Broadband-RLAN technologies – *similar or slightly deeper* – as in the case of the application of any CDLMR technology to maritime in the protected VHF Maritime frequency band – or *larger* – as

[‖‖‖]The LPWAN family of technologies is omitted here, because their technologies are essentially designed with the target to minimize energy consumption at (remote) sites without any supporting infrastructure, resulting in a few short transmissions per relatively long periods of time and are thus not suitable for voice communications.

in the case of applying CDLMR technology to the protected Maritime VHF and also – as originally intended in [35] – maritime MF/HF frequency bands. All options would require demonstrations of their capabilities in the maritime domain for voice transmission, still.

To introduce any of the earlier technologies capable of voice transmission into IMO recognised usage, i.e. as a replacement of present analogue technology's usage in the long run, an amendment of SOLAS (beyond the scope of GMDSS!) would be required as well as an elaborate, long-term migration path for phasing in digital voice technology while concurrently phasing out analogue voice technology eventually.

7.4.3 Intermediate conclusion for Maritime RATs

From the previous considerations, one could draw third *intermediate conclusions* as follows: while the default terrestrial RATs of the IMT-2020 RAN cannot be ignored by the maritime domain – at least not in the long run – due to their superior characteristics as introduced earlier, even with their advent there is and will remain margin for application in the maritime domain of at least *some* of the previous candidate Maritime RATs.

But how many? Just a few of the earlier for each distinctive field of maritime application or even just one? Or will they serve as bridge technologies for a period of time (maybe longer as originally intended/anticipated, as the history of maritime telecommunications shows).

7.5 Satellite component

So far, the discussion has focussed on the terrestrial component of the IMT-2020 RAN with its default RATs on one hand and on the potential Maritime RATs on the other, also being terrestrial means. The previous sections demonstrated that both default terrestrial IMT-2020 RATs and terrestrial Maritime RATs, when combined with a (terrestrial) HetNet-Maritime (compare Figure 7.1), will be capable of carrying the bulk of present and future maritime voice and non-voice communication needs in 'under shore' scenarios. The remaining scenarios, in particular considering 'open ocean', would require the satellite component.

Not surprisingly, therefore, a satellite component has already been introduced both to the IMT-2020-framework and to at least one of the Maritime RATs introduced earlier, namely to the VDES.[****] These satellite components were introduced with the intention that the respective specific terrestrial component and the specific satellite components of IMT-2020 and VDES, respectively, interoperate seamlessly to the extent that the terrestrial component actively supports the operation of the satellite component as in the case of the VDES.

[****]For the satellite component of IMT-2020 compare [38]; note that IMT-2020 comprises IMT-Advanced. And for the satellite component of VDES compare [19].

With the improvement of presently existing communication satellite systems, which were deployed independently of the IMT-2020 and VDES, and in particular with the advent of new satellite constellations covering the full maritime domain, the satellite component – combined with the functionalities introduced by IMT-2020 – may pose a *criticality* for the justification of maintaining dedicated Maritime RATs, *if and as soon as* the resulting satellite communications are truly ubiquitous, sufficiently reliable, and cheap (to the extent that the price for individual maritime user's satellite communications may be even neglected).

On the other hand, requirements such as

- distributed control over communication means (as opposed to a few satellite system operators) and
- systemic redundancy, i.e. the necessity to have fallback arrangements due to single mode of failure (despite any promises of technology reliability)

will need to have a weight in the balance between satellite and terrestrial RATs in the overarching maritime radio communication architecture (compare Figure 7.1, again).[††††]

This leads to another intermediate conclusion: while these are arguments to maintain terrestrial, the undeniable advent of massively improved satellite communications in the future – combined with terrestrial IMT-2020 capabilities – will add pressure on the plurality of candidate Maritime RATs within the HetNet-Maritime.

7.6 Shipboard equipment and shore-based deployment – the touchstone for any real application

So far, the present author has considered the promised IMT-2020 features in operational terms and then focussed on the envisaged IMT-2020 RAN – as a HetNet of default terrestrial IMT-2020 RATs *supplemented* by candidate Maritime RATs, *overlayed* with a powerful emerging satellite component. The potential applications in the maritime domain focussed purely on the applicability of technological features for maritime operational scenarios – i.e. it was a feasibility consideration in principle. In doing so, the present author has touched upon certain regulatory aspects already at some points, such as recognition by IMO. Now, these need to be elaborated fully.

It is now necessary to turn towards another key factor which can be called without exaggeration the 'gate keeper' of any technology set-up's real entry into the maritime domain, namely towards *deployment.* That is how will any technology set-up's user equipment[‡‡‡‡] – be it for a terrestrial and/or a satellite component – be actually implemented and actually operated onboard vessels with the number of installations that is

[††††]The same line of considerations lead the international maritime community to recognise that in the field of radionavigation for shipping, both Global Navigation Satellite Systems and terrestrial radionavigation systems are required, at least as backup solutions.

[‡‡‡‡]'User equipment' is construed here as the required *set* of shipboard equipment (such as detached antenna installation, and cabling) together with its integration of the bridge equipment at large, i.e. not only as the user frontend equipment, but including it.

meaningful and sufficient for the maritime application under consideration – ***ship/ship and ship/shore–shore/ship***? What mechanism will drive the shipboard part of the deployment process? (This applies to both the terrestrial and satellite components.)

Or to put it in different words, when will a technology set-up's fully operational user equipment permeate the fleet of maritime vessels to a degree that shore-based organisations will complement it by deploying relevant shore-based applications building on that fact? Will a shore-based organisation take the risk of the deployment of the shore network of the terrestrial RAT in question and thus incur the costs of operating this shore network over a long period of time (i.e. at least one or two decades) when the complementary onboard user equipment is not implemented, yet?

Clearly, this is a classic chicken-and-egg problem. It must not be underrated, because – as the history of maritime technology shows – several promising technologies have failed to pass by the earlier 'gate keeper' and disappeared from consideration over time.

Developing a strategy to resolve that chicken-and-egg problem would lie beyond the scope here. For the purpose here, it might suffice to name some relevant factors as follows:

- *For the deployment of the default terrestrial IMT-2020 RATs, i.e. those which carry the bulk of the big promise as introduced earlier*, a good start might be to consider maritime applications where these default terrestrial IMT-2020 RATs will be deployed most likely independently of the maritime domain, i.e. in close proximity to land, namely in port-within city situations.
- For harbour approaches where shore is clearly visible, such as in estuaries, archipelago passages, canals, and inland waterways, the challenges for deployment of service to the maritime community may become already severe because the much lesser density of participants in applications may, in turn, decrease the commercial incentive for the deployment of the default terrestrial IMT-2020 RATs along the full stretch of a waterway to achieve seamless coverage.
- The extended usage of default terrestrial IMT-2020 RATs may be generated as a spin-off by any of the following factors, however:
  - if the waterway as such would be IoT monitored or IoT activated for reasons other than shipping, and a default terrestrial IMT-2020 RAT would be deployed to that end;
  - if a constant vessel-to-waterway infrastructure interaction (ship/shore–shore/ship) using default terrestrial IMT-2020 RATs would allow for a trade-off with more expensive, potentially outdated Aids-to-Navigation technologies, or at least in a reduction of their number;
  - if the advent and/or efficient operation of remotely controlled vessels and/or MASS would require the features of the default terrestrial IMT-2020 RATs, at least in those waterways where they operate; and
  - if particular importance of that approach for the coastal and/or port state may be reflected in digitalisation policies accordingly, resulting in a preparedness for equipment without immediate return, i.e. preparedness for an 'investment into the future' decision.

- In the past and until today, it were/are predominantly maritime safety administrations which operated and still operate coastal wide, terrestrial networks of maritime radio communication technologies – analogue radio communication technologies like VHF and digital radio communication technologies in the order of 1G like AIS.
  - Therefore, they may be considered of being predestined to provide some of the initial deployment of default terrestrial IMT-2020 RATs to 'commercially less interesting' sea areas and waterways and thus trigger consecutive additional deployment. But will they be capable and/or prepared to jump to the operating demands of an IMT-2020 Core Network plus several terrestrial RATs?
  - Since their concern traditionally lies with coastal-wide deployments of safety-related radio communication systems, they might be inclined to accept the notion of a HetNet-Maritime, including the default terrestrial IMT-2020 RATs, provided that they can be assured to deploy the ***best HetNet-Maritime***.
  - Since they are required to support legacy systems, too, an international consolidation process for phasing out legacy systems on one hand and phasing in the best HetNet-Maritime on the other might be required. That consolidation process clearly would be a task of the competent international body, namely the IMO, and the guiding question at hand would be: which best HetNet-Maritime is going to determine the future shipboard user equipment to become mandatory equipment and which IMO member states are required to provide the supporting (shore) infrastructure?
- (National) prestige should not be underrated as a driver: This may introduce progressive technologies much faster than by ordinary acceptance processes or even by mandatory carriage requirement(s). And it may suffice that some particularly relevant countries follow that road to introduce the technology as-a-matter-of-fact.

Whatever strategy might be selected eventually, it is essential that trust in new systems is allowed to grow over time taking into account also systemic reliability and fall-back arrangement considerations (beyond the single-technology reliability which is part of the big IMT-2020 promise).

7.7 Conclusions

From the earlier considerations, the present author draws the following conclusion:

- The promises of 5G/IMT-2020 are sufficiently trustworthy for the maritime community to seriously venture into considering potential applications of it in the maritime domain.

- Is the maritime domain 'ready' – in terms of understanding, strategic consideration, user requirements, use case development and acceptance, migration path considerations and the like – for IMT-2020? *Not yet.*
- The *IMT-2020 cannot be ignored*, however, by the maritime domain, because it has the potential *to cover a ground floor of maritime ship–shore/shore–ship applications* – in particular when shore is close or when vessels are in harbours, *by its terrestrial component alone.*
- More specifically, the maritime community should recognise and employ the convergence of data processing and tele-/radio communication technologies in IMT-2020 as was already started by vanguard initiatives like the Maritime Connectivity Platform. Or to put it in other words, take all benefits of IMT-2020 because it can offer unprecedented functionalities to the maritime domain.
- This may necessitate the acceptance of the system architecture(s) developed by IMT-2020 as the architectural and functional framework for any future application of any radio communication technology to the maritime domain.
- The maritime community should start by getting acquainted, by using and by introducing the already standardised terminology and concepts such as RATs and HetNet, in particular. While – necessary to stress at the same time – not discarding the overarching service-oriented architectures developed in the maritime domain during the past decades, such as the overarching e-navigation architecture [39–41], and the Common Shore-based System Architecture (CSSA) [42]. This requires to fit the IMT-2020 into their architectures.
- With the proliferation of a high quality of user experience in the private lives of maritime users due to the advent of IMT-2020, the maritime community needs to proactively address the topic of introducing and maintaining an equally or at least reasonably high QoE for – in particular, mandated – shipboard equipment, thus mitigating the risks created by an ever increasing 'QoE gap'.
- It must be recognised, though, that the big challenge for the terrestrial IMT-2020 RAN when applied to the maritime domain is the availability of the default terrestrial IMT-2020 RATs for substantial parts of the maritime domain, i.e. for all those ocean areas beyond line-of-sight to the nearest land infrastructure site.
- Hence, can the candidate Maritime RATs introduced here be easily discarded due to the advent of IMT-2020? *Certainly not*, as the earlier considerations have shown.
- A fusion of maritime-specific technologies as individual RATs into the terrestrial IMT-2020 RAN, i.e. the creation of a ***HetNet-Maritime*** with an IMT-2020 core and umbrella, might be the optimum solution for the availability issue, when considering terrestrial RATs alone.
- This would imply also the feedback of IMT-2020 concepts into the domestic set-ups of the Maritime RATs: this may result not only in technology creep in detailed functionalities, but also in improved approaches of setting up shore-based networks for the Maritime RATs.

- The satellite component to IMT-2020 as well as of communication satellite systems independent of IMT cannot be ignored, too: they provide the only communication means, both voice and non-voice, where there is no coverage from any terrestrial RAT – default IMT-2020 RATs or Maritime RATs – but they may constitute a challenge also to the justification of individual terrestrial RATs, in particular, in the long run. Relying solely on the satellite domain would not be an option for the maritime domain either.
- Only if and when the full (potential) import of IMT-2020 as applied to the maritime domain is understood, should long-term decisions as to (mandatory) carriage requirements of any Maritime RAT be made – the risk to have to support a system mix which is outdated even on the conceptual level already upon its launch as a mandatory system is too high.
- But because radio communications applied to maritime are needed, this should be an incentive to have the maritime community venture into strategic considerations as to the optimum mix of technologies – i.e. for the ***best** HetNet-Maritime* – ***soon***.
- IMO – following the example of the MASS scoping exercise – may be in the best position to initiate a scoping exercise for that best HetNet-Maritime with the support from other international organisations such as ITU, IHO, and IALA.

7.8 Some afterthoughts

Some necessary considerations were left out deliberately at this stage as follows:

- *Detailed, specific or even formalised statements of 'user needs' and/or 'user requirements':* While this is required at any more detailed level of consideration, in particular when it actually comes to the suggested international scoping exercise for the best HetNet-Maritime, for the point of this overview it suffices to *assume* that the promised capabilities of IMT-2020 as sketched earlier truly satisfy *all* presently known user needs and/or user requirements. Once the suggested international scoping exercise for the best HetNet-Maritime would start, the user needs and/or user requirements as described in more detail in the relevant IMT-2020 documentation would be a good starting point.
- *Incomplete list of candidate Maritime RATs:* While striving for the best possible oversight on the candidate Maritime RATs, it may be the case that one or another may have been overlooked. Should such a case be detected, the reader is kindly asked to apply the logic of evaluation developed here to that overlooked candidate Maritime RAT, or even better, bring it to the attention of the body which will run the suggested scoping exercise on the best Maritime-HetNet.
- *Cybersecurity:* It was not addressed because this is a prerequisite for any and all of the IMT-2020 components, including the candidate Maritime RATs. Cybersecurity thus is a part of the 'level playing ground' for the previous considerations: could there be demonstrated, upon more detailed investigation, that there are significant weaknesses regarding cybersecurity, this might lead the affected IMT-2020 component and/or candidate Maritime RAT to be excluded from further consideration.

References

[1] International Telecommunication Union. 2015. *Recommendation ITU-R M.2083-0. IMT Vision – Framework and overall objectives of the future development of IMT for 2020 and beyond*. Geneva.

[2] Eluwole, Opeoluwa Tosin, Nsima Udoh, Mike Ojo, Chibuzo Okoro and Akintayo Johnson Akinyoade. 2018. 'From 1G to 5G, What Next?' *IAENG International Journal of Computer Science*. 45(2018) 3. www.iaeng.org/IJCS/issues_v45/issue_3/IJCS_45_3_06.pdf; accessed 04 February 2020.

[3] ITU. 2018. *Report ITU-R M.2441-0. Emerging usage of the terrestrial component of International Mobile Telecommunication (IMT)*. Geneva.

[4] ITU. 2018. *Report ITU-R M.2373-1. Audio-visual capabilities and applications supported by terrestrial IMT systems*. Geneva.

[5] ITU. 2018. *Report ITU-R M.2440-0. The use of the terrestrial component of International Mobile Telecommunications for narrowband and broadband machine-type communications*. Geneva.

[6] ITU. 2015. *Report ITU-R M.2370-0. IMT traffic estimates for the years 2020 to 2030*. Geneva.

[7] ITU. 2014. *Report ITU-R M.2320-0. Future technology trends of terrestrial IMT systems*. Geneva.

[8] ITU. 2015. *Report ITU-R M.2375-0. Architecture and topology of IMT networks*. Geneva.

[9] ITU. 2015. *Report ITU-R M.2376-0. Technical feasibility of IMT in bands above 6 GHz*. Geneva.

[10] ITU. 2019. *Recommendation ITU-R M.1036-6. Frequency arrangements for implementation of the terrestrial component of International Mobile Telecommunications in the bands identified for IMT in the Radio Regulations*. Geneva.

[11] https://portal.3gpp.org/relevant reference documentation for 5G is contained in cumulative 'Releases' #14, #15 and #16 so far. Accessed 11 August 2020.

[12] ITU. 2019. *Report ITU-R M.2480-0. National approaches of some countries on the implementation of terrestrial IMT systems in bands identified for IMT*. Geneva.

[13] Jahn, Carlos and Sebastian Saxe. 2017. *Digitalization of Seaports – Visions of the Future*. Stuttgart: Fraunhofer-Verlag.

[14] ITU. 2017. *Recommendation ITU-R M.2101-0. Modelling and simulation of IMT networks and systems for use in sharing and compatibility studies*. Geneva.

[15] https://maritimeconnectivity.net/; accessed 11 August 2020.

[16a] ITU. 2016. *Radio Regulations*. Geneva.

[16b] ITU. 2020. *World Radiocommunication Conference 2019 (WRC-19). Final Acts*. Geneva.

[17] International Association of Marine Aids to Navigation and Lighthouse Authorities (IALA). 2017. *Guideline G1117, Edition 2. VHF Data Exchange System (VDES) Overview*. St-Germain-en-Laye, France.

[18] ITU. 2015. *Recommendation ITU-R M.2092-0. Technical characteristics for a VHF data exchange system in the VHF maritime mobile band*. Geneva.

[19] ITU. 2018. *Report ITU-R M.2435-0. Technical studies on the satellite component of the VHF data exchange system*. Geneva.

[20] ITU. 2015. *Report ITU-R M.2371-0. Selection of the channel plan for a VHF data exchange system*. Geneva.

[21] ITU. 2018. *Report ITU-R M.2443-0. NAVDAT Guidelines*. Geneva.

[22] ITU. 2014. *Recommendation ITU-R M.2058-0. Characteristics of a digital system, named navigational data for broadcasting maritime safety and security related information from shore-to-ship in the maritime HF frequency band*. Geneva.

[23] ITU. 2019. *Recommendation ITU-R M.2010-1. Characteristics of a digital system, named Navigational Data for broadcasting maritime safety and security related information from shore-to-ship in the 500 kHz band*. Geneva.

[24] ITU. 2010. *Report ITU-R M.2201-0. Utilization of the 495-505 kHz band by the maritime mobile service for the digital broadcasting of safety and security related information from shore-to-ships*. Geneva.

[25] International Maritime Organization (IMO). 1998. *Resolution MSC.74(69) (adopted on 12 May 1998) Adoption of New and Amended Performance Standards, Annex 3, Recommendation on Performance Standards for Universal Automatic Identification System (AIS)*. London.

[26] IMO. 2004. *SN/Circ.236, 28. May 2004. Guidance on the Application of AIS Binary Messages*. London.

[27] ITU. 2019. *Recommendation ITU-R M.2135-0. Technical characteristics of autonomous maritime radio devices operating in the frequency band 156–162.05 MHz*. Geneva.

[28] Jo, Sung-Woong and Woo-Seong Shim. 2019. 'LTE-Maritime: High-Speed Maritime Wireless Communication Based on LTE-Technology.' *IEEE Access*. 7(2019). DOI 10.1109/ ACCESS.2019.2912392. https://ieeexplore.ieee.org/document/8694989; accessed 08 December 2019.

[29] ITU. 2019. *Report ITU-R SM.2153-7. Technical and operating parameters and spectrum use for short-range radiocommunication devices*. Geneva.

[30] ITU. 2018. *Report ITU-R SM.2423-0. Technical and operational aspects of low power wide area networks for machine-type communication and the Internet of Things in frequency ranges harmonised for SRD operation*. Geneva.

[31] ITU. 2019. *Report ITU-R M.2479-0. The use of land mobile systems, excluding IMT, for machine-type communications*. Geneva.

[32] ITU. 2010. *Recommendation ITU-R M.2202. Maritime broadcast wireless mesh networks*. Geneva.

[33] ITU. 1997. *Report ITU-R M.2010-1. Improved efficiency in the use of the band 156-174 MHz by stations in the maritime mobile service*. Geneva.

[34] Republic of Korea. 2014. Consideration on available services from the perspective of shipboard users of e-navigation. Submitted to the International Maritime Organization, Sub-Committee on Navigation, Communications and Search and Rescue, 1st session, as document NCSR1-INF.17, 25 April 2014. Can be retrieved publicly at www.IMO.org.

[35] ITU. 2013. *Report ITU-R M.2288-0. Digital voice communication system on MF/HF radio channels of the maritime mobile service for shore-to-ship/ship-to-shore applications*. Geneva.

[36] ITU. 2019. *Report ITU-R M.2474-0. Conventional digital land mobile radio systems*. Geneva.

[37] ITU. 2012. *Recommendation ITU-R M.1084-5. Interim solutions for improved efficiency in the use of the band 156-174 MHz by stations in the maritime mobile service*. Geneva.

[38] ITU. 2012. *Report ITU-R M.2176-1. Vision and requirements for the satellite radio interface(s) of IMT-Advanced*. Geneva.

[39] IMO. 2018. *MSC.1/Circ.1595, 25 May 2018. E-navigation strategy implementation plan – Update 1*. London.

[40] IALA. 2015. *IALA recommendation e-NAV 140 on the architecture for shore-based infrastructure 'fit for e-navigation, Edition 2*. St-Germain-en-Laye, France.

[41] IALA. 2015. *IALA Guideline 1113 on design and implementation principles for harmonised system architectures of shore-based infrastructure*. St-Germain-en-Laye, France.

[42] IALA. 2015. *IALA Guideline 1114 on a technical specification for the Common Shore-based System Architecture (CSSA)*. St-Germain-en-Laye, France.

Chapter 8
A data-driven methodology for maritime Patterns of Life discovery

Konstantinos Chatzikokolakis[1], *Dimitris Zissis*[1,2], *Giannis Spiliopoulos*[1] *and Marios Vodas*[1]

In recent years, mostly due to the increased availability of maritime-related data, the interest in maritime analytics has grown and several academic works have started to gradually exploit this wealth of data for tasks, including frequent pattern discovery, trajectory clustering, classification, forecasting and outlier detection. A precursor to all these works is building a data-driven model of normal behavior or commonly traveled routes that can then act as a baseline for numerous real-world applications.

In this chapter, a data-driven methodology suitable for discovering maritime traffic patterns and revealing "roads of sea" is discussed. The proposed solution exploits the MapReduce paradigm to perform parallel distributed processing of large vessel tracking datasets, collected through the automatic identification system (AIS), and analyzes vessels' navigational patterns in a computationally efficient and accurate way. Unsupervised machine learning algorithms are employed to distinguish the spatio-temporal characteristics of vessel routes and distill global maritime "Patterns of Life." A "Pattern of Life" emerges through an aggregated analysis of spatiotemporal directed port-to-port connections bearing thus the burden of post-processing analysis. Within this context, Patterns of Life are perceived as an enabler for numerous maritime applications.

This work presents an overview of the ROute exTrActor (ROTA) approach, while focus is given toward the technical design aspects and challenges of this scheme. The merits of this work can be helpful to a wide spectrum of maritime services such as understanding and predicting vessels' activities, evaluating shipping's impact on the environment, detecting dangerous situations and providing risk reports, assessing the vessel's navigation performance and performing voyage optimization.

8.1 Introduction

Maritime transport is vital to the global economy with more than 80% of the world's trade being carried by sea. It is by far the most cost-effective way of

[1]Department of Product and Systems Design Engineering, University of Aegean, Greece
[2]MarineTraffic, Athens, Greece

transporting massive amounts of supplies (i.e., goods, raw material, liquids such as oil or gas) around the world [1]. The commercial shipping fleet follows specific paths when sailing across the oceans, widely known as shipping routes or "sea roads." These paths form a global sea transportation network connecting major ports. Shipping routes are regularly overcrowded, and in many cases, they are quite narrow (often just a few kilometers wide) having various navigation constraints due to physical obstacles (e.g., reefs). They are usually traversed by enormous vessels that perform risky maneuvers under constantly changing environmental conditions (e.g., wind, sea currents and limited vision due to fog). For most connections, the shipping route between a departure and a destination port is neither a direct line nor the shortest path between them. Vessels typically follow "climatological routes" along which higher speeds can be achieved due to the existence of currents or the prevalence of wind, sea or swell, making it possible for vessels to reduce their fuel consumption while meeting delivery deadlines.

Contrary to road networks, sea roads are not paved in concrete. In fact, sea transportation network is a highly dynamic environment that its spatial characteristics (i.e., connection paths) can significantly change over time, due to supply and demand trends that influence trade and carrier patterns, to large infrastructure investments (e.g., canal expansions, construction of new ports and offshore facilities), to climate changes (e.g., ice melting due to global warming), to seasonal effects (i.e., weather affecting the wind speed and direction leading to storms and hurricanes), to traffic regulations enforcements (e.g., emission control areas (ECA)) or due to political events (e.g., imports/exports embargos) and other international incidents (e.g., increase of piracy in specific areas). Accurately modeling this network and the spatiotemporal characteristics of ship traffic patterns is critical for enhancing maritime decision-making and for applications such as traffic optimization, navigation, environmental impact assessment and policy making to name a few. Particularly for navigation, accurate definition of the sea transportation network improves safety and security at sea, making possible the detection of vessels that deviate from normalcy.

In the last years, interest in data analytics has grown significantly. Data analysis falls within the broader scope of knowledge discovery from data (KDD) and data mining that are not new topics. KDD process is a methodology of transforming low level and raw data to more abstract, compact and useful information for the end user [2]. Data mining is a fundamental step at the core of the KDD process consisting of applying data analysis and discovery algorithms that, under acceptable computational efficiency limitations, produce a particular enumeration of patterns over the data [33]. Hence, extracting patterns also means fitting a model to the data, finding structures, or in general any high-level description of a set of data. The fitted models play the role of inferred knowledge [2]. Nowadays, KDD is hindered by the "4V" challenges stemming from big data (i.e., volume, velocity, variety and veracity of data) produced by numerous distributed real-world sensors and systems.

Several academic works have started exploiting the wealth of data made available for tasks, including frequent pattern discovery, trajectory pattern clustering, trajectory classification, forecasting and outlier detection. However, a precursor to all these works is building a data-driven model of normal behavior or commonly traveled routes that can then act, for example, as a baseline for detecting critical deviations, forecasting arrival times and weather routing. In [3], "Patterns of Life" for maritime applications are defined as a graph where nodes are semantically annotated areas of vessel activity and edges are connections between the nodes holding statistical information of the path followed such as speed statistics and traveling time.

The most commonly used tracking system is the AIS, since International Maritime Organization (IMO) made AIS transponders compulsory to carry for all vessels with more than 299 gross tonnage [1,4]. AIS is a self-reporting, collaborative system through which vessels transmit periodically their sailing information (i.e., GPS location, speed over ground, course over ground, heading, navigational status and rate of turn) originating from on-board sensors and devices, their identification data (i.e., their Maritime Mobile Service Identity (MMSI), name, vessel type and dimensions) and their voyage characteristics (i.e., destination, estimated time of arrival and draft) [5]. These data can be received by other vessels equipped with AIS transceivers, as well as by on the ground or satellite-based sensors.

With the vast proliferation of data available for analysis and the uncertainty of the data itself, the effectiveness and efficiency of traditional processing systems and algorithms are rapidly diminished. AIS datasets demonstrate some unique attributes and peculiarities. For instance, the transmission interval of AIS devices is correlated to ships' navigational behavior and not fixed as it would be expected in data produced with positional trackers. More specifically, a vessel is transmitting its data every 2 s when sailing with more than 14 knots, or when changing its course, while the transmission period increases to 3 min when the vessel is traveling with less than 3 knots. Hence, the collected positions follow a nonuniform spatial and temporal distribution. This is further affected when vessels are sailing on the edge of coastal-based receivers' network coverage, causing communication gaps that range from several minutes to many hours. Particularly when sailing in areas that have limited coastal reception (e.g., in open seas), vessels can use satellite AIS communication, in which case, communication gaps delays of several hours are typical. Most of the existing studies focus on traffic pattern analysis near coastal areas or in ports, where communication gaps do not occur, failing thus to tackle the data veracity issue [6].

To address this, we showcase a distributed data-driven approach suitable for discovering maritime traffic patterns and defining "roads of sea." The maritime data modeling methodology introduced, named maritime ROTA, can extract "origin to destination" connections at a global scale and identify their spatial characteristics (e.g., course and speed along the various segments of each path), without dependencies on a priori knowledge or any additional information sources (e.g., nautical maps), addressing thus the challenge of automatically generating an accurate and data-driven representation of maritime traffic. ROTA overcomes big data challenges (i.e., processing huge volumes of uncertain data that arrive at unprecedented speed in the system) and transforms tracking data into a representative model of maritime traffic patterns. In the related literature, "sea routes" are represented as thin lines connecting ports causing significant information losses. On the contrary, ROTA follows standardized route modeling principles [7] to construct contextual routes with spatial variations depending on the ship traffic dynamics, revealing variations in traffic patterns that are unseen using the global and uniform scope (e.g., routes followed by different vessel categories or shipping markets under specific conditions). Thus, ROTA reinforces the maritime traffic understanding, distinguishing "Patterns of Life," a feature that is currently missing from the majority of navigation charts [8], unlocking thus, several potential applications that this methodology can be used for, such as understanding and predicting vessels' trajectories, monitoring areas of interest and protecting the environment, performing routing optimization or detecting dangerous situations and providing risk reports [9].

The structure of this chapter is as follows: we first review the related literature and identify shortcomings in Section 8.2, before presenting in Section 8.3 the proposed knowledge discovery approach and the distributed method for maritime data modeling, capable of automatically extracting ship routes and their spatial characteristics at a global scale. Following this, potential maritime applications that would benefit from ROTA methodology are presented in Section 8.4, before concluding this chapter in Section 8.5.

8.2 Related work

Defining "Patterns of Life" and understanding the mobility patterns of vessel activities are crucial steps for the deployment of complex applications such as trajectory forecasting and anomaly detection. "Patterns of Life" are understood as observable human activities that can be described as patterns in the maritime domain related to a specific activity (e.g., fishing) taking place at a specific time and place [10]. Essentially, vessel-based maritime activity can be described in space and time while classified to a number of known activities at sea (e.g., fishing). The spatial element describes recognized areas where maritime activity takes place, including ports, fishing grounds, offshore energy infrastructure and dredging areas. The transit paths to and from these areas also describe the spatial element (e.g., commercial shipping and ferry routes), while the temporal element often holds additional information for categorizing these activities (e.g., fishing period and time of year). The development of an accurate network for modeling these activities enables the deployment of relevant descriptive and predictive analytics that are critical for improved situational awareness.

Understanding the behavior of a vessel and its surrounding are essential elements of Maritime Situational Awareness (MSA). MSA is defined as the effective understanding of activities, events and threats in the maritime ecosystem that could affect global safety, economic activity or have an environmental impact. Some empirical studies attempt to strengthen MSA by generating heatmaps or density maps (as the one depicted in Figure 8.1) of the data for operator engagement and to support stakeholders' decision-making [11–13]. However, large data volumes overwhelm the end user and dispute the established analysis workflows, thus

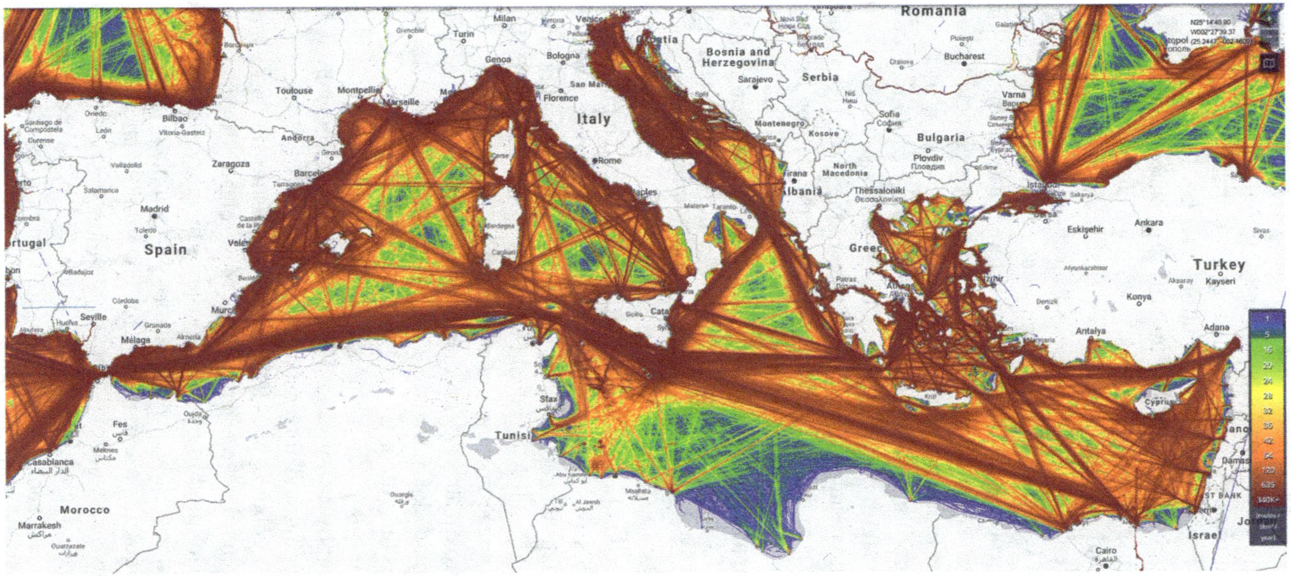

Figure 8.1 Vessels' density map in the Mediterranean Sea

requiring more complex approaches to handle the data volume and veracity [8]. Heatmaps, for instance, although they are considered a useful visualization tool, may result in representations heavily affected by the communication gaps that are inherent in the AIS data, missing thus, important connections (i.e., commercial routes). A number of publications rely on statistics to produce vessel traffic analytics [14,15]. More complex solutions can be divided into (i) methods that use vector representations of traffic and (ii) grid-based solutions.

In the first category, vessels' positions are initially sorted chronologically, compressed and then grouped per vessel so as to form a set of connected waypoints (formally referred to as the vessel's trajectory). High compression rates make the processing of vessel motions in large areas (including global scale) possible [16,17]. Using the trajectory spatiotemporal structure for modeling the kinematics of vessels, a Bayesian vessel prediction algorithm based on a particle filter was applied in [18]. Traffic Route Extraction and Anomaly Detection is a methodology that employs DBSCAN algorithm to detect anomalies automatically and provide trajectory forecasts for short-term planning [19]. DBSCAN is also used in [20] for clustering trajectories and to reduce the processing time of the second step which is an Ant Colony algorithm used to find the optimal path from the starting turning node to the ending turning node. Similarly, a combination of Douglas–Peucker algorithm and kernel density estimation is used in [16] to perform processing on simplified trajectories and thus strike a balance between computational time and performance. The authors of [14] present a method that is based on big data technology stack and that performs traffic analysis in Singapore port using AIS data. Graph-based solutions have also been introduced in [21,22] to automatically learn and represent commercial maritime traffic. Graph's nodes represent clusters of waypoints, while graph edges are the navigational legs that connect the nodes. However, this work is applied in the Iberian Coast and the English Channel, which is a limited geographical area with good terrestrial coverage and minimum uncertainty in the dataset. The approach in [23] aimed to approximate vessel routes between ports and defined route boundaries. The dataset used had significant time span (i.e., 10 years of data) but was limited geographically to the Atlantic Coast of the United States. The analysis produced continuous but not smooth boundaries and in some cases the boundary lines were changing abruptly at transects. Finally, in [24] a methodology that performs trajectory classification and forecasting is presented. The longest common subsequence algorithm is combined with DBSCAN to measure similarity and group similar trajectory segments into clusters. Then, these clusters indicate the vessels' mobility patterns and are used in a long short-term memory network to forecast subsequent vessels' positions. However, the algorithm is applied in a small geographical area, and the prediction is accurate only for few minutes ahead.

In grid-based solutions, the geographic area of interest is split into cells, creating a spatial grid [19,25–30]. Each cell preserves motion properties of the crossing vessels, such as (average, minimum and maximum) speed, course and number of stops within the cell area to name a few. A paradigm of such grid is depicted in Figure 8.2 where the area of interest is split into hexagons using Uber's hexagonal hierarchical spatial index [31]. Learning the distribution of kinematic

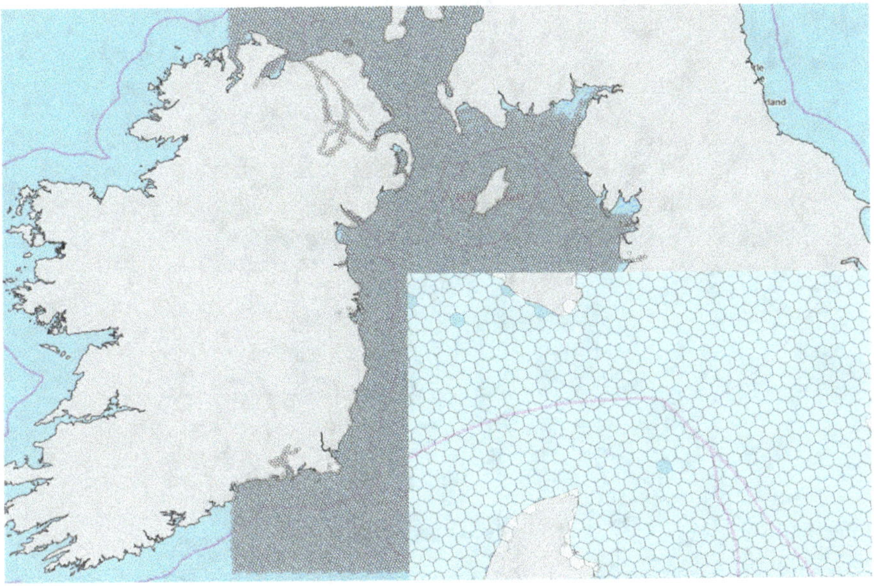

Figure 8.2 Spatial grid using hexagonal spatial indexing

variables (e.g., by applying kernel density estimation method [28]) in each cell is possible and can be used for more complex data-driven analysis, such as estimating vessel arrival times in port areas [25], or performing grid-based anomaly detection using algorithms such as Fuzzy ARTMAP [26], Holst Model [27,28], Support Vector Machine [32], GeoHash encoding [24] and others [29]. The main limitations of grid-based methods is that they lack scalability and are effective only for a limited geographic area [19], requiring also a priori selection of the optimal cell size. In areas with dense maritime traffic, like intersecting sea lanes, the resulting multimodal behavioral description would lead to complex algorithms to perform anomaly detection. A grid-based method for computing shipping density, fast enough to be performed at a global scale, is presented in [29], in which 33 months of data (i.e., from August 2012 to April 2015) were processed and produced global monthly ship density, traffic density and AIS receiving frequency maps in 56 h.

8.3 Approach

ROTA methodology builds upon the traditional data mining process [33] involving algorithms to extract, mine and discover meaningful information patterns and knowledge in maritime data. As also shown in Figure 8.3, ROTA methodology consists of the following steps:

- Data selection
- Data transformation

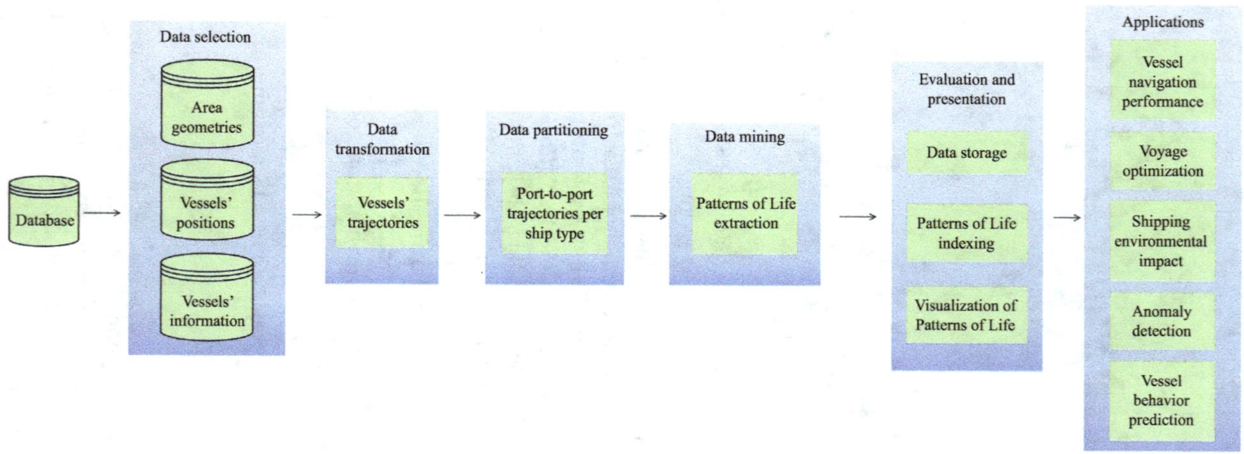

Figure 8.3 ROTA methodology

- Data partitioning
- Data mining
- Evaluation and presentation

During the data selection step, the sources that will help the data analytics process for the domain-specific problem that ROTA methodology is trying to solve are chosen. More specifically, the selected data includes the following:

1. Vessels' positions: This dataset consists of position reports collected through the AIS. Such data should have sufficient spatial and temporal coverage (i.e., global coverage for at least 2 years) so as to produce a generalized data analytics model that will not be affected by regional or seasonal effects [9].
2. Area geometries: These data (similar to the node definition provided in [3]) include definitions of geographic areas of interest. Depending on the application, areas of interest could include port and anchorage geometries, offshore structures, fishing areas, exclusive economic zones, ECA zones and piracy-affected areas.
3. Vessels' information: This dataset includes "static" information reported from vessels through AIS and consists of vessels' identification information and characteristics (i.e., MMSI, vessel's name, type and dimensions) and voyage information (i.e., reported destination and estimated time of arrival). Such information is fundamental for understanding the navigational patterns of vessels (e.g., passenger vessels repetitively performing the same itinerary or liner vessels transiting along regular routes on fixed schedules), or for reasoning about alternative routes selected from vessels performing the same itinerary.

Ship journeys can be split into segments linking two geographical areas. A ship journey begins and ends at a seaport, but in between it could include multiple stops. For instance, a fishing vessel could sail from a port to a fishing area and transit from there to one or more other fishing areas before returning to a port (which could also be the same with the departure port). Vessels' positions included in the selected data are not linked with a specific journey. During the data transformation step, departure and destination information are linked to the positional data. This can be done, using the destination port reported through the AIS; however, this information is manually inserted by the vessels' crew members without following a specific standard and thus it is prone to errors. Furthermore, intermediate stops are often neglected (e.g., passenger vessels visiting a sequence of ports during a day report only the final port as destination) leading thus to misconception of route information. To alleviate this issue, ROTA performs spatial analysis during this step, assessing intersections between stored area geometries and positional data. All the positions that intersect with an area geometry are characterized by that area's unique identifier (i.e., area id). Then, sorting chronologically the positions of each vessel leads to visiting area events (i.e., entering or exiting an area), and grouping the data between consecutive port call events (i.e., entering or exiting a port geometry) enables the creation of a sequence of visited areas during a vessel's journey. Furthermore, journeys are curated before proceeding to the data partitioning phase to remove unintended crossings of vessels with area geometries. This is performed through creating in-port stay sessions as defined in [9]. Finally, in data transformation step positional data are cleaned from incomplete entries, spoofing events and empty, erroneous and inconsistent (e.g., ship characteristics such as ship dimensions, changing during a voyage) or conflicting fields (e.g., ship name changing during voyage).

After the data transformation phase, all the positional data are linked with a specific journey (journey id) defined by two consecutive port events (i.e., a port departure followed by a port arrival) and grouped by their departure and their destination port as well as their vessel type. Such grouping will allow performing route analysis per port-to-port connection considering the special characteristics of different ship types. For instance, cargo vessels and tankers are limited by their size and cannot sail in areas with shallow waters, following thus different routes than lighter and smaller vessels. Furthermore, separating the data of each port-to-port connection enables distributed storage and parallel processing during the data mining step, as extracting the "Patterns of Life" can be performed separately for each connection without the need of any other data. In ROTA, this is modeled as a MapReduce scheme executed on a spark cluster, but other distributed processing techniques can also be exploited using the same methodology in which positions are transformed into key-value pairs (with key being an identifier for the route and the value being the position and the vessel details) during the map phase and then in the reduce phase, they are grouped in lists, each having all the positions of a route. As mentioned afore, each journey may include one or multiple intermediate stops, and performing similar data partitioning process for journey segments instead of journeys as a whole is trivial and would produce groups of journey segments with the next steps in the proposed methodology being exactly the same.

With the data segmented into a multiple partitions where keys indicate unique departure–destination connection per vessel type and values include all the AIS positions of those journeys, the next step is creating a model for each connection that describes the motion patterns of vessels performing the same journey. This can be achieved through clustering algorithms that reduce the volume of the initial dataset and generate a normalcy model from the vessels' trajectories. Following the clustering algorithms' analysis given in [34], ROTA does not rely on density-based clustering but performs a dynamic variant of the K-means partitioning clustering algorithm, at which the number of clusters' parameter is calculated per route and is dependent on the number of points of each route. To increase computational efficiency and stability of the data mining technique and normalize the geometries of the produced clusters, the k farthest points of each route are selected to be the initial centroids instead of random selection [35] and the rest of the positions are clustered by calculating the haversine distance of their coordinates from the route's centroids. Further details on the clustering algorithm are beyond the scope of this chapter and can be found at [9].

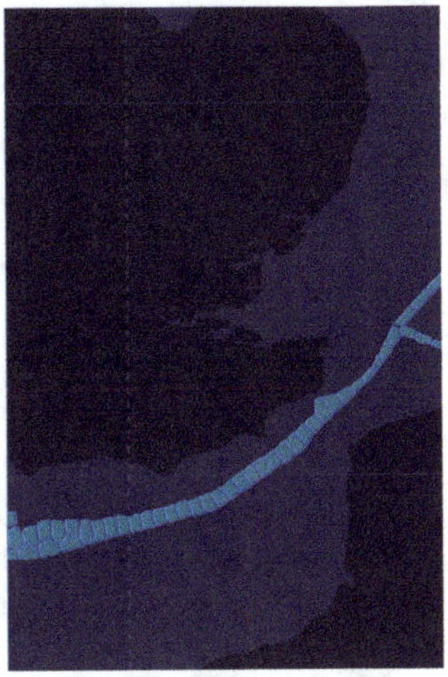

Figure 8.4 Spatial clusters in the form of convex indicating Pattern of Life for a port-to-port connection between South West England and the Netherlands

The result of the data mining step is a collection of clusters calculated from the positions of vessels with the same ship type that performed the same journey (i.e., having the same departure and destination ports). This dataset is passed in the evaluation and presentation step of the ROTA methodology. In this phase, each cluster is represented by a polygon formed by the convex hull that includes all the positions belonging to the cluster. A convex hull is the smallest polygon (w.r.t. covered area) formed by the set of points belonging to a cluster. They are the most accurate geometric shapes that can be used to represent the area that each cluster of points occupies. Figure 8.4 illustrates an example collection of convex hulls for a port-to-port connection between South West England and the Netherlands.

Each convex hull is also semantically annotated with statistical information extracted from the positions that formed the cluster, such as speed statistics (i.e., average, median, minimum and maximum speed when traversing the convex hull), direction statistics (i.e., heading and course over ground statistics), number of turns and stops in the convex hull and entry/exit points. The collection of those high-level and semantically annotated geometries constitutes the global network of "Patterns of Life." All this information is stored in a database system and can be retrieved for analysis depending on the application that it is used for. For instance, in real-time applications data can be cached in memory, or only a subset of the polygons could be retrieved from the database depending on the specific journey segment. R-tree is exploited to index the polygons on the basis of their spatial distribution and reduce rapidly the query response time. Finally, the K-means variant assigned each position to a cluster, without prion outlier elimination. Therefore, some convex hulls included noise and were subject to further analysis. ROTA applied a density indicator (presented in [9]) and maintained convex hulls with high densities that indicate areas of high concentration of vessel traffic. Using this density metric and a threshold, provided by the user, ROTA methodology selected the convex hulls with low density that should be removed. The "Patterns of Life" that ROTA produces can be used in multiple maritime applications as will be further described in the next section.

8.4 Applications

8.4.1 Vessel's navigation performance and voyage optimization

AIS data can be used from multiple maritime stakeholders such as ship operators and freight forwarders to monitor and analyze fleet performance and manage vessels' voyages. AIS provides information with a much higher data density than other vessel reporting systems (e.g., vessels' noon reports), which can be exploited to derive various key indicators. For instance, speed variability during deep sea passage, average sea passage speed or nautical miles per day can be produced on the basis of a vessel's speed. However, such indicators lack context in the sense that they are not compared against the performance of other vessels and could be misinterpreted by analysts.

ROTA provides statistical analysis of historical voyages, which can be used to compare a vessel's navigation performance against historical tracks of the same or other vessels sailing under similar conditions. So far ROTA has exploited vessel information, area geometries and vessel positions to extract Patterns of Life. The data selection step can be extended with weather data, sea bathymetric data, vessel engine information and vessels' cargo capacity (i.e., gross tonnage, draft and deadweight) to perform path planning, fine-tune route analysis and improve the decision-making for maritime professionals.

Path planning is a time-consuming process that is manually conducted by experienced navigators, often leading to a suboptimal route selection. Ship operators can use such data in conjunction with voyage statistics from ROTA to find the safe route that has a minimum cost with respect to voyage time, distance traveled, vessel maneuvers or fuel consumption taking into consideration all the potential influencing factors such as weather conditions, traffic density or vessels' load. Freight forwarders can use ROTA statistics to improve the accuracy of the estimated time of arrival (ETA) of vessels. In conjunction with port congestion analytics (i.e., anchorage waiting time and time in port), ETA can be exploited to plan operations for multimodal cargo transportation, warehouse storage optimization and accurate package delivery to customers.

8.4.2 Shipping environmental impact

Long-term observations and earth monitoring over the past two centuries indicate that human activities have aggravated climate change. Industrial revolution and economic prosperity led to tremendous growth in transports and trades globally leading to an intensified usage of sea pathways for transferring commodities and supplies around the world [36]. Consequently, the carbon footprint and the engine noise of vessels sailing across the oceans have increased to such levels that pollution in marine ecosystem has become evident, leading to destruction and even extinction of species. In an effort to make marine and coastal resources sustainable, European Union (EU) and IMO have set new regimes to reduce the carbon footprint of vessels. More specifically, EU issued in 2017 monitoring, reporting and verification regulation that requires ships to monitor and report their CO_2 emissions, fuel consumption, transport work and average energy efficiency starting from 2018. Similarly, IMO in its third greenhouse gas study [37] published in 2014 presented the organization's estimates of shipping emissions during the 2007–12 period and provided maritime emissions projections till 2050 taking into account the expected market growth, the market-driven efficiency changes and the regulation changes. IMO followed a bottom-up approach for estimating the emissions by combining fleet activity data derived from AIS observations with the global fleet technical data. The fuel consumption estimates of vessels were grouped by vessel category to provide statistics for different sectors of the market, describing activity, energy use and emissions for each of the years from 2007 to 2012. The bottom-up model relates hourly measures of the vessel's speed and draft to fuel consumption using equations similar to those deployed in the second IMO GHG Study 2009 and the wider naval architecture and marine engineering literature. In all the business as usual scenarios presented in the report, shipping emissions are expected to increase from 50% to 250% till 2050 depending on the economic growth, and even the most significant improvements in efficiency modeled do not result in a downward trend. This study became the principal reference for the organization's Marine Environment Protection Committee to consider further measures appropriate for the regulation of international shipping emissions at the global level. IMO strategy on vessels' emissions reduction was further reinforced in 2018 when IMO dictated the use of a data collection system on fuel consumption, with data collection starting from 2019, for all ships of 5,000 gross tonnage and above. The aggregated data are reported to the flag state after the end of each calendar year and the flag state, having determined that the data have been reported in accordance with the requirements, issues a statement of compliance to the ship.* The vessel reports should include the following information:

- period of calendar year for which the data are submitted,
- distance traveled,
- amount of each type of fuel consumed in total,
- hours underway under own propulsion, and
- DWT to be used as cargo proxy.

ROTA methodology can assist ship navigation to reduce emissions. For each port-to-port connection, ROTA produces multiple routes, including statistics on

*http://www.imo.org/en/OurWork/Environment/PollutionPrevention/AirPollution/Pages/Data-Collection-System.aspx

speed, distance traveled and underway time. This information can be used during voyage planning to select a path that will be compliant with the EU and IMO regulations on fuel consumption without violating any delivery deadlines.

Apart from the carbon footprint, the vast increase of vessel traffic in rivers and seas results in acoustic emissions of extremely high power that propagate over several square kilometers. Aquatic animals are sensitive and vulnerable to vessels' underwater radiated noise, and the increase of shipping activities near coast (e.g., commercial vessels anchored near ports, high-speed yachts and fishing vessels) can affect the health of marine and river fauna. In this context, IMO's Committee on marine environment protection has set up a process for designing Particularly Sensitive Sea Areas (PSSAs) and adopting associated protective measures. Several international and regional instruments promote the protection of marine areas but enforcing PSSA can affect navigation in areas designated as such. In such cases, the results of ROTA can be combined with animal migration patterns to quantify and thoroughly investigate the behavioral effects of sound on river and marine animals. Understanding the effect of ship noise has on animals' behavior and taking also into account regional vessel traffic patterns can lead to enforcing protective measures in PSSAs that in one hand will reduce the vessels' noise to acceptable levels and on the other hand minimize the impact to shipping activities.

8.4.3 Anomaly detection

In the shipping industry, anomaly detection is perceived as a set of methods, tools and services that supports MSA by indicating vessels and situations that, in some sense, deviate from the expected, known or "normal" behavior. Increasing automation, through advanced machine learning and AI, beyond simple rule-based systems, can enable the system to spot complex situations by correlating various events from all surveillance sensors and classify them into important incidents.

ROTA is an essential pillar for increasing MSA as the outcome of the proposed methodology is a model of normalcy that can assist in understanding complex maritime environments. Patterns of Life and their related statistical analysis of voyage behavior can be used to detect route deviations in vessels' voyages in real time and provide early warnings and avoid potential accidents. A route deviation happens when a vessel is sailing out of the Pattern of Life it should follow, based on its vessel type, the departure and destination port. Route deviation may also occur when a vessel is sailing inside the limits of its Pattern of Life, but its behavior is unusual compared to the path's statistical analysis. For instance, its speed, course or traveling time may deviate from the "normal" behavior and this could be an early indicator of a potential anomaly in vessel's voyage. Using bathymetric data and vessels' cargo information in combination with the Patterns of Life can help in vessels' navigation and trigger warnings when a vessel is sailing in shallow waters and avoid running aground. In many cases, sailing in shallow waters would also imply that the vessel is traveling out of its corresponding Pattern of Life and thus, route deviation events would also be triggered before the vessel ran aground.

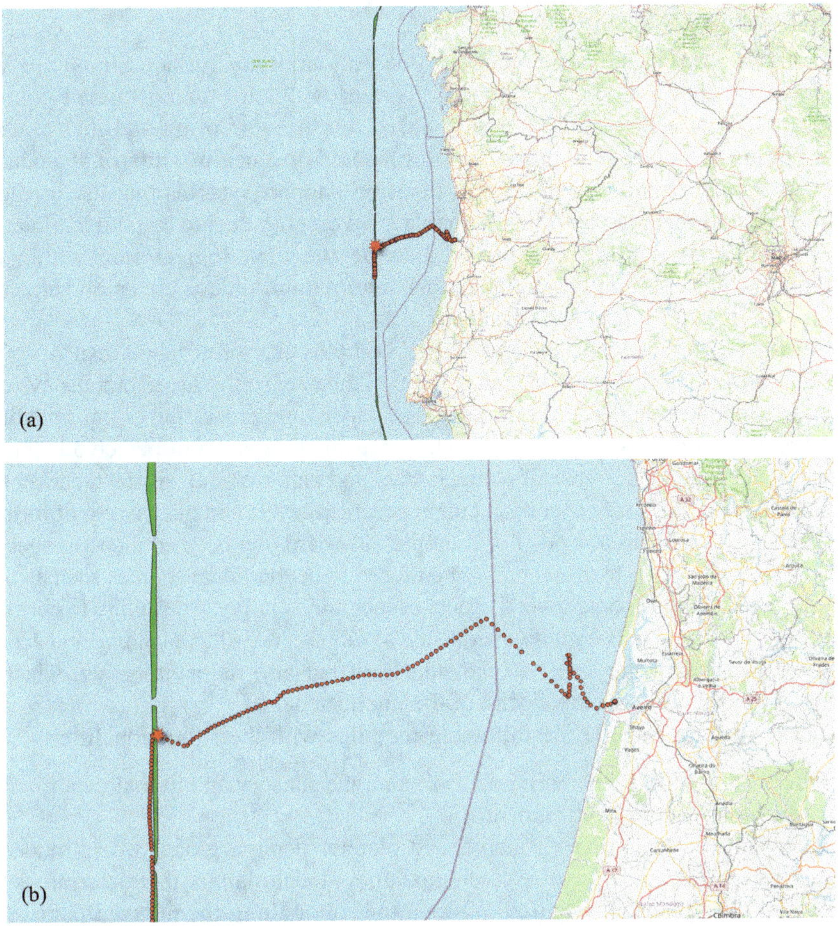

Figure 8.5 YM Pluto route deviation while sailing from Ceuta to Rotterdam shown at two different zoom levels (a) and (b)

For instance, Figure 8.5 illustrates the route deviation of the YM Pluto vessel that was traveling from Ceuta, Spain to Rotterdam. During its voyage, the vessel faced adverse weather conditions, and its master was severely injured while trying to stop a water leakage. The vessel changed its course and headed toward the port of Aveiro. The Pattern of Life that connects Ceuta to Rotterdam is shown in green color and the YM Pluto voyage is shown in red color.

Our previous work in [9] demonstrates further the relevance of ROTA to this application scenario and validates the hypothesis that defining normal vessel behavior can assist in detecting deviations by testing the ROTA methodology against data from real-world maritime incidents. The results show that ROTA successfully detected 93.4% of those incidents.

8.4.4 Autonomous ships

Autonomous ships (also known as Maritime Autonomous Surface Ships) are seaborne vessels that can sail in navigable waters with little or no human control. There are different levels of autonomy based on the monitoring and the (remote) control of the vessel from a human operator located on a nearby ship (or at a control center at shore), or through intelligent decision support systems that use artificial intelligence and machine learning to enable the vessel to decide its course of action [38]. Autonomous ships are considered to be the next logical step within the shipping industry, aiming to maximize task automation, reduce crews on ships and improve navigation safety and efficiency.

Autonomous ships can be perceived as complex maritime robots that have the capability of sensing and perceiving data through their sensors. These complex systems rely on machine learning and AI technologies, analysis and processing of information at all stages in a loop of sense, plan and act. For a robotic system to perceive its environment, it needs to make sense of complex data and patterns at an abstract, conceptual level in real time. Data are generated at unprecedented levels and in a variety of formats from numerous on-board sensors. For example, on-board unprocessed video imagery is produced at a rate of up to 500 mb/m, when fused with additional sources such as lidar volumes can quickly become several orders of magnitude larger. The ability to perform, as quickly as possible, this definitive sequence of robotics (sensing, planning and acting) is not only critical to the safety of the ship itself but also its environment, which is hampered by the volume and velocity of the implicated data.

These systems include the following modules with their functionalities:

- Perception module: Its main goal is to analyze sensory and global data to build a local map of the environment.
- Path planner: It involves a map-based planner using a global world model to generate safe paths toward a given goal and a local planner that determines the details of the motion required to move the robot along the planned paths.
- Execution monitoring module: It plans a collision-free course by simulating the displacement of the robot across along the planned course.

Solving the motion planning problem for self-guided vessels that are moving in a dynamic environment is considered as a highly complex and challenging task, highly dependent on building first an accurate world view. To plan an effective collision-free path for an autonomous vessel through numerous moving and static obstacles, contextual information and available safe paths, such as those defined by ROTA, is critical.

Autonomous ships will strain the waterways utilization, making thus imperative the need to ensure safe and efficient sailing. Shore-based control centers will adopt automatic monitoring of vessel traffic which currently is used by vessel traffic services. Such services are strongly dependent on linear prediction techniques. Such approaches can assess the environment and predict the near future and possible impending critical situations (e.g., collision risk or grounding). However, due to their narrow time window applicability, they will fail to predict long-term goals such as avoiding situations that affect the navigation efficiency of a voyage.

ROTA methodology can be extended with bathymetric data, weather data and weather forecasts and offer options for safe and efficient routing of the ship. Route statistics retained across the Patterns of Life and their spatial characteristics (e.g., sea depth and number of turns) will be taken into account during the decision-making process, either that is performed by a human operator at a shore control center, or by the vessel in cases of fully autonomous vessels.

8.4.5 Vessel activities identification

ROTA methodology exploits vessel trajectories to produce Patterns of Life for port-to-port connections and various vessel types. However, some shipping operations do not conform to the typical "departure"–"destination" connection. More specifically, supply vessels, fishing vessels, tugboats and pilot vessels generally perform port operations inside or near port. Furthermore, in some shipping operations such as transshipments or bunkering multiple vessels are involved and such operations usually last for many hours. ROTA methodology should be further extended to support the identification of vessel activities by employing mobility pattern classification techniques. This way it is possible to process vessels' mobility patterns, compare them against a set of known mobility patterns (each one mapped to a specific activity type) and estimate whether a vessel is involved in a specific operation. Accurately detecting piloting, tugging and bunkering operations will help stakeholders of port ecosystem such as port authority, terminal operators and ship suppliers as well as ship owners to analyze statistically those operations and thus improve the planning of operations. In the occasion of a fishing activity, such a service can be used for the identification of illegal, unreported and unregulated fishing activities in prohibited areas or nature protection areas, or even in distinguishing fishing activities with specific gear such as trawling that have been linked to the decrease of marine fauna.

8.5 Conclusions

In this chapter, a data-driven methodology suitable for discovering maritime traffic patterns and revealing "roads of sea," ROTA, has been presented. The proposed solution analyzes vessels' navigational patterns in a computationally efficient and accurate way and extracts maritime "Patterns of Life" globally through machine learning techniques. Furthermore, ROTA can be employed to a wide spectrum of maritime applications that have been discussed in this chapter.

References

[1] International Maritime Organization profile, https://business.un.org/en/entities/13 [Accessed: 15-May-2020].

[2] U. M. Fayyad, G. Piatetsky-Shapiro, P. Smyth, and R. Uthurusamy, Eds., Advances in Knowledge Discovery and Data Mining. Menlo Park, CA, USA: American Association for Artificial Intelligence, 1996.

[3] A. Itani, C. Ray, A. El Falou, and J. Issa, "Mining ship motions and patterns of life for the EU common information sharing environment (CISE)," in OCEANS 2019, Marseille, France, 2019, pp. 1–6, doi: 10.1109/OCEANSE.2019.8867219.

[4] SOLAS chapter v – regulation 19 – Carriage requirements for shipborne navigational systems and equipment, http://solasv.mcga.gov.uk/regulations/regulation19.htm [Accessed: 15-May-2020].

[5] M.1371: Technical characteristics for an automatic identification system using time-division multiple access in the VHF maritime mobile band, https://www.itu.int/rec/R-REC-M.1371/en [Accessed: 15-May-2020].

[6] D. Nguyen, R. Vadaine, G. Hajduch, R. Garello, and R. Fablet, "A multi-task deep learning architecture for maritime surveillance using AIS data streams," 2018.

[7] DTI, Methodology for assessing the marine navigational safety & emergency response risks of offshore renewable energy installations. Department of Trade & Industry (DTI), 2013.

[8] L. Cazzanti, A. Davoli, and L. M. Millefiori, "Automated port traffic statistics: From raw data to visualisation," in Big Data. IEEE, 2016, pp. 1569–1573.

[9] D. Zissis, K. Chatzikokolakis, G. Spiliopoulos, and M. Vodas, "A distributed spatial method for modeling maritime routes," IEEE Access, vol. 8, pp. 47556–47568, 2020, doi: 10.1109/ACCESS.2020.2979612.

[10] M. R. MacLeod and W. M. Wardrop, "Operational analysis at combined maritime forces," in Proc. 32nd Int. Symp. Mil. Oper. Res., 2015, pp. 1–14.

[11] L. Goldsworthy and B. Goldsworthy, "Modelling of ship engine exhaust emissions in ports and extensive coastal waters based on terrestrial AIS data – An Australian case study," Environmental Modelling and Software, vol. 63, pp. 45–60, 2015.

[12] M.-C. Tsou, "Discovering knowledge from AIS database for application in VTS," Journal of Navigation, vol. 63, no. 3, pp. 449–469, 2010.

[13] N. Willems, H. van de Wetering, and J. J. van Wijk, "Visualization of vessel movements," Computer Graphics Forum, vol. 28, no. 3, pp. 959–966, 2009.

[14] L. Zhang, Q. Meng, and T. F. Fwa, "Big AIS data based spatial-temporal analyses of ship traffic in Singapore port waters," Transportation Research Part E: Logistics and Transportation Review, 2017.

[15] Q. Meng, J. Weng, and S. Li, "Analysis with automatic identification system data of vessel traffic characteristics in the Singapore strait," Transportation Research Record: Journal of the Transportation Research Board, vol. 2426, pp. 33–43, 2014.

[16] Y. Li, R. W. Liu, J. Liu, Y. Huang, B. Hu, and K. Wang, "Trajectory compression-guided visualization of spatio-temporal AIS vessel density," in WCSP. IEEE, 2016, pp. 1–5.

[17] M. Fiorini, A. Capata, and D. D. Bloisi, "AIS data visualization for maritime spatial planning (MSP)," International Journal of e-Navigation and Maritime Economy, vol. 5, pp. 45–60, 2016.

[18] F. Mazzarella, V. F. Arguedas, and M. Vespe, "Knowledge-based vessel position prediction using historical AIS data," in Sensor Data Fusion: Trends, Solutions, Applications (SDF), 2015. IEEE, 2015, pp. 1–6.

[19] G. Pallotta, M. Vespe, and K. Bryan, "Vessel pattern knowledge discovery from AIS data: A framework for anomaly detection and route prediction," Entropy, vol. 15, no. 6, pp. 2218–2245, 2013.

[20] S.-k. Zhang, G.-y. Shi, Z.-j. Liu, Z.-w. Zhao, and Z.-l. Wu, "Data-driven based automatic maritime routing from massive AIS trajectories in the face of disparity," Ocean Engineering, vol. 155, pp. 240–250, 2018.

[21] P. Coscia, P. Braca, L. M. Millefiori, F. A. N. Palmieri, and P. Willett, "Multiple Ornstein–Uhlenbeck processes for maritime traffic graph representation," IEEE Transactions on Aerospace and Electronic Systems, vol. 54, no. 5, pp. 2158–2170, 2018.

[22] L. M. Millefiori, P. Braca, K. Bryan, and P. Willett, "Modeling vessel kinematics using a stochastic mean-reverting process for long-term prediction," IEEE Transactions on Aerospace and Electronic Systems, vol. 52, no. 5, pp. 2313–2330, 2016.

[23] S. A. Breithaupt, A. Copping, J. Tagestad, and J. Whiting, "Maritime route delineation using AIS data from the Atlantic coast of the US," The Journal of Navigation, vol. 70, no. 2, pp. 379–394, 2017.

[24] W. Li, C. Zhang, J. Ma, and C. Jia, "Long-term vessel motion predication by modeling trajectory patterns with AIS data," in 2019 5th International Conference on Transportation Information and Safety (ICTIS), July 2019, pp. 1389–1394.

[25] A. Alessandrini, F. Mazzarella, and M. Vespe, "Estimated time of arrival using historical vessel tracking data," IEEE Transactions on Intelligent Transportation Systems, 2018.

[26] N. A. Bomberger, B. J. Rhodes, M. Seibert, and A. M. Waxman, "Associative learning of vessel motion patterns for maritime situation awareness," in FUSION. IEEE, 2006, pp. 1–8.

[27] R. Laxhammar, "Anomaly detection for sea surveillance," in FUSION. IEEE, 2008, pp. 1–8.

[28] B. Ristic, B. F. L. Scala, M. R. Morelande, and N. J. Gordon, "Statistical analysis of motion patterns in AIS data: Anomaly detection and motion prediction," in FUSION. IEEE, 2008, pp. 1–7.

[29] L. Wu, Y. Xu, Q. Wang, F. Wang, and Z. Xu, "Mapping global shipping density from AIS data," The Journal of Navigation, vol. 70, no. 1, pp. 67–81, 2017.

[30] G. Wang, J. Meng, and Y. Han, "Extraction of maritime road networks from large-scale AIS data," IEEE Access, vol. 7, pp. 123035–123048, 2019.

[31] I. Brodsky, "H3: Uber's hexagonal hierarchical spatial index," 2018.

[32] D. O. D. Handayani, W. Sediono, and A. Shah, "Anomaly detection in vessel tracking using support vector machines (SVMs)," in 2013 International Conference on Advanced Computer Science Applications and Technologies, Dec. 2013, pp. 213–217.

[33] M. H. Dunham, Data Mining: Introductory and Advanced Topics. Prentice Hall, 2003, pp. 9–10.

[34] G. Spiliopoulos, K. Chatzikokolakis, D. Zissis, *et al.* "Knowledge extraction from maritime spatiotemporal data: An evaluation of clustering algorithms on Big Data," in 2017 IEEE International Conference on Big Data (Big Data). IEEE, 2017.

[35] T. F. Gonzalez, "Clustering to minimize the maximum inter cluster distance," Theoretical Computer Science, vol. 38, pp. 293–306, 1985.

[36] BDVA, Whitepaper big data in earth observation, 2017. http://www.bdva.eu/sites/default/files/TF7%20SG5%20Working%20Group%20-%20White%20Paper%20EO_final_Nov%202017.pdf [Accessed: 28-May-2020].

[37] T. W. P. Smith, J. P. Jalkanen, B. A. Anderson, *et al.* Third IMO GHG Study 2014: Executive Summary and Final Report, 2014.

[38] B. Batalden, P. Leikanger, and P. Wide, "Towards autonomous maritime operations," in 2017 IEEE International Conference on Computational Intelligence and Virtual Environments for Measurement Systems and Applications (CIVEMSA), Annecy, 2017, pp. 1–6, doi: 10.1109/CIVEMSA.2017.7995339.

Chapter 9
Real-time information with ITS technology

Ioannis Filippopoulos[1]

9.1 Introduction

Thousands of vessels are navigating the ocean and making port calls at any one time. With information technology becoming a central pillar in all industries, advance automation processes are also entering the maritime industry, which is slowly integrating technological systems in the business environment. Included in this process are intelligent transport systems (ITSs), which are becoming part of the vessel's ecosystem and so-called smart ports, creating a data streaming circle between the vessel, port and managing company. The literature on the topic has been very slow to develop, partly as a result of the gap between academia and the industry.

In light of the lack of a comprehensive literature on the integration of ITS technologies in the maritime industry, the article describes how ITS technologies can be used as decision support systems in the hands of seafarers as well as operators to maximize efficiency, which in turn, supports the financials of the companies.

In order to do this, the chapter is structured as follows: The first part of the chapter introduces the concept of ITS by defining what it is and the role it plays in the transport sector with an example drawn from the automotive industry where the concept was first applied. The section then continues to narrow down the focus by discussing the application of ITS technologies in the maritime sector more specifically. The subsequent section of the chapter considers the smart vessel concept in more detail by highlighting the elements associated with the vessel and those associated with operations onshore. The chapter then moves on to consider the 'smart port' concept and the intelligent maritime systems (IMSs), namely systems applied in the context of the land–marine interface. Finally, the chapter analyses the impact of these systems on business outcomes, before concluding.

9.2 ITS technologies and their application in transport

Automation has been part of transportation systems for a long time. However, there has been a recent surge in the use of words 'smart' or 'intelligent' to describe

[1]Cyprus International Institute of Management (CIIM), Nicosia, Cyprus

technological solutions in the industry. This has emerged alongside the increasing attention on the Internet of Things (IoT) and the ability to transfer information over the Internet between two physical objects. Given the large role played by transport systems both from a personal and business point of view, it was almost inevitable that IoT would rapidly expand in the field of transportation. The automotive industry is one of the first sectors using IoT. Back in 1970, the first electronic control units were introduced in commercial vehicles, and the automotive companies started to program and control the fuelling and electrical equipment of a car. Today there are vehicles with around 100–120 sensors, and this number is expected to increase more as the technology evolves. The pending question for this sector as well as the maritime sector, nevertheless, is, what does it mean to call vehicles smart and intelligent, and what specific problems this technology means to address?

Answering this question first involves defining ITS in the contest of the maritime sector. ITS is a system that aims at collecting data from transportation mean, transferring them to a source, and then analysing this data to use the results for various operations. The possibilities are endless and vary according to the industry. For example, we can use ITS to control traffic management in a city, to get information on timetable and current location for public transport, to prepare port actions for a vessel arrival for maritime, to schedule a parking spot in an airport for an airliner or get information about seat availability on a railway. Recently Ducati, Audi and Ford along with Qualcomm presented car-to-bike communication technology using ITS and performed safety scenarios for vehicle-to-vehicle, vehicle-to-pedestrian and vehicle-to-infrastructure showing the potential of using such systems for safety.

Emerging ITS technologies can also be implemented in other areas that will benefit a greater number of users. For example, an eco-friendly approach in smart cities where traffic management and congestion control can have a huge impact on the reduction in carbon footprint. Later is an example of how an ITS can be applied in a city. A passenger requests traffic information that occurs from ITSs in vehicles. At the same time, live information is forwarded to emergency services and public transportation services that are used for preparing and operating optimum facilities for civilians. Citizens may also request public transportation information and emergency services from mobile devices. From this example it is evident that ITSs have the potential to support the development an ecosystem where all members can interact and be informed from live information (Figure 9.1).

The way in which ITS technologies have been applied in the maritime industry differs in part due to the scope of the services involved. Since ancient times, transportation by vessel was the most popular way of carrying people and goods from and to different places. For many years, ships were the only means of mass transportation. Communicating with the shore was always a challenge. The first merchant ships would communicate with each other with special flag signals or other visible signals. Communication with the shore was only possible in the same way, assuming that the weather conditions permitted. At the same time on-board

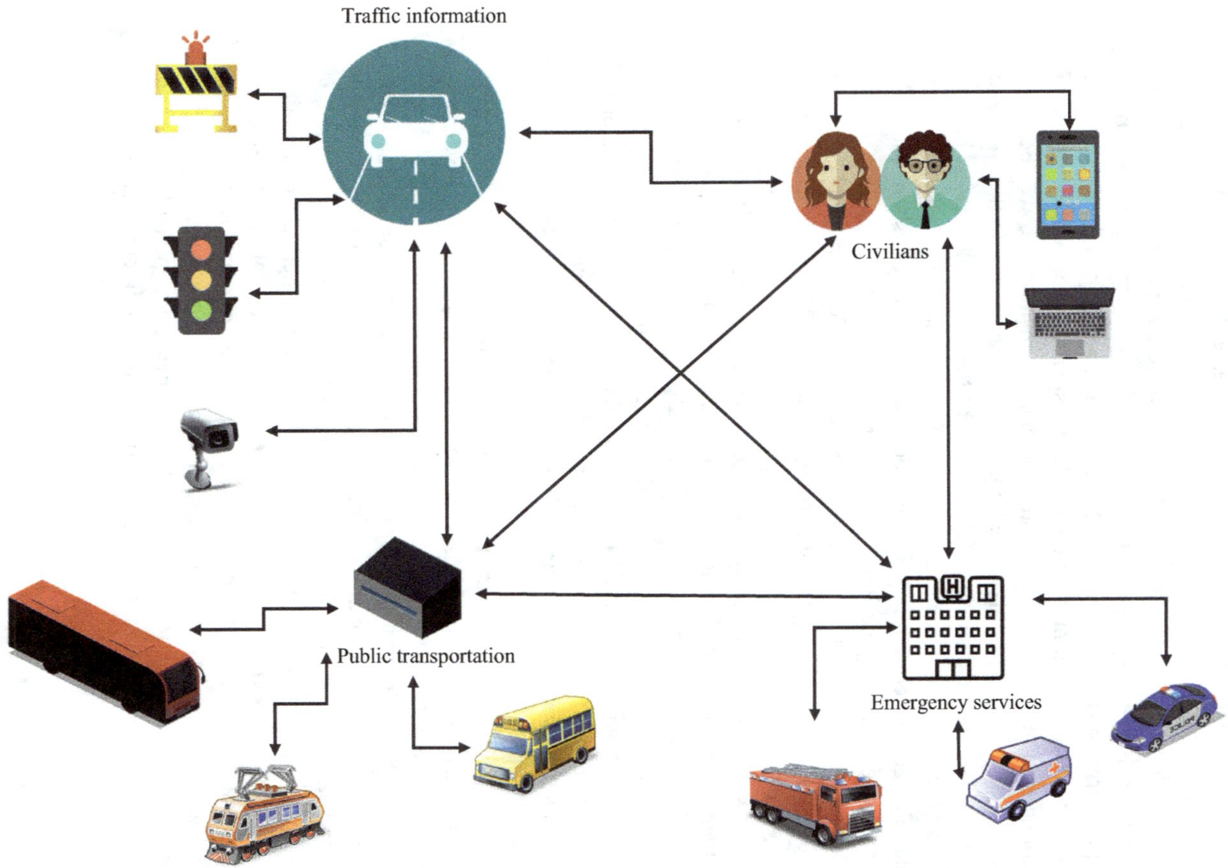

Figure 9.1 Intelligent transport system. Source: Author

the vessel seafarers were using the ship's bell for indicating the time and performing other tasks on-board.

It was many years later when a noon report was invented, and it was named 'noon' because noon time was the only time that the vessel knew accurately due to different time zones, therefore, could use to crosscheck with previous days and calculate its position in the ocean along with other information. A noon report is a form that the captain sends everyday to the shore and contains all information concerning the navigation, cargo, position of the vessel, weather and other operational data. The first noon reports were communicated to the shore using radio communication and telex and contained only the most essential data such as position and weather conditions. Nowadays, noon reports may take hours to be completed from the crew and are composed of various data across different areas of the vessel such as the bridge, engine rooms, cargo compartments. At the same time, satellite communications evolution has changed the communication between the vessel and the shore and real-time date and sensors can provide information on request.

Does this mean that vessels can be considered as ITSs? Not completely. Due to the nature of merchant shipping which usually finds 20–25 people on-board a vessel for several months, seafarers live in an ecosystem that is ruled from traditions and sea laws and most of their work is based on sea experience. This situation influences the use of automated procedures and data exchange because in most cases their decision may have large financial and safety impacts, so they prefer to trust their own judgement [8]. It is also worth noting that maritime industry's traditional practices are widely accepted from all involved parties of maritime business such as owners, traders, ports, oil majors, international organizations, charterers and agents, which makes the evolution of smart systems slower but that does not mean it is not there.

Just like any other industry, maritime is also part of the digital revolution era. Most vessels that have been built the past 5 years are delivered from the yards to their owners with ready-to-play sensors in every single part of the vessel. All parties involved want to ensure that their profit margin remains high and at the same time organizations and governments are paying high attention to environmental regulations and safety; therefore, they are all having 'sneak peeks' to technology and pushing things towards intelligence systems. As of today, there are companies that can see real-time vessel information and use performance data for immediate actions and prediction from their smart vessels and there are smart ports that are able to identify port traffic and priority as well as perform automated loading and discharging of cargo. Further to that, technology is becoming part of seafarer's training so people on-board can feel and interact with intelligent systems with confidence.

9.3 Smart vessel concept

When it comes to the maritime industry, connectivity and data exchange, there are two main pillars to consider. The first is the vessel and the second is the port or shore-side operation. To consider the vessel as an intelligent system, we focus on the 'smart' vessel concept.

ITS is part of the IoT platform but in the shipping industry, IoT can be defined as the Internet of Vessels (IoV) [1]. IoV is a system in which a vessel can produce real-time data from various sensors and services which can eventually be used both on-board and from the shore from all different shareholders of the vessel for various cases. Earlier, automatically generated information and data was derived from engines as live readings from the machinery with reference to basic information like engine output, temperature, thrust and power and concerned only the crew on-board the vessel that was using them for vessel operational activities. Lately, with the increasing capability to receive information from sources that are not necessarily on the vessel, more information can be received and combined to perform actions of high complexity and difficulty (Figure 9.2).

The earlier one is a graphical representation of IoV. Information comes directly from the vessel such as remaining on-board (ROB) item for gas, oil, food and others, as well as information from various sensors from different vessel compartments like shaft torque, propulsion, rudder angle, slip and trim. Additionally, there are services that can provide general information like latitude, longitude, emissions standards (NO_x, CO_2, SO_2), air and sea temperature.

The first and main objective in collecting vessels' data is a Global Geographic Information System (GIS) monitoring platform. The platform can give all the accurately needed information about vessel's condition in real time, and the vessel's responsible management officer (operator or superintended) can command and control. However, the most important part is the combination of those and the ability to make predictions. Having an engine running lean for many hours can produce danger prediction on engine heating well before the event occurs. For vessel performance, which is a rapidly expanding field in the maritime industry, the system can combine real-time data from speed and drought to produce optimum trim and alert the captain to correct the vessel's bow in up or down movement. The ability to perform these tasks may save thousands of dollars in consumption given the fact that the vessel's hull will move in accordance with its speed and cargo volume (Figure 9.3).

Planned maintenance is another costly operation for vessel operators and the accurate prediction on the time that a part is closing to the end of its operational life is of high importance. Having access to extended information on maintenance, a maritime company can predict when the vessel will need to be in dry dock, and thereby book in advanced for the services, which can impact on the ability of the owner to achieve economic scale.

Optimal route planning or route optimization is also a very popular concept associated with IoV and ITS. This application gathers information from data on-board the vessel and correlates these with data related to factors external to the vessel such as wave height, sea currents, chlorophyll in the water, risk zones and special fuel zone areas and typhoons, just to name a few. Using various algorithms and formulas, the system creates the optimal routing of the vessel from point A to point B. It is very interesting that although a straight line is the fastest way to go

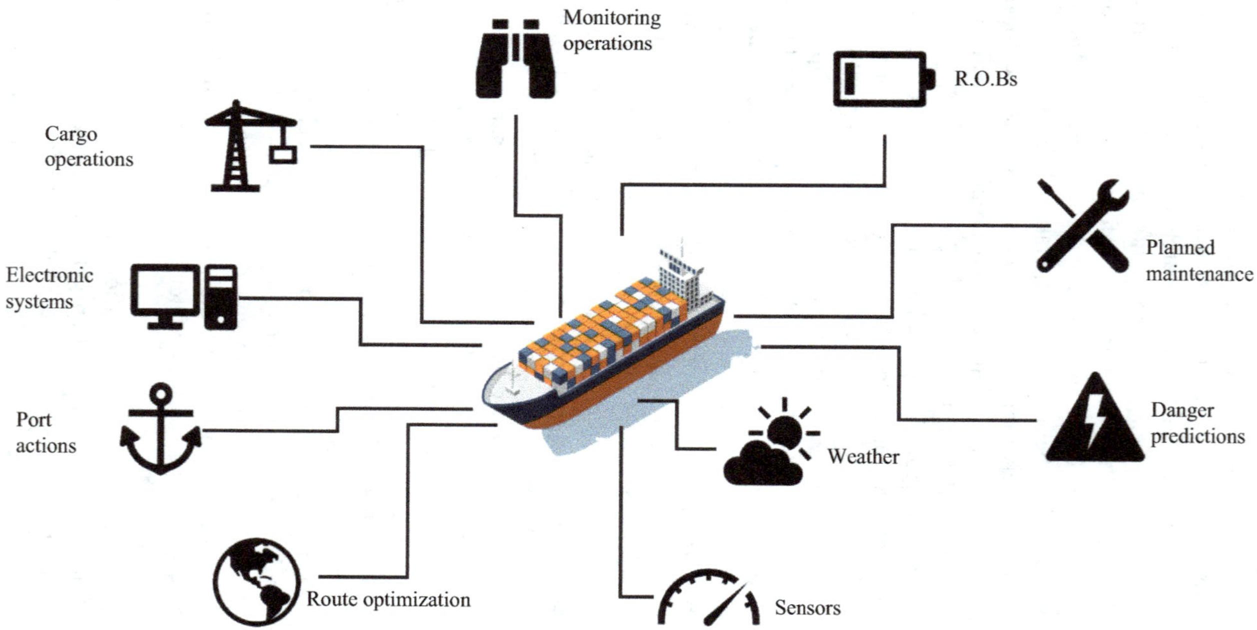

Figure 9.2 Internet of Vessels [1]. Source Author

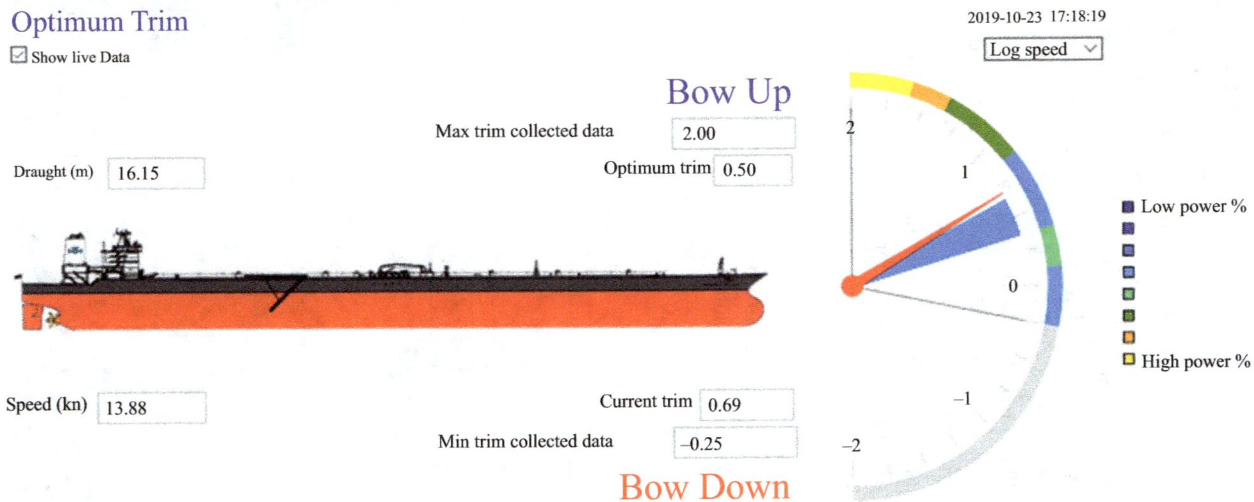

Figure 9.3 Optimum trim application [2]

from one point to another, in maritime a deviation from the route that will take the vessel outside an emission control area (ECA) or sulfur emission control area (SECA) may be beneficial for the management company given the fact that outside these zones the vessel may have higher consumption in fuel, but the captain will not be forced to change to low sulfur fuel which is more expensive. As a result, traditional maritime practices are being questioned and real-time information is being used to achieve this. Ultimately, these practices transform a vessel into a hub. An increased number of 'smart' vessels will mean the development of a global oceanic information stream from vessel-to-vessel connections that can have a huge impact on worldwide trading since over 90 per cent of the volume is transferred from ships [8].

In view of these developments, data collection from sensors on vessel equipment can be considered less advanced than what can be achieved today. Legacy or modern satellite communication systems can instead be used to transmit information to a centralized data storage. Using different methods, this information can be analysed and presented to users. All the earlier can be done almost in real time and at a relatively low cost. The collecting data process can be easily described with the following diagram (Figure 9.4).

If sensors are used, data can be saved into lightweight vessel storage. If the sensor is analogue, this should first be converted from the analogue measurement to digital. All the stored data can be presented to the vessel's crew in a simple overview of the measurement. The next stages are data encryption and compression.

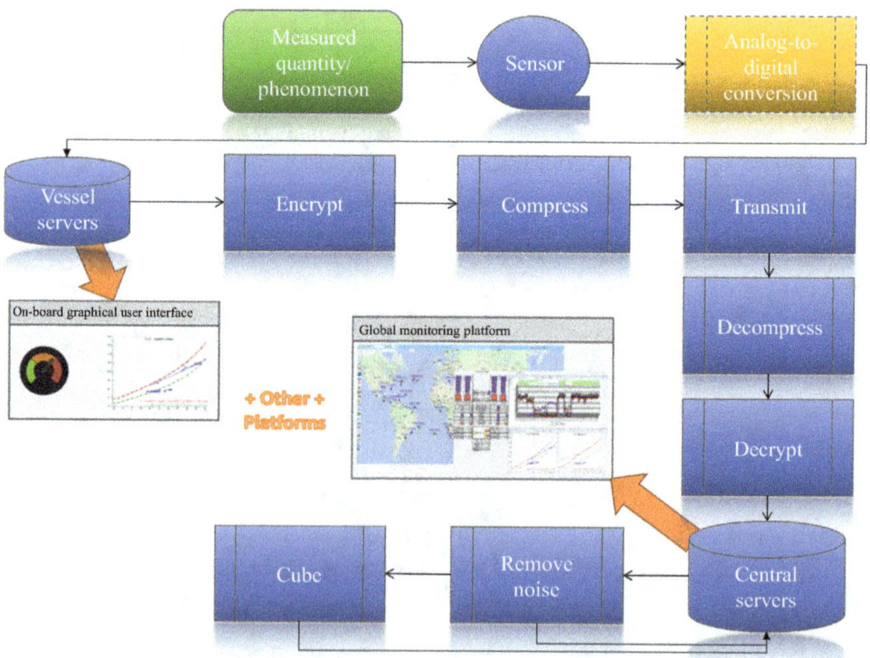

Figure 9.4 Collecting data process [1]. Source: Author

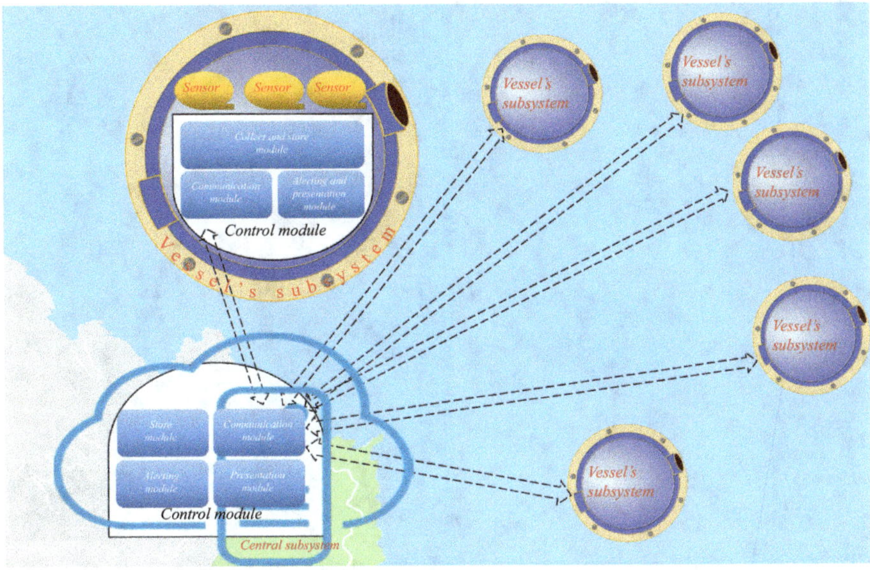

Figure 9.5 Internet of Vessels system diagram [1]

The outcomes are transmitted to the premises where the information is decompressed and decrypted. The received row data is stored into a relational database management system (RDBMS). The noise in the data is cleared by using filters. Finally, the information is placed in cube, if needed, and presented to the users.

Two clearly separated and identified subsystems are synthesized to create the described IoV platform: the vessels' and the on-premises subsystems. These two subsystems are composed of several modules as shown in Figure 9.5.

Concerning the vessel's subsystem, there are lots of sensors on-board that can be used to collect data. A very common example is a Coriolis mass meter with two tube sections connected by a coupling extending between connection points of the tube sections wherein the coupling extends a distance between the connection points which exceeds the shortest perpendicular distance between the first and the second tube sections. The coupling facilitates less stiffness of the tubes against Coriolis oscillation and thus provides a mass meter with a stronger and potentially more reliable measuring signal [3].

Indicatively some of the sensors that are used to collect data include anemometer, gyrocompass, speed log, depth ammeter, rundle angle indicator, Global Positioning System (GPS), flow meters and shaft torque. In new buildings, sensors are digital but in older vessels many are analogue. For analogue sensors, analogue-to-digital converters (ADC) must be used to collect the measurement. The 'collect and store' module is responsible for ADC. This module is responsible for communicating with the sensors and collecting the measurements and data (Figure 9.6).

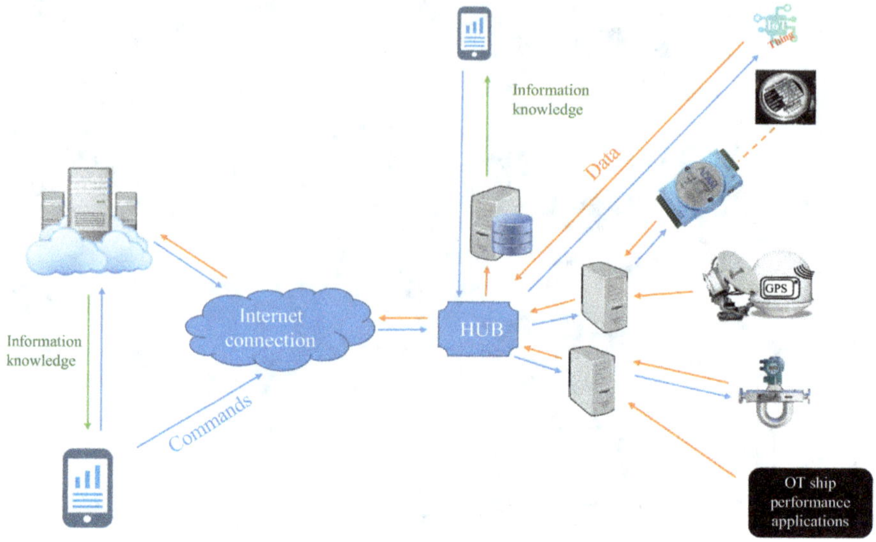

Figure 9.6 IoV connections and data flows [1]

If the sensor is analogue, then this module must convert the analogue measurement to digital using a convertor (ADC). An ADC is a device or component of a larger device that receives an analogue signal and converts it into a digital quantity with given accuracy and resolution. The analogue signal input is compared to a given reference signal, and the difference between signals is used to compute the digital quantity indicated by the analogue signal. The reference signal is normally equivalent to the maximum value of transmitted data [4] (Figure 9.7).

The protocols that are used to collect data from sensors are either the National Marine Electronics Association (NMEA) or MODBUS. NMEA [5] 0183 Interface Standard defines electrical signal requirements, data transmission protocol and time, and specific sentence formats for a 4,800-baud serial data bus. Each bus may have only one speaker but many listeners. This standard is intended to support one-way serial data transmission from a single talker to one or more listeners. This data is in printable ASCII form and may include information such as position, speed, depth, frequency allocation [6]. MODBUS and MODBUS TCP/IP is an application-layer messaging protocol, positioned at level 7 of the OSI model. It provides client/server communication between devices connected to different types of buses or networks. The de facto industrial serial standard since 1979, MODBUS continues to enable millions of automation devices to communicate. Today, support for the simple and elegant structure of MODBUS continues to grow. The Internet community can access MODBUS at a reserved system port 502 on the TCP/IP stack. MODBUS is a request/reply protocol and offers services specified by function codes. MODBUS function codes are elements of MODBUS request/reply PDUs. This protocol specification document describes the function codes used

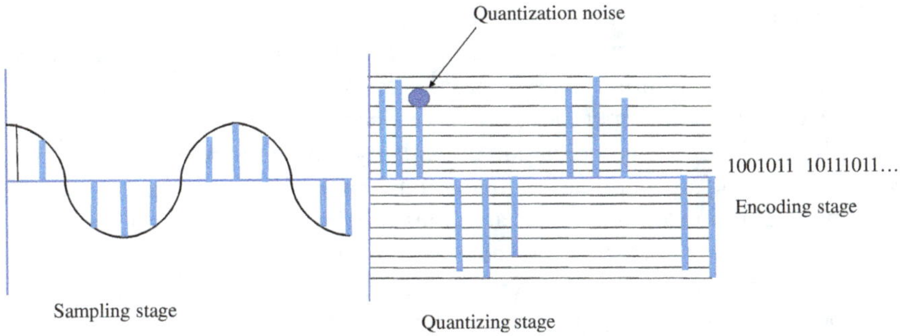

Figure 9.7 ADC [1]

within the framework of MODBUS transactions [6]. In other proprietary ASCII protocol, the module after retrieving the measurement from a sensor, and ADCs if needed, stores it into a 'light-weight' database engine (XMLs, SQLite, Firebird, H_2, express version of well-known RDBMS, etc.). The sample rate is configurable by the System Administrator or the owner and is being controlled by the Control Module.

Finally, the communication module sends the stored measurements back to the on-premises' subsystem. Using the stored administrator's preferences from the control module, the system 'decides' which data must be transmitted (from which sensor), transmission rate, the minimum and maximum size of transmitted packages, etc. Communication module creates the packages, encrypts, compresses and sends them back to premises. If there is Internet connection, the module sends them back using one of the protocols, TCP, UDP, FTP or web services, depending on the implementation. If there is slow Internet connection the module uses the SMTP protocol as the last solution. The encryption uses the triple DES algorithm and for the compression, a well-known zip class. The communication module receives administration or commando packages for the on-premises' subsystem mostly for control module configuration and for software updates.

In reality, the smart vessel concept already exists. All technologies are available, and time will only improve on the data-flow exchange performance. It does not matter anymore if you build the largest or fastest ship since the intelligent vessel is expected to be safer and more efficient. Building them is not a vision of a few idealists but is the way to go in the shipping industry.

9.4 Smart port concept

'Smart ports are the only ports that will survive', argues Olaf Merk, the Administrator for Ports and Shipping at the ITF of the OECD, 'Smart means to me: no waste of space, time, money and natural resources' [7]. This is a valid statement. Smart port is a concept in which technology and innovation are used to automate processes and reduce the time for port actions. In order for a port to be smart, it should use IoT data to its full potential. Indicatively, the Port of Rotterdam which is one of the biggest hubs in Europe, for the first 6 months of 2019 achieved a throughput of 240.7 million tons [9]. Besides the volume of cargo in and out of a port, there are many other operations that take place for every port call that involves many parties and must be completed in a relatively short amount of time. Suppliers, crew members, port agents, loading operations, bunkering operations, repairs and spare parts operations are just a few activities of what we call as port actions.

Being 'smart' a vessel should be able to know all actions need to be done upon arrival long before the port call. As a matter of fact, arrival date and time are part of the ITS between vessel and port because being as precise as possible can correlate with real-time port traffic data and estimation of slot availability in order to reduce or eliminate waiting time. The port is using position sensors and knows exactly the number of vessels that can serve and how many vessels are having discharge operations at any given time. The first procedure takes place when the vessel automatically triggers an alert to the port that contains information for many sub-systems of the port as a single list of tasks need to be performed. The port receives, among other information, the drawings of the vessel, type of cargo and volume and alerts specific tugboat and pilotage services according to vessel's size in order to guide the vessel in the dock. Alternatively, using a GIS system creates a geofence and communicates to the vessel longitude and latitude coordinates, with sea depth analysis, that the vessel will be able to anchor until a slot is empty. The complicated part starts when the vessel docks on the dedicated area where all operations must be done fast, efficient and most of the time at the same time (Figure 9.8).

The agent that acts on behalf of the vessel uses an application on their mobile device that provides data for crew changes on the vessel. The vessel has access and

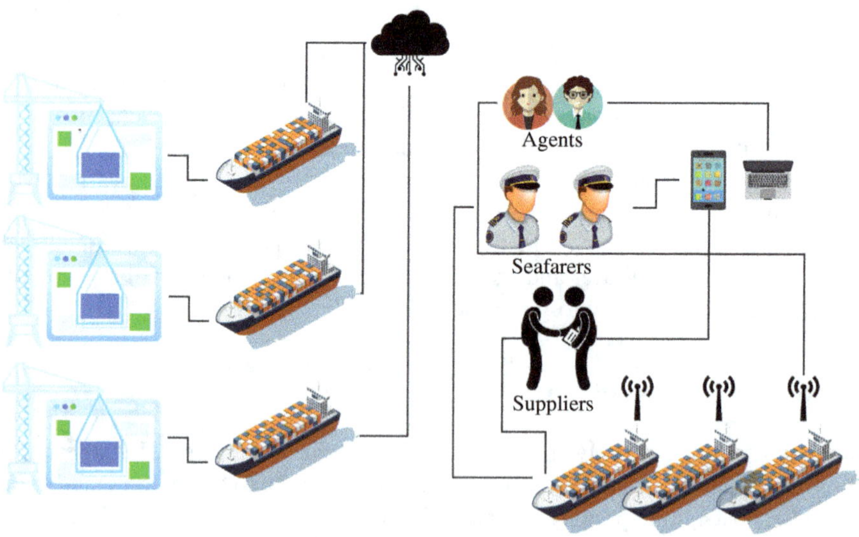

Figure 9.8 Smart port operations. Source: Author

is able to communicate with the web portal, which is a single access point of the port. This allows for the transfer of information on seafarers that are about to disembark by providing information such as passport, flight destination and luggage volume. The agent selects the companies for which he or she is the representative of and can arrange taxi service and bulk bookings for more than one person. This reduces the waiting time for the seaman and has cost reduction for the agent company.

Logistics optimization is another feature of the smart port that allows for suppliers and vendors to track goods, using the same platform, in an accurate and efficient way which leads in keeping up with the demand for vessels concerning procurement, bunker and spare parts operations. To achieve this task the smart port is able to use big data and IoT for support making decisions, which as the traffic increases the system learns to handle situations better. Bunkering operations which are very costly can be scheduled depending on the vessel docked or the demand at any given time. The volume that is required from a vessel has been communicated to the web portal of the port either prior of arrival or when the vessel has been ‘plugged in’ in the dock which means that smart sensors of the port interface with the sensors on-board and know the actual amount required for oil and gas.

Smart port and ITS does not concern only the relationship between the vessel and the port. Ports have grown to be like cities in population with all related issues we meet in a society such as economic, energy, environmental, transportation, medical and others. In the Port of Antwerp, there are more than 60,000 people working every day; therefore, problems like water pollution, traffic and food distribution are more than critical. In comparison with a smart city, there are also additional challenges such as heavy machinery moving around public roads or operating in restricted areas, and there are the vessels that berth and unberth every single day for 24 h. As a result, an ITS in a port will perform the same way as in a smart city but having as a pillar the vessel and the operations related to the vessel. In addition, the system will interact with all involved members of the ecosystem for everyday operations.

Smart port development is becoming an easier concept to implement as the market becomes more mature. The implementation in such case would be based on technology that already exists and hardware that is being used by many companies around the world. To be more specific, cloud technologies will play a major role in a smart port. Nowadays large data centres and local installations are becoming obsolete. All physical characteristics and operational activities of a server are now based on a web datacentre. As a result, ports experience high data availability, latest safety protocols and minimal effort in updating or upgrading physical infrastructure.

Concerning hardware devices, a smart port must be connected in a constant state for ITS and IoT to work. This is possible if there are connections and infrastructure like Local Area Network, Wireless Local Area Network, Wide Area Network (WAN), Radio Frequency Identification, GPSs installed. To get into the ecosystem of the port both physically and virtually, biometric sensors and two fact authentication mechanisms exist for fingerprint recognition and access the systems.

Sensors that are low powered and connected by Low Power WAN are scattered all around the port area and provide measurements like environmental and weather. Other sensors are deployed on vehicles or on workers and provide location information and safety incidents. Finally, sensors on every dock can provide information on the availability of the spot and using specialized algorithms can also alert for potential collision due to the mishandling of the vessel. The whole network, including docked vessels, is connected with Software-Defined WAN which provides high-performance secure data transmission. These networks can power automated vehicles like drones that are used to monitor traffic and cargo operations as well as assisting vessels with navigational manoeuvres inside the port area. Additionally, sensors can control aquatic drones that are used for underwater inspections and filming and cranes for heavy cargo operations.

Economic deals are taking place and key performance indicators (KPIs), as well as data analytics, are available to all shareholders at request. At the same time, smart sensors on the streets release information about the traffic and heavy operations that are taking place at current or future times. Using smartphone applications workers, civilians, seafarers, agents and suppliers are connected to a system flow that ensures each party will follow tasks that are related to specific jobs thus minimize time and maximize efficiency. The port authority is monitoring waste distribution, water supply and power supply through sensors that monitor demand from on-shore personnel and the vessels and is using predictive mechanisms, decision support tools and live reports to plan upcoming supply and demand. As a result, the port can also monitor and control the environmental footprint. Simultaneously, health and safety services are standing by to assist in the case of an accident.

Given the volume that a large port is handling, stakeholders involved in operations and the economic scale of such a system, smart ports are an asset for local societies or even countries. Based on the fact that a port can exist only in a specific geographical location, countries that have operational ports are imminent to invest in smart technologies and capitalize on smart port growth. It is not a risk anymore. It is vital for the survival of port operations.

9.5 Intelligent maritime systems

A whole system is born. The vessel, port and headquarters joined in a single entity using ITS. IMS as a unit is, basically, the collection, distribution and management of data flow between three different entities with different characteristics but one common purpose (Figure 9.9).

The key driver for this technology is the management of the business imperatives of the vessels. With the use of cloud infrastructure, all personnel can work in web-based applications, remotely, with high availability and security standards. This practice reduces maintenance server time, increases update frequency and practically eliminates downtime. At the same time, all data are stored in a data warehouse where data scientists can produce real-time analytics and KPIs for assisting in making decisions. Eventually, a virtual cockpit is created where

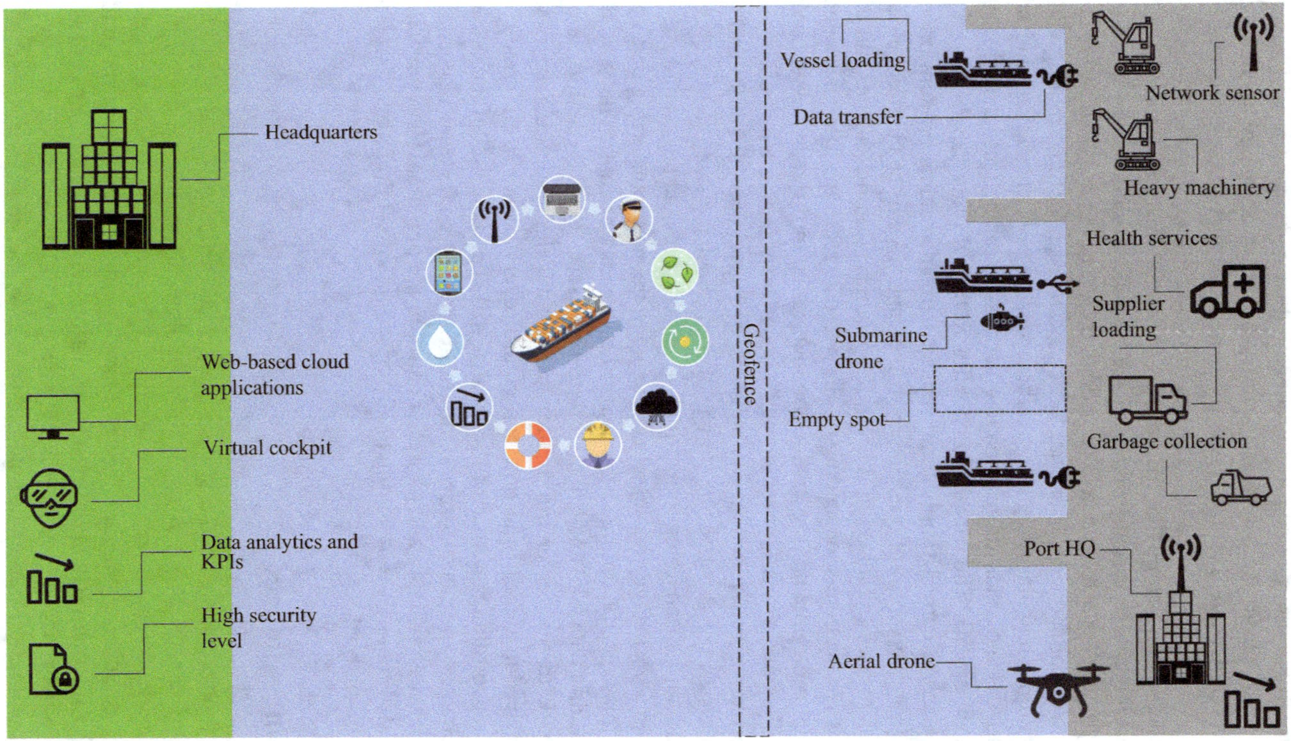

Figure 9.9 Intelligent maritime system. Source: Author

onshore personnel can monitor in real time all vessel actions. Virtual reality can be used in a virtual cockpit for seafarer training that will reproduce precisely vessel departments. In the future, onshore personnel and trained seafarers will be able to control the vessel from land that will lead in the era of autonomous vessels. All stakeholders of a voyage, including oil majors and charters, are able to monitor the progress of a voyage or a voyage leg in real time, transparently avoiding insurance disputes and contract claims that increase the cost and create frustration.

The vessel relates to various services as well as other vessels and the headquarters and at the same time acts as data beckon. Additionally, she controls her own data flow through her internal network. Therefore, officers can create optimal routing with the use of algorithms and parameters like weather, consumption, charts, load and others. She also retrieves and sends information to other vessels sailing close thus creating an oceanic stream of data. The captain can control ROB and order supplies in the next port call or analyse current supplies and request items depending on item price and availability in any given location. Moreover, the captain with the chief engineer can improve performance by adjusting the various parameters on the engine, hull, thrust and propeller among others, depending on the live data received. The interconnected vessel is able to use mobile devices for communication in different vessel compartments as well as perform task checking while being away from a terminal.

The port has a geofence. The geofence acts as a geographical border and when the vessel crosses the geo-line the port is alerted about the size the vessel, cargo carried, load and discharging operations required, crew changes and supplies requested. On a later stage, the vessel is plugged in the port transferring data to port authorities for capacity optimization and other various actions. The port's web platform acts as a data lake and can alert third-party services on the actions required and send back to the vessel information concerning dock spot and tasks to be performed. Headquarters of the port can control every corner of the facility that extends from controlling submarine drones that will check the hull of the vessel for reported damages and aerial drones for traffic control and vessel navigation to goods transfer, water supply and garbage control in the port area. Heavy machinery can be operated remotely and automatically based on vessel drafts and cargo to be loaded and discharged.

9.6 Business impact

In view of the developments and systems described earlier, it is clear that ITS for the maritime industry is far more than technological progress and information technology achievement. If we see the maritime economy from a statistical point of view, there are some significant figures that show the extent of marine trading. According to International Maritime Organization, more than 90 per cent of the world's trade is carried by sea as it is by far the most cost-effective way to move mass goods and raw materials around the world. At the same time, based on statistical analysis from IHS Markit as of 31 December 2018, the world fleet of propelled sea-going merchant ships of not less than 100 gross tonnages now stands at

118,525 ships of 1,333.6 million gross tonnages and an average age of 22 years. Completions during the year amounted to 2,400 ships of 57.8 million gross tonnages. Provisionally, ships out-going from the world fleet as losses or disposals totalled 981 ships of 19.1 million gross tonnages and an average age of 33 years. The world's cargo carrying fleet is 59,687 ships of 1,263 million gross tonnages (1,883.7 million dwt) and an average age of 18 years. During the year, completions totalled to 1,475 ships of 56.2 million gross tonnages (79.2 million dwt). As a result, if automation and ITS reduce the workload and man hours, it will have a very large impact on scale of economy.

As far as the cost of operations is concerned, a shipping company must calculate crew costs, repair and maintenance costs, marine insurance and P&I, stores, spares and lubrication oil costs. The current bunker price for a cargo ship is between $300 and $350 per ton. A cargo ship can use anything from 1 to 16 t/h for the largest ships travelling at a maximum of 23 knots. Therefore, it is safe to assume that a large ship of 300–350 m could cost as much as $5,440 per hour or $130,560 per day of steaming and this figure is for main engine fuel only. As a result, shipping companies are digging deeper into technology innovations and digital evolution to try and decrease costs. The amounts are so large that if for example a company is using performance monitoring from sensors on-board a vessel and manage to save 1 or 2 per cent of fuel consumption due to proper speed monitoring in correlation with weather and cargo data, the vessel is looking at cost-saving of thousands of dollars in a single voyage. Using real-time data for every operation through a voyage or a port call what impact of economic scale will have in a shipping company. Moreover, ITS applications can assist in other aspects of the maritime economy that are not calculated from OPEX but have a large effect on operations of economic scale such as avoidance of piracy zones or ECA and SECA zones. Finally, a company becoming 'green' gains a strategic advantage over the competitors for charterers and customers. According to World Shipping Council, International Maritime Shipping accounted for approximately 2.1 per cent of annual global greenhouse gas emissions in 2012. Container ships account for approximately 25 per cent of that amount while moving roughly 52 per cent of maritime commerce by value.

When an oil major or charterer decides which company will use to transfer goods all those things are taken into consideration. Additionally, during a contract, it is more efficient for the client to monitor the progress of the voyage using real-time data instead of the traditional position list and noon reports that come once per day from the seafarers. It is more effective for insurance companies to cross-check incidents and make decisions on a dispute if they are provided with information directly from the vessels and it is far more productive for the seafarers and operators to monitor their task lists and minimize the idle time of the vessel.

9.7 Conclusion: an eye to the future

What does the future hold for the shipping industry in relation to the technological revolution we are experiencing? What we know is that the global fleet is increasing in numbers. We also know that like airlines, shipping companies are looking to alternative lightweight materials and structural designs that will lead to the creation of mega-ships with improved hydrodynamics, capable of transferring an enormous volume of cargo faster and more efficiently. At the same time, alternative fuels like liquefied natural gas for commercial shipping will play a major role in operations, in energy efficiency for the associated emissions and environmental protection. But this is just from a technical point of view. Is it enough? From a technological point of view, digital transformation will elevate automation and with the developments in satellite communications that will eventually boost ship connectivity, vessels will become sea hubs and data generators.

Technology will eventually lead to autonomy and autonomous vessels, and the contrary to the general belief seafarers will play a major role in autonomous vessels. Being a seafarer is a challenging profession that requires being months away from family and friends facing difficult and stressful situations. On the other hand, it is a very exciting business that requires high skills and focuses. So why not keep those better aspects of the profession? With the use of virtual cockpits or better virtual bridge seafarers will be able to control the vessel via remote control and make decisions based on decision support systems that stream and analyse real-time data. Furthermore, we expect trading patterns to change because of a high availability of automated data. Web platforms and portals are to be created that will stretch from information concerning the type of cargo trading in certain areas and ports to global markets of carbon dioxide distribution. Blockchain is expected to bring a revolution to the maritime industry by speeding up cargo trading adding security, transparency, tracking, improved document flow and faster customs checking, and it will force maritime to re-evaluate business models.

Other trends that will dominate the market until 2030 and will be adopted from the maritime industry are robotics, emissions storage, powering and cybersecurity. Robots will carry out complex and high-risk actions on the vessel and while connected to three-dimension printers will be able to create and install spare parts during a voyage. As the global fleet increases hybrid propulsion systems will be installed on the vessels and electrical power will reduce consumption and global carbon footprint.

Finally, as the industry is moving towards digitalization, there will be an increasing need for cyber attach prevention and proactive actions and probably we will experience the creation of cybersecurity forces [10]. The maritime industry and people around the marine ecosystem are ready to experience the most exciting era in shipping after the Age of Exploration.

References

[1] Ioannis Filippopoulos and Georgios Stamoulis, "Collecting and Using Vessel's Live Data From On Board Equipment Using 'Internet of Vessels (IoV) Platform'". IEEE / SEEDA-CECNSM Conference 2017, May 2017. DOI: 10.23919/SEEDA-CECNSM.2017.8088242.

[2] Ioannis Filippopoulos, George Mitsopoulos Zoran Lajic, Alexandros Senteris, and Mark Pearson, "Multi-Sensor Data Fusion for the Vessel Trim Analyzer and Optimization Platform". IEEE/3rd International Conference on System Reliability and Safety ICSRS 2019, Rome, Italy, 20–22 November 2019.

[3] Ioannis Filippopoulos, Georgios Stamoulis, and Ioannis Sovolakis, "Transferring Structured Data and Applying Business Processes in Remote Vessel's Environments Using the InfoNet Platform". IEEE/South Eastern European Design Automation, Computer Engineering, Computer Networks and Social Media Conference (SEEDACECNSM), Kastoria, Greece, 22–24 September 2018, DOI: 10.23919/SEEDA-CECNSM.2018.8544918.

[4] THE ANALOG-TO-DIGITAL CONVERTER. http://electronicstechnician.tpub.com/14091/css/The-Analog-To-Digital-Converter-315.htm.

[5] National Marine Electronics Association – NMEA. https://www.nmea.org/content/standards/nmea_0183_standard.

[6] U Modbus Messaging Implementation Guide. http://www.wireless-iot.beanair.com/files/UM-RF-08-ModBus-Messaging.pdf.

[7] What is a Smart Port? https://www.porttechnology.org/news/what_is_a_smart_port/.

[8] Zoran Lajic, Ioannis Filippopoulos, Alexandros Senteris, and Mark Pearson, "Adaptive GLR Change Detector for Increasing Reliability of Vessel Performance System". IEEE/3rd International Conference on System Reliability and Safety ICSRS 2018, Barcelona, Spain, 24–26 November 2018, DOI: 10.1109/ICSRS.2018.8688875.

[9] Port of Rotterdham. https://www.portofrotterdam.com/en/our-port/facts-figures-about-the-port.

[10] Stergios Oikonomou, Ioannis Filippopoulos, and Alexandros Voliotis, "An Integrated Maritime Cyber Security Policy Proposal". ISER – 637th International Conference on Science, Technology, Engineering and Management (ICSTEM), Crete, Greece, 8–9 August 2019.

Chapter 10

Digital supply chain and port information modeling (PIM)

Georgios Kapogiannis[1], Jiequn Guo[2,3] and Craig Matthew Hancock[4,5]

This chapter presents how digital transformation could inspire innovation within the supply chain in ports. It consists of three core themes: digital supply chain transformation, port information modeling, and innovation and integrated supply chain management. In particular, it aims to aid in understanding the background of why digital transformation within ports is vital and how this transformation could be implemented. The chapter enhances the understanding of what digital supply chain transformation (digital SCx) is and how the implementation of a collaborative culture could enhance the added value of a port by using an innovative and integrated environment. A key lesson presented is that digital transformation, though originated from technology, is equivalent neither to information technology nor to information systems. More importantly, the reader should understand that the journey through the digital SCx and its integration with the management of a port will be achieved through port information modeling (PIM) and the hierarchy of human understanding, both of which are derived on the basis of the building information modeling (BIM) paradigm.

10.1 Introduction

The operation and asset management of a port is a domain that has increased in interest over the last few decades. Port asset management, divided into the subsections of infrastructure management, capital management, and utility management, has become of particular importance. As a result of the increased importance of global logistics, there is a need to use integrated solutions that incorporate access

[1]Geospatial and Hazard Research Group, The University of Nottingham Ningbo (China), Ningbo, China
[2]Ningbo Institute for Supply Chain Innovation – MIT, China
[3]School of Architecture, Building and Civil Engineering, Loughborough University. UK
[4]Geospatial Engineering, University of Nottingham Ningbo China, Ningbo, China
[5]Geospatial Research Group, University of Nottingham Ningbo China, Ningbo, China

to data that could provide valuable information to facilitate the improvement of efficiency within ports (supply chain management). An improvement in efficiency will, in turn, lead to more efficient use of assets and therefore reduced costs and shorter delivery times. Due to recent advances in technology, including the creation of cheaper but more reliable sensors, improved data transfer technology and cheaper and more accessible digital storage provide the opportunity to collect and utilize "big data." As a means of improving, management within ports has become more feasible. The collection and storage of data also allow for the modeling and visualization of such data in real time or in simulated environments that can aid in better decision-making. Over the last 10 years, BIM has been introduced as a paradigm to improve the design, construction, and operation of assets. By combining BIM with big data and analytics, including artificial intelligence (AI), it is possible to create a more integrated approach to port management that has the potential to improve efficiency across the supply chain. The implementation of BIM in ports and the development of an integrated intelligent system will be discussed in detail in this chapter starting with digital SCx, then proceeding to port information modeling and management, and finally concluding with digital supply chain strategy.

10.2 Digital supply chain transformation

10.2.1 What is digital transformation?

Since the 1960s, the technological development of integrated circuits—computers, and later, personal computers, the Internet, and smart phones—has proliferated and changed the society and business processes around the world in a revolutionary way. Corporate managers are constantly besieged with the latest development in digital technology and have taken steps to extend digitalization to formerly analog transactions and operations. Whether a firm can grow, or even survive, in large part depends on its adaptability to meet structural shifts in markets and information technology in particular. A research was conducted by 500 companies listed in Standard & Poor's to show that the average corporate lifespan of leading companies has fallen from 61 years in 1958 to 18 years in 2011, largely due to the pressure exerted from digitalization according to a McKinsey report [1]. However, more recently, Fitzpatrick in 2020, due to COVID, highlighted that *A digital future lies ahead. By acting early and being bold and decisive, CEOs can accelerate their digital transformation and reach the next normal sooner.*

Thanks to the computer revolution, or digital transformation, most organizations, including government, business corporations, NGOs, and so on, have gained tremendous power to acquire, store, and access data. This can be considered a direct product of digital technology. The business community, in particular, has experienced an explosive growth of data over the recent decade, which, if used appropriately, can provide meaningful insights about the most important objectives of an organization. However, many companies do not have sufficient capacity to cope with the large amounts of data, let alone to perform a smart in-depth analysis of the data they collect. While everyone believes that data are valuable assets, it is, in reality, the information derived from the data that are the most useful. In other words, merely hoarding data without ideas on how to put the data to use is meaningless. In addition, some data have a feature that is similar to money, i.e., time value; in other words, these data may become irrelevant or outdated over time, and any future reference based on these data would be incorrect or misleading.

A more pressing issue, the rate at which data is collected and stored, has now become faster than ever. According to a Deloitte report [2], by 2020, data created would be 44 times more than the total data volume of 2010. Interactive Data Corporation (IDC) predicted a compound annual growth rate of 61% from 2018 to 2025. Therefore, the way to handle data in the era of digitization has to be fundamentally changed from how it used to be. Business leaders need to recognize that digital transformation is a process relying on advanced technology to unlock value in a business, to efficiently operate the business, and to give value to customers through recognizing their needs. Due to the competitive nature of business, companies that ignore the importance of digital transformation may struggle to stay afloat, as businesses that embrace this change gain a competitive edge.

Venkatraman [3] classified digital transformation into five levels sorted by the degree of impact on business that are generated from the transformation process. A schematic representation of the five levels within the context of innovative 3D Information Modelling (re-elaborated by the authors, elaboration based on Venkatraman's paper—Figure 10.1) applied has been presented.

On the lowest level (level one), only individual business activities are affected by the deployment of certain isolated IT functionality; level two is an extension of level one and attempts to integrate internal existing systems and business processes; level three emphasizes the redesign of business processes rather than superimposing new IT functionality on the existing processes; level four goes beyond transformation on a single organization and focuses on the redesign of the business network among multiple participants; on the most comprehensive level (level five), business models, processes, and new strategies can be generated via a strategic implementation of digital transformation.

Despite the hype about digital transformation taking place in recent years, according to a McKinsey report [1], the success rate of digital transformation among organizations is lower than one-third. A 2019 Harvard Business Review [4] article shows that out of the $1.3 trillion spent on digital transformation in 2018, about $900 billion was estimated to be ineffective. A key lesson is that digital transformation, though originated from technology, is equivalent neither to information technology nor to information systems. Digital transformation can be defined as a strategy that requires a holistic and systematic approach from all business functions within an organization to act together. The same report listed several actions, including the following to ensure a successful implementation: using a rigorous approach, communicating effectively, leading actively, empowering employees, and creating an environment of continuous improvement [24].

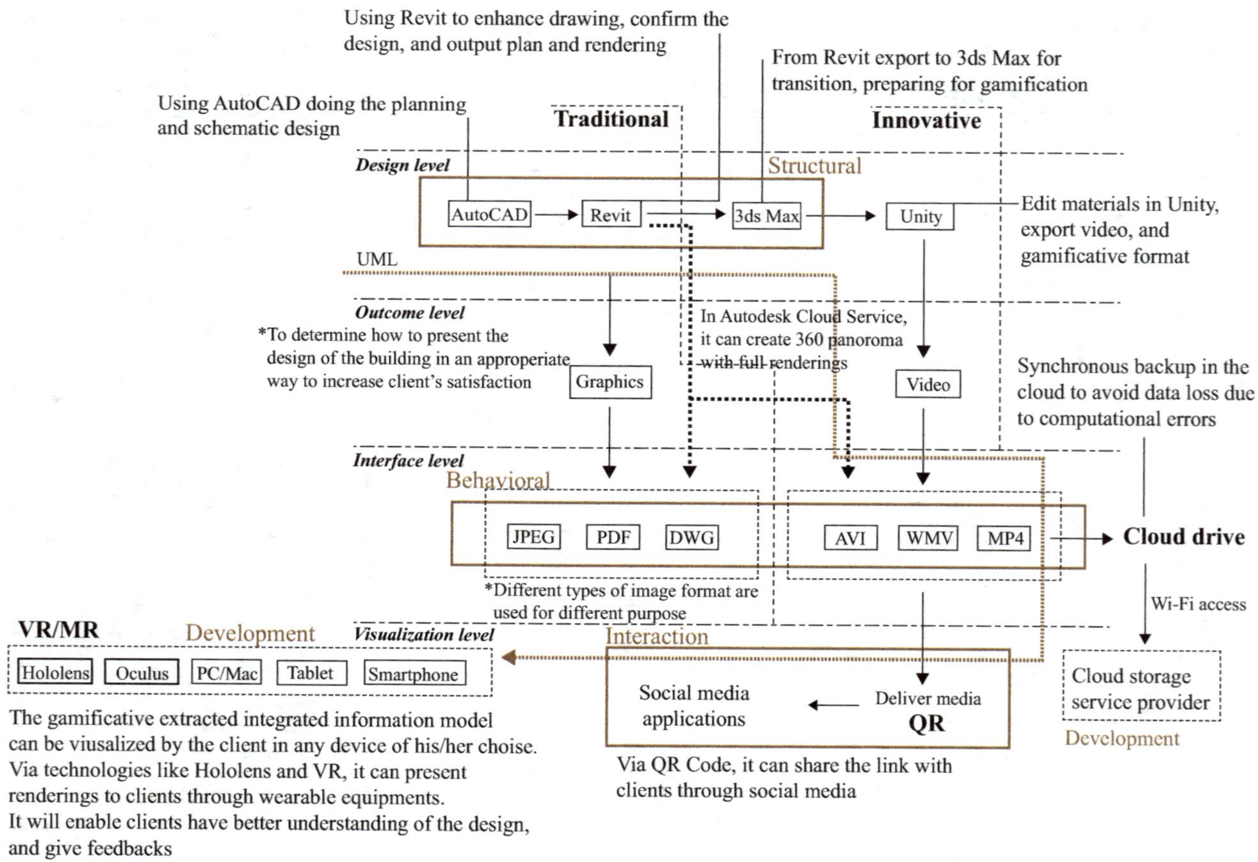

Figure 10.1 Five levels of IT-enabled business transformation within the context of innovative 3D Information Modelling. Source: authors production based on Venkatraman [3]

10.2.2 What is the supply chain?

There are many ways to look at the supply chain like the following:

1. From a structure perspective, the supply chain can be defined as a network. For any company, the upstream chain connects the company with its supplier(s); the downstream chain connects the company with its distributor(s).
2. From a functionality perspective, the whole chain represents steps of a product from initial material extraction (from the ground) to final disposition. The company purchases material from suppliers for production, moves the material to manufacturing facilities, and creates final products to deliver to customers. Certainly, the suppliers may also have their suppliers, and customers may also have their customers.

Because the network involves so many activities, such as product development, procurement, logistics, finance, operations, marketing, distribution, customer services, and so on, and so many actors, such as producers, vendors, warehouses, transportation companies, distributors, retailers, and so on, the final product actually cumulates efforts from all organizations along the chain (those organizations can also be collectively referred to as the supply chain). The organizations that make up the supply chain are connected via information systems that manage physical, information, and financial flows, in which physical flows refer to the movement of goods and/or materials and are the most touchable and visible parts of the supply chain.

The purpose of a company developing a supply chain is to reduce its costs (maximize its profits), be agile enough to create innovative products or services, and stay ahead of its rivals. In supply chain management, specific skills and expertise are required to connect entire activities along the chain, to break the business boundaries of each organization to ensure an overall lower cost and faster production cycle, and to achieve a sustainable competitive business advantage. Because of the continuity of reshaping the business environment and development of technological factors, supply chain management has experienced multiple phases since its establishment in the nineteenth century.

10.2.3 What is digital supply chain transformation (SCx)?

In this study having introduced the concepts of digital transformation and supply chain, digital SCx can be viewed as a natural extension or combination of the two. The birth of digital SCx reflects the needs of the business world and the availability of technology advancement. It allows the acquisition of supply chain data in a convenient digital format and the conduction of deep analysis via the latest algorithms, such as AI and machine learning. Henceforth, digital SCx is a revolution across the supply chain, not a migration on a specific point on the chain.

Being used as an analogy, industry 4.0 is defined as *information exchange between machines and processes within industrial production* [4]. Without newly created technologies, it is impossible for human beings to process in real time the amount of data generated by sensors. In the modern supply chain, the capacity of handling the explosive growth of business data is presented as a common phenomenon in the information era. Thanks to dramatically powered integrated circuits and significantly lowered costs, it is possible to synergize information technology and operation technology with a holistic approach on the data. Digital SCx, by leveraging digital and advanced physical technologies such as drones and automated guided vehicles that collect data using a variety of sensors, in vast amounts, and by using powerful analytical systems, can improve real-time access to information, enhance cross-organizational business collaboration, streamline quantitative analysis to generate new insights, provide accurate and quick responses to react to market fluctuation, and facilitate effective decision-making processes.

In short, digital SCx allows a perfect combination of the digital information flow with the physical flow and the financial flow and fundamentally changes the way that products are developed, sourced, produced, and delivered. This is expected to bring a major improvement in business supply chain management.

To further ensure the understanding of digital SCx, there are three important points to make:

1. Technology used in digital SCx plays only a supporting role. One needs to keep in mind that it is the business need that is the most important objective. Just like playing music, one can have a very expensive instrument; but without notes, there is not anything meaningful. So, for instance, in a competitive business community, it does not make any sense to implement AI algorithms just because the competitors are using AI. It makes perfect sense to implement AI if one's business has a specific need of this technology, and AI can deliver added value that none of the other existing technologies can provide. In other words, even though we have many digital technologies that can be deployed, the purpose of supply chain is the most important. Otherwise, the business is wasting limited resources.
2. Digital SCx is not equivalent to automation. Although the confluence of digital technology, physical technology, and data enables the possibility to automate the supply chain, digital SCx will not likely neglect human intervention. There are a few reasons, including an increasingly complicated business environment and growing regulatory compliance requirements, that make any stereotypical approach of automation unable to satisfy the required needs. One may automate a process today but it will certainly fail to serve one's evolving purposes in an evolving business environment in the future. In addition, digital SCx is not a self-contained, or a well-defined, or a delineated package that companies can simply plug into their business models. Digital SCx is always changing and tailored to specific businesses. Therefore, a company must evaluate all options carefully before selecting what best suits its purposes. Furthermore, digital SCx goes beyond automated processes and creates added value to enterprises.
3. Implementation of digital SCx is a long journey. As digital SCx is not a fixed package with functionalities, one has to build a core digital infrastructure first. With the needs of coordination among diverse stakeholders, resources devoted to replace legacy systems and to create the digital holistic systems, and sustaining their current business without interruption, therefore, a long journey seems to be guaranteed. Digital SCx is a perpetual process and evolving in a business environment that is full of disruption and dynamics.

10.2.4 Digital SCx—an illustration in port management

In the global supply chain, ports play an important role. Not only do they impact the coordination of cargo flows through services, such as storage of material and in-flow services, but also do they deliver value-added supply chain services, such as the implementation of postponement strategies. More importantly, among global trades, about 90% by volume and over 70% by value is carried by sea, for which a port is a must-have from a logistics point of view [5]. But with so many ports and the changing business environment, there is a high competition among them, meaning that a successful port today does not guarantee its future status. Ningbo port of China, for example, which was a lesser known port several years ago, is now one of the largest ports in the world in terms of import and export container volumes [6]. Maersk, the largest container shipping company of the world, has made significant investment [7] in the area, including the establishment of automated warehouses and an international supply chain platform. In 2018, Ningbo port handled 3.35 m TEU (twenty-foot equivalent unit) in import and export container volumes for Maersk with an increase of 20% than 2017 [8].

Port development has experienced three stages: loading/unloading port, industrial port, and logistics/supply chain port. This is followed by the fourth stage: smart ports that utilize integration of digital transformation. Through development, ports have gradually transferred from physical service providers and regulators to logistic cluster managers and information service providers. During the past half of the century, digital transformation has made invaluable contributions. Heilig [9] classified modern port digital transformation into the following three phases:

1. Phase 1, from the 1960s to the 1980s, included the time during which ports expanded their focus from cargo flows to information flows by utilizing electronic data interchange systems, terminal operating systems, etc. However, during this period, most digital transformation is represented by isolated information technology systems with limited functionality. It was not until the 1980s that major advances in enterprise resource planning systems promoted the idea of data integration from different business activities taking place within terminals. During this period, paper-based processing was greatly reduced.
2. Phase 2, from the 1990s to the 2000s, automated handling equipment and began to be integrated with terminal operating systems by using laser technologies, vessel traffic services, and other newly developed technologies. During this period, to foster collaboration among shipping companies, Global Alliance and Grand Alliance, strategic global container liner shipping alliances were formed in 1994 by integrating information systems among participating members.
3. In Phase 3, from the 2010s till date, when the integration of traffic and infrastructure management is continued by utilizing real-time data technology, machine learning and blockchain facilitate ongoing interaction among related parties in port activities. All data collected are processed in a central information system to support decision-making processes, for instance,

interterminal transports and optimization of route technologies. Therefore, this phase emphasizes information analyzing and smart decision-making to enhance the efficiency of port operations by using digital transformation, further improved from data pooling and exchanging in Phase 2.

10.2.5 Multidimensional collaboration in digital SCx

One of the important characteristics of industrialization is the division of labor. The production, since the start of industrialization, has been gradually divided into different roles and specialty areas, which means that collaboration for mass production is required. In the information era, when an organization, such as a port, digitizes its supply chain processes and digitalizes its value chain, collaboration can take multiple forms and becomes a key driver of competitive advantage. A study by IDC Financial Insights [10] on retail and corporate banks found key elements in creating enterprise-wise digital transformation, including a collaborative culture and a focus on a digital core. Kapogiannis [11] shows that integrated collaborative technologies have an impact on team collaboration between stakeholders, including clients, in the planning, design, and construction stage.

So, in an organization, the approach for digital SCx has to be top-down with a consensus at the high level and a clarity on bringing all stakeholders within organization along the journey. The Chief Digitization Office of Cisco once said during an interview with MIT Sloan Management Review,* "we must orchestrate this cross-functionally with the executive leadership team, sales, engineering, human resources, finance, and all of the rest." He continued: "The challenges and opportunities of digital transformation just can't be done inside of marketing or inside of your supply chain. They must involve stakeholders from around the company to make it stick and to create scale."

While it is easy to understand that collaboration among enterprises is required to successfully implement digital SCx effectively, most companies focus too much on their own profit targets, undermining collaboration for maximizing the supply chain value. Lee [26] stated that "*great companies take care to align the interests of all the firms in their supply chain with their own*" [5]. He further pointed out to develop collaborative relationships with suppliers and customers to respond to market change quickly and smoothly. In the digital information era, collaboration becomes even more necessary, as the technology has stretched across distance and physical boundaries, becoming pervasive along the entire supply chain.

From another perspective, the multidimensional feature of digital SCx reflects on its involvement with customers, technologies, suppliers, systems, and tools; it has both short-term objectives and long-term goals. With so many moving parts integrated inside a dynamic and uncertain business environment, it surely is complex and challenging.

The digital SCx within ports is still developing. Despite recent progress, the overall impacts remain limited. With advanced technology proliferating among ports, the smart process will eventually align the interests of all participated

*https://sloanreview.mit.edu/article/how-cisco-encourages-innovation-and-digitization/

members (including importers, exporters, shipping companies, and port authorities), improve port operations, and achieve the efficiency as well as the sustainability of the logistic process. Architecture, engineering, and construction industry through BIM paradigm could be an inspirational mode for digital transformation in supply chain management that could potentially impact also the asset management of the port.

In particular, according to Eastmans [12], BIM can be applied to any built environment project, including ports. Following on from Eastmans [12], it can be said that when BIM is applied to ports, we can call it PIM. The large number of external stakeholders in ports, when compared with other built environment projects, lends importance to the implementation of the ISO19650 standard for better operation of the port through better utilization of team members collaborating and sharing information among external port stakeholders distributed, collocated, and mobile. The EU-funded CoSpaces project (2012) aimed to develop three clusters: methods, technologies, and workspaces. Through this process, research findings have shown stakeholders' capacity to develop the understanding of human and organizational requirements and constraints for collaborative workspaces by using integrated collaborative environment (cospaces.org). The purpose of the technology cluster is to collaboratively develop underlying technologies necessary for supporting collaborative workspaces, namely, the foundation platform layer and collaboration tools layer for supporting reconfigurable collaborative work environments for design and engineering. The purpose of the collaborative workspaces cluster is to implement the three generic collaborative workspaces so as to support various collaboration types identified as important for the engineering sector.

Despite the fact that there is not much research conducted on Port projects, this paper tries to build the backbone of how port information modeling—from now on will be used as a term (PIM)—could be integrated with digital SCx in order to develop a Multidimensional Immersive Collaborative Culture. This unique culture will comply with the international standards—ISO19650 [13]—to achieve digital transformation successfully.

10.3 Port information model, modeling, and management

As discussed in Section 10.2, integration and collaboration within the supply chain is a key element for the digital transformation of ports. According to the international standards, ISO19650 BIM aims to integrate people, processes, and technology [12]. In particular, according to ISO19650 that was published in 2019, BIM is separated into the following three main strands:

1. modeling the geometrical 3D representation of an object,
2. modeling referring to the construction of an object, and
3. information project processes and management of the information flow between stakeholders.

It is of vital importance that this information is secure and consistent when shared among stakeholders and shareholders. As stated in Section 10.2.1 of this chapter, data from digital assets play a significant role in the enhancement of the supply chain process and can be expensive to collect, store, and analyze, although this cost is reduced, as sensors and storage become cheaper. Therefore, it is vital to visualize and archive these data securely and in an organized way to defend against malicious attacks and to facilitate the use of such data in future decision-making processes. In order to achieve the SCx as discussed in Section 10.2.4, a collaborative culture has to be developed; so, the information is available to stakeholders easily from anywhere in the world at any time. To facilitate this kind of collaborative culture, a common data environment (CDE) is required. This has to be secured considering the latest security technologies, e.g., blockchain. This will ensure secure project data transformation in an on/off-line mode in the port.

Additionally, geometrical and spatial data about the port environment are also of vital importance while attempting to increase collaboration and efficiency through the use of PIM. Good-quality geometrical and spatially correct models can be used in virtual and/or augmented environments to aid in the improvement of the supply chain management processes. Geospatial data, collected using a variety of techniques such as laser scanning and photogrammetry, can be used to create and build 3D models of ports in a semiautomated way [14,15]. These models can be used to update design PIM models with more accurate information. Geospatial data can also be used for maintenance and management purposes in conjunction with PIM models and geographical information systems (GIS) in maritime environments. Information that has been created for these virtual models can be important during the whole building of life cycle and can be updated on a regular basis to obtain up-to-date information about spatial and geometrical relationships within the port. As a result, port asset management and its supply chain mechanism can be improved significantly.

Considering the importance of ISO19650 and its impact on port asset management, a detailed analysis of Port Information Model–Modeling–Management is presented in Sections 10.3.1–10.3.3.

10.3.1 Port information model (3D representation)

Considering Eastman [12] and ISO19650 [7], PIM could be considered as a 3D geometrical representation of the port environment, including information about digital and physical assets, underpinned by the creation, collation, and sharing of spatially correct 3D models that are linked to useful metadata. A 3D spatially correct representation of the port environment can be created using design models and preliminary survey data or from new spatial data collection with the aim of building an accurate up-to-date model often known as "as-built." There are plenty of alternative and complementary technologies that allow the collection of high-quality spatial data that can be processed into accurate and useful information. In particular, for spatial data collection, 3D laser scanners are often used in order to digitize a physical asset. This digitization can be useful when there is no existing BIM model or the current BIM model is out-of-date due to updates to the built environment. In some cases, it is possible to use existing drawings and combine them with spatial data collected from laser scanning or similar technology to create a geometrical representation of any built environment. Models built using spatial technology can be of very high quality due to the level of precision achievable from survey grade laser scanners [25]. An accuracy of spatial data, and perhaps more importantly, integrity and reliability are major considerations in a port due to the continuous movement of parcels and containers as well as unique weather conditions that might affect the space/layout.

It is therefore important to understand the relative positions of all assets within the port environment if efficiency is to be achieved. Additional geospatial technologies can be utilized to optimize the workflow in and around ports by providing position and navigation information. Geospatial technologies, such as Global Navigation Satellite Systems (GNSS), inertial navigation systems, ultra-wideband, integrated with photogrammetry and laser scanner technology in simultaneous location and mapping algorithms can be used for enhanced location and navigation. GIS tools, integrated with BIM, can be used for visualization of object locations and for spatial analysis, leading to further optimization of supply chain and management processes. Based on the asset management and supply chain management needs, additional data might be required to complement spatial data (e.g., weather, camera, and sound) in order to increase the level of security on site and aid better decision-making. To do this, additional sensor networks can be installed.

Data from these sensor networks can then be stored as metadata within a digital 3D information model. The actual data needs to be stored in a particular CDE and/or blockchain [27], but the decision to implement such as strategy is something that has to be made at an early stage of the design/redesign of a port with the aim of digital transformation. As a result, in the first instance, there is a need to understand and plan which data are the most important and what infrastructure is required to collect these data. Also, the design of the data transfer and storage strategy is also of vital importance. If designed and carried out properly, the collection of spatial data from geospatial technology and design plans combined with additional useful data networks and a clear strategy for data storage and transfer will create a PIM that can be used during various stages of the project life cycle.

10.3.2 Port information modeling (construction information)

In Section 10.2.1, *what things are produced* was discussed, whereas this section will discuss *how things are produced*. The added value of the construction information process acquired from the design, construction, and operation of a port incorporating certain challenges for the port owner and its clients will also be discussed. In particular, the information production process plays a significant role in understanding the port information project process. Figure 10.1 shows the process of various technologies and digital tools conducted during a case study at the University of Nottingham Ningbo China. The idea was to develop a process that will improve the design workflow through engaging and incorporating inputs from both end users and investors. In fact, the process started with the use of 2D drawings in AutoCAD and 3D modeling of the port in Revit, and then by using 3ds Max and Unity to process a gamificative environment (virtual reality). This kind of virtual-reality environment can be used to support engagement and collaboration among different stakeholders. The presented workflow (Figure 10.2) shows the possibility of creating several different presentation media formats such as images or videos that can aid in the facilitation of client discussion in virtual environment and/or in CDE. Access is given directly to clients using QR codes from a terminal, a phone, or a tablet. In this setup, storage cloud and/or social media platforms such as WeChat have been used as a simple CDE (but could be replaced with blockchain, if needed by the client).

In order to make the system more interoperable, interactive, and innovative, the integration of these technologies can be visualized through a mixed reality environment using a tool such as the Microsoft HoloLens platform. This unique approach allows the client to be within a dynamic and interactive environment in which the information can be visualized in a 3D representation of the real-world environment. The HoloLens 3D environment allows the client to walk through the virtual representation of the building and allows him to give feedback directly to the design and construction teams while being in the environment [16]. ISO9650 was used for information generation, management, and flow purposes within the project. This concept can be utilized to achieve SCx in a more comprehensive, intelligent, and innovative way.[†]

10.3.3 Port information management

PIM can also refer to the management of the port environment that can aid the supply chain efficiency through a better design of ports and better port operation on the basis of higher quality information that can be more easily shared, thus creating better collaboration. Within the modeling part of PIM (that has been described in Sections 10.3.1 and 10.3.2), information is gathered and stored. This information generation is related to time (4D), cost (5D), and other factors (*n*D), which could add value to the decision-making process of the port design, redesign, or operation. Time or 4D is important for the prediction of the completion time of a construction project in contrast to 5D that is important for those data related to costs (bill of quantities) associated with the construction process. Other factors (i.e., *n*D) are related to the operational processes of port referring to data that could add value to any digital asset and their operation, e.g., capital expenditures and operational expenditures, energy performance, position information, and flood simulation, to name a few.

Recently, the Internet of Things (IoT), AI, and digital twinning are becoming synonymous with digital transformation because they allow users to integrate data

[†]It has to be noted that this process was submitted and awarded first prize at the 2019 National Intelligent Construction Technology and Innovation Competition, with the project entitled "Architectural Design Intelligence based on Gamification." The team consisted of members from the University of Nottingham Ningbo China jointly with an industrial partner (Zhejiang Jingwei Engineering Design Co Ltd).

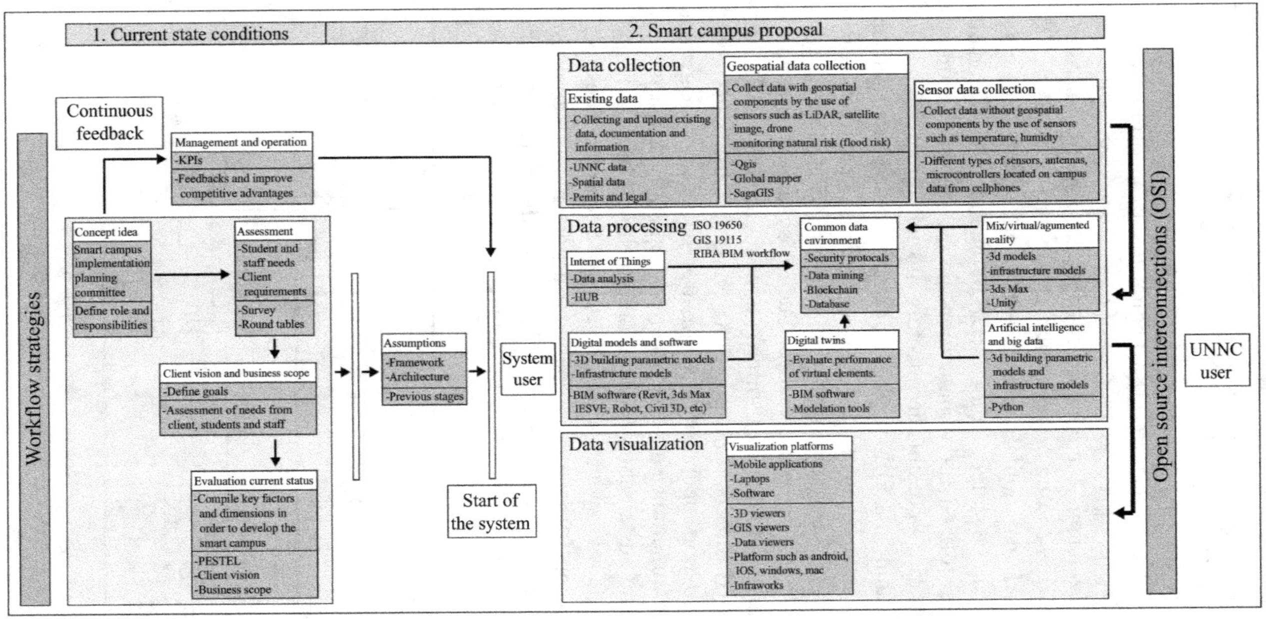

Figure 10.2 Presented workflow to produce new coordinated information in a construction project

and reproduce new information that is more "accurate." IoT can be defined as "*a system of interrelated computing devices, mechanical and digital machines, objects, animals or people that are provided with unique identifiers (UIDs) and the ability to transfer data over a network without requiring human-to-human or human-to-computer interaction*" [17].

AI is defined as improved devices or algorithms that could help a system successfully achieve user objectives [18]. Digital twins, according to IET, is considered as *a realistic digital representation of assets, processes, or systems, in the built or natural environment. The complexity of that representation, and degree of connectedness varies depending on maturity.*[‡]

Utilizing IoT, AI and digital twining within an online platform linked to geospatial data that can be stored within a CDE in the industrial foundation classes (IFC) format can be utilized for visualization, prediction, and optimization purposes. Figure 10.3 shows a possible architecture of such a platform and attempts to illustrate how geospatial data, BIM, and other data, from dedicated sensor networks stored in the cloud, can be integrated into a single system. Within the project management life cycle, key performance indicators (KPIs) are used by project managers to improve project performance. Such a system architecture could be used to evaluate and communicate information about these KPIs. If such an architecture is applied directly to a port environment, these KPIs could include aspects connected to port operation performance or asset management performance. Some of the important KPIs are time, cost, quality, business performance, health and safety, and sustainability. Therefore, accuracy of data is an important factor in assessing and reviewing whether KPIs have been met. These KPIs will be used for designing, constructing, and operating a port and in keeping up-to-date with technological advances in the area. Therefore, the more accurate the data acquired, the better quality of model can be created, and it follows that the better

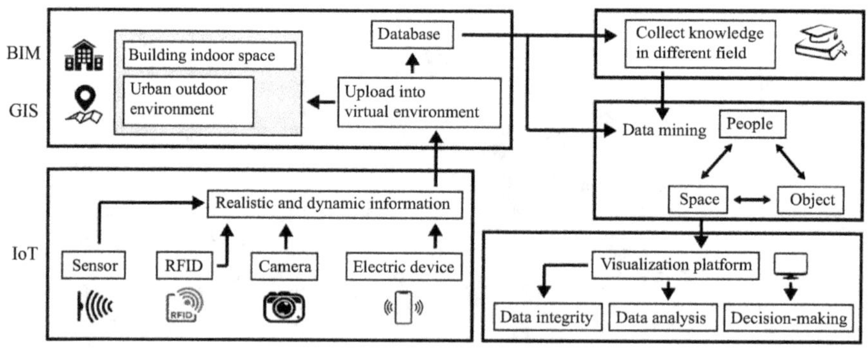

Figure 10.3 Integrating BIM, GIS, and IoT for sustainable buildings

[‡]https://www.theiet.org/impact-society/sectors/built-environment/built-environment-news/digital-twins-for-the-built-environment?origin=short-url

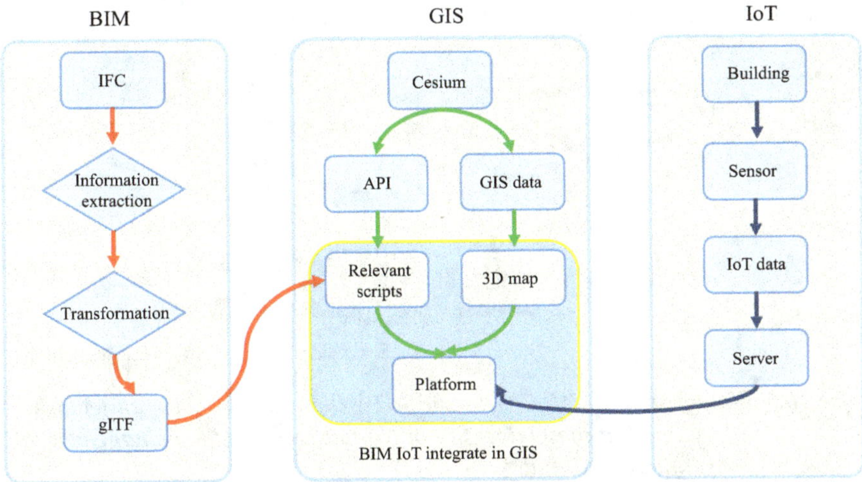

Figure 10.4 General System Architecture developed at the University of Nottingham for the management of a sustainable building

quality the model, the more efficient the design, operation, and construction of the port will be in the context of the supply chain.

As a specific example of a system architecture similar to the one in Figure 10.3, an example is given from the work done in 2017 by the BIM innovation team at the University of Nottingham Ningbo China, which developed an integrated digital architecture to gather and visualize information about the built environment.

The overall design of the system architecture can be seen in Figure 10.4. This system was designed as a proposed solution to the management of a sustainable green building through the integration of BIM, GIS, and IoT (supported by dedicated sensor networks) [19]. The project involved a creation of as-built 3D models of buildings on the university campus. This was done by combining data from existing design drawings (paper- and CAD-based) with laser scanning and photogrammetry data that had been registered using GNSS coordinates. The purpose of using real-world coordinates is served so that the models and associated data can be viewed in a real-world environment through GIS. This means that data from different buildings can be visualized and analyzed with respect to their absolute spatial coordinates. The figure shows that the system can be divided into the server block (where data are stored and analyzed) and the browser block (where data are visualized and queried). Figure 10.4 also shows that the server is again subdivided into three other blocks: the sensor network (IoT), the BIM, and the GIS.

In Figure 10.5, the details of the BIM, IoT, and GIS blocks are shown in more detail. Each block performs an important task within the system architecture to help one to achieve an efficient management system of the sustainable building. Each step performed in each block will be described next:

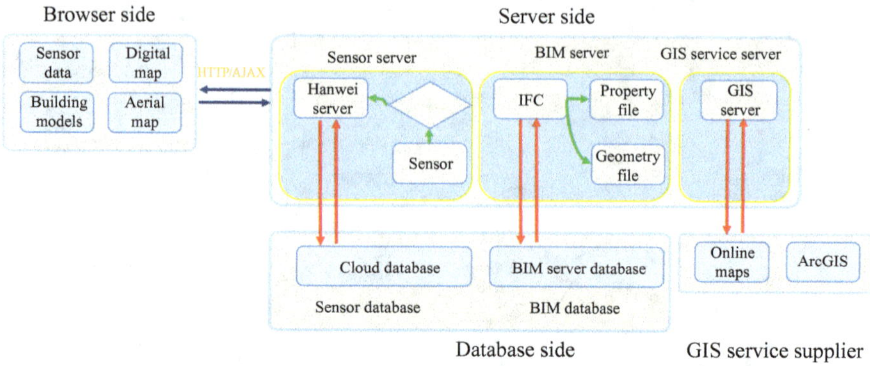

Figure 10.5 Integration of BIM, GIS, and IoT within the system architecture developed at the University of Nottingham for the management of a sustainable building

1. BIM block
   Step 1: The industry standard IFC data file is created as a part of the BIM creation process.
   Step 2: The semantic and geometry information are extracted from the BIM.
   Step 3: The geometry information is converted to a Web Graphics Library that is used for rendering the buildings in the web browser.
   Step 4: Semantic information is collected and stored in a bespoke data format.
2. IoT block
   Step 1: Sensor network is installed inside the building.
   Step 2: Sensor locations are linked to the BIM geometry data.
   Step 3: Sensor network data are uploaded and stored in the cloud in near real time.
3. GIS block
   Step 1: Open source GIS web service is developed for displaying data via a web browser.
   Step 2: Using the AJAX/HTTP protocol facilitates the display and querying of sensor data at specific spatial locations within the model (see Figure 10.6).

Sensors measuring air quality, illuminance, humidity, temperature, sound pressure level, formaldehyde, and carbon dioxide were installed around the building. The sensors were connected to the cloud via a wireless Internet connection, and data were uploaded to the cloud in near real time. Figure 10.6 shows information from the sensors being displayed through the web browser. For facilities and asset management, such a system offers the advantage to managers and decision makers being able to query real data connected with geometry and spatial information that will aid the decision-making processes and provide evidence for any proposed changes to the building on the basis of sustainability or the comfort of the building users. This type of system architecture also means that many people can query and

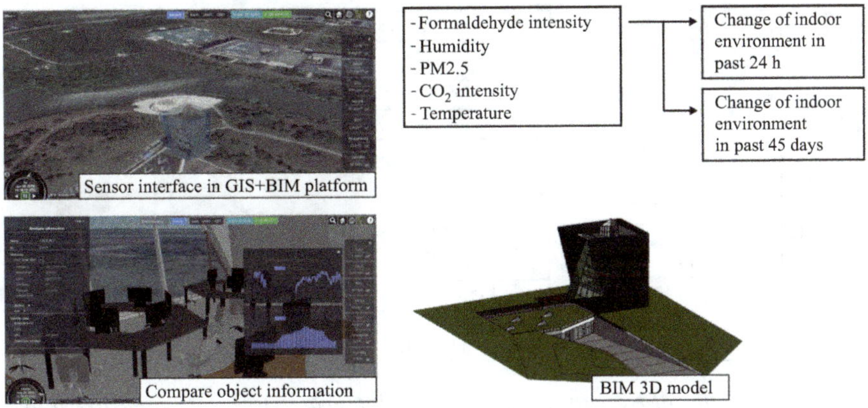

Figure 10.6 Smart campus UML

visualize the same data from different locations at the same time, thus increasing usability and convenience. In addition, warnings and alarms could be built into the system to automatically notify managers or users of problems that may arise in the building.

A similar architecture concept could be easily applied to digital ports where GIS, BIM, and IoT can be integrated to provide a more collaborative sustainable solution that can be utilized to improve port operations. The general concept of this is shown in Figure 10.7. The added value of integrated collaborative solutions is that certain conflicts, due to limitations in information sharing processes, can be avoided and corrected synchronously [20]. The implementation of Unified Modeling Language (UML) can aid in this process. Moreover, the need to use a *general-purpose, developmental, modeling language in the field of software engineering that is intended to provide a standard way to visualize the design of a system*[§] could add value. An example workflow of how a similar strategy might be implemented in a port can be seen in Figure 10.7.

10.4 An integrated strategy toward SCx and PIM

A comprehensive discussion on how ports could integrate PIM and digital supply chain to improve its operations is contained in this section. According to the Deloitte report on smart ports [21] and also according to Mussomeli [22], there is a need for port terminal integration, port-city integration, integration of ports, and the wider supply chain. This section will outline a digital transformation strategy in order for ports to become integrated through a collaborative culture so that they can work toward becoming smart ports in the future.

[§]https://www.uml.org/

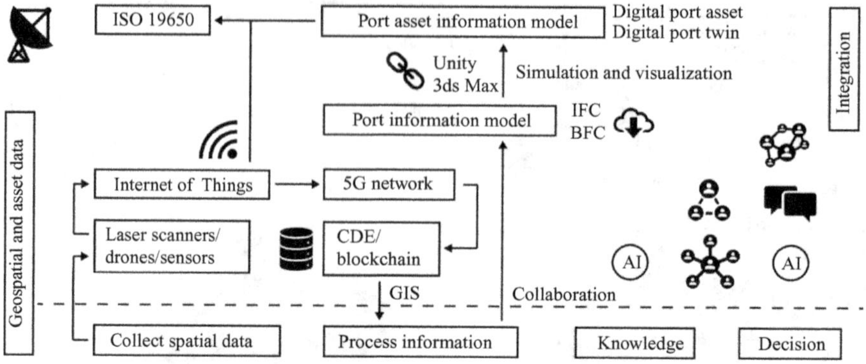

Figure 10.7 General System Architecture to integrate BIM, GIS, and IoT that could be applied to a port environment

10.4.1 Development of multidimensional immersive collaborative culture

A collaborative culture can be developed in construction enterprises and implemented into a variety of projects [11]. As an example, the Scottish Government developed an approach in order to continuously provide instant and constructive feedback for *n*D models[¶] within construction projects. The approach developed by the Scottish Government consisted of three distinct stages: (1) before review meeting, (2) during review meeting, and (3) post review meeting. This approach has been tested and shown to be successful in various different scenarios and projects and therefore this same approach could viably be applied within ports with small adjustments. The outcomes of this project have also highlighted that "accurate" data in the early stages of a project can significantly help and improve the review process of a model [11]. Therefore, it is important to collect high-quality data from the beginning of the project.

Ultimately, following a similar working method to the approach used by the Scottish Government could help one to achieve improved collaboration and digital transformation. By using such an approach, data and model and associated information are stored digitally in the cloud and therefore are accessible to the web using various CDE platforms allowing access to data anytime and anywhere. However, there are still several uncertainties associated with such port operational data that can affect the implementation of digital transformation. These data are linked to economic, social, and environmental (sustainability) characteristics and are likely to be subject to change over time, the precise nature of which cannot be predicted [20].

By integrating UML as a part of the digital port transformation strategy, stakeholders will have a better understanding of uncertainties during the operation of a port, because a port by default is heavily affected by uncertainties such as weather

[¶]https://bimportal.scottishfuturestrust.org.uk/level2/nhs-scotland/stage/5/task/16/model-review-meetings

and on/off-load containers. An example of the use of UML was used for the modeling process of the smart campus of the University of Nottingham Ningbo China and can be seen in Figure 10.2.

10.4.2 Designing a digital port transformation strategy

A port as an asset consists of numerous information that can be generated from a variety of data sources that will affect the decision-making process. Sensor networks can be used throughout the port to gather data that can be available in real time (e.g., position data for people and/or containers, temperature, equipment, and traffic). Advanced networks can be used to support the transfer, using 5G, and analysis, using AI algorithms of collected data. The products derived from these data can be used for identification and prediction of delays, problems, and also for making suggestions about appropriate solutions to these problems using current assets. These data are defined as digital assets that have a unique ID number and that, through IoT, can be traced and tracked by the suppliers. The digital assets are stored in a 3D format by the manufacturer and could be shared with the asset management team to integrate them into a new Asset Information Model (AIM). ISO19650 1–2 will be used for the creation, development, storage, and maintenance of the new information. Each time an AIM is updated, it will be easier, using AI, to teach the model how to predict arrival times of containers, to automatically detect any error in the system and automatically report this to a senior level. ISO19650 3–4 will help one to advance in security and asset modeling in a closed and private network such as a blockchain, so that the asset management team can run simulation and visualization exercises (prediction) by identifying optimum solutions for the suppliers. Through this approach, the port can be managed in an intelligent way because all mechanisms and technologies are integrated in an interactive environment and culture that helps decision makers think proactively and react, whenever required, when standards are in place. Figure 10.8 shows a graphical representation of such a digital port transformation strategy.

Furthermore, if any port plan changes, such as when expansion and maintenance are required, the owner will be in a better position because they have access to the right information (following a high volume, valuable, velocity, variable, and veracity amount of (non)structured data) from anywhere at any time. In principle and according to a 2025 construction report by the UK Government, the implementation of information modeling could help one to reduce project costs (design and construction) by 33%, contribute to faster delivery of projects by 50%, reduce emissions by 50%, and improve exports by 50% [23].

Moreover, this approach will contribute to improving information coordination, team collaboration, detection of clashes (design, construction, and maintenance), increased customization, flexibility, scheduling (4D), cost (5D) optimization, and easier maintenance of the building/asset life cycle (6D) while keeping the quality high. In addition, conflicts among team members can be reduced and transparency can be increased. In turn, this has the potential to improve cash flow due to the

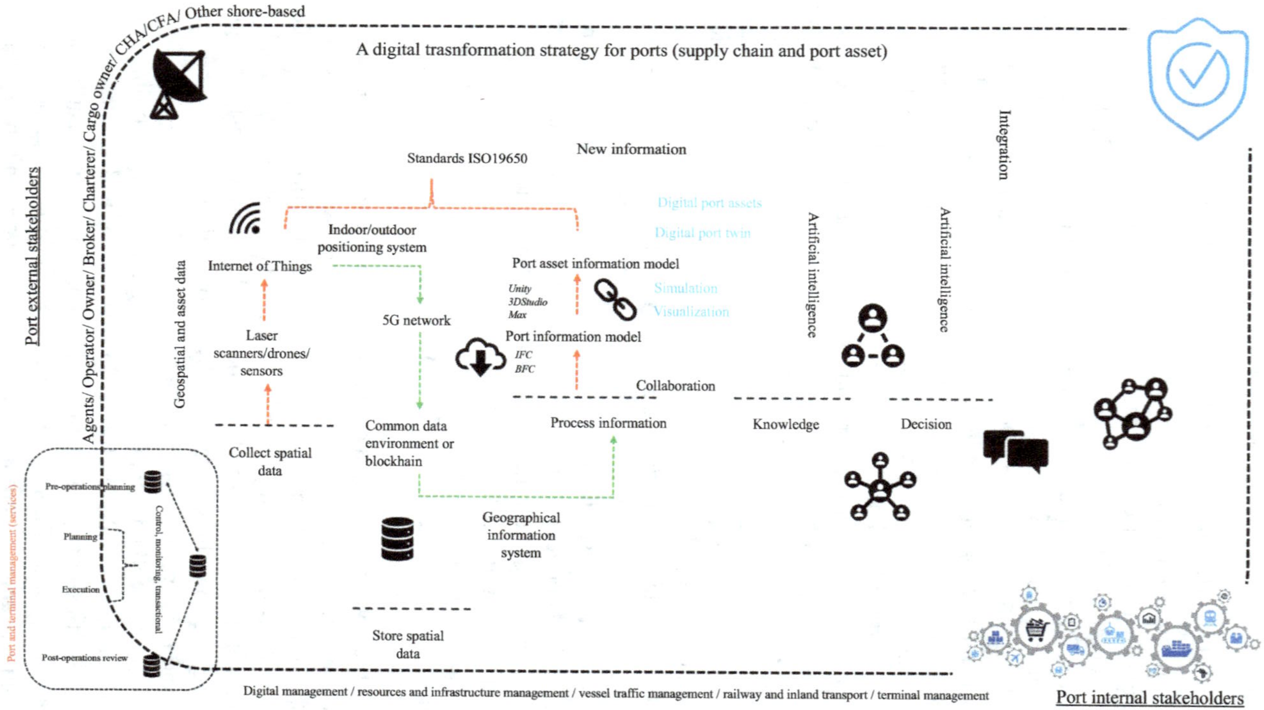

Figure 10.8 Digital port transformation strategy

provision of more "accurate" data related to a number of factors that can be visualized and analyzed automatically either by experts or by algorithms.

10.4.3 Impact of SCx and PIM on the design and operation of ports

Preliminary research has been conducted to identify whether there is an impact of SCx and PIM on the design and operation of ports at Ningbo Port, Zhejiang Province, the Republic of China.

Based on the preliminary research conducted by the University of Nottingham Ningbo China/Geospatial Research Group in collaboration with Massachusetts Institute of Technology (MIT)-Ningbo Supply Chain Institute, communication, transportation, and financial flow are the drivers behind the digital transformation in the port of Ningbo (Figure 10.8). As digital transformation accelerates, and all ports move toward digital transformation, the effect on the supply chain linked to other port operations could be massive, as it has the potential to ensure consistency, effectiveness, and efficiency.

In addition, collaboration will be increased by the development and deployment of an integrated and innovative system. However, despite the implementation of such innovative systems designed to improve collaborative environments, there might be challenges in understanding who has the responsibility for different tasks and thus to obtain a common ground of knowledge among external/internal stakeholders, this can be especially difficult across different disciplines.

Knowledge sharing may also be increased by digital transformation and might lead to decisions that will decrease risks and thus optimize asset management and port operations. Furthermore, findings have shown that the capacity of the use of an integrated innovative solution to the port asset and operation management could add value to senior facilities/asset team, on the basis of developing certain proactive antecedents that could impact its management process and thus could pre-identify problems or react in a timely manner, whenever a problem occurs.

It is important for stakeholders to be aware of their resources in order to maximize, use, and thus move toward smoother operation. Therefore, there is a need to develop a collaborative culture in the port followed by the design of a digital strategy and development of an integrated system that will allow the Ningbo Port Authority to have better control of its operations and assets, including the improvement of its relationships with international trade partners. This idea complies with the Deloitte report [21] on the high demand of port digitalization to achieve integration and, in particular, port terminal integration, port–city integration, and wider supply chain integration. It has to be noted that Ningbo International Marine Science and Technology City has developed the strategy to link Meishan Island and the port with the city of Ningbo and the rest of the planet through several core ports, e.g., Greece, Rotterdam, Frankfurt, Italy, India, and Africa. This aligns with the Chinese Government—One Belt—One Road Initiative [20].

In order to reach a mature level of digitalization and thus improve the supply chain, there is a need for a high level of modeling development (level of detail (LOD) 400). With this level of maturity, the digital information model can be linked with digital assets by using an integrated solution that incorporates people, processes, and technologies. This integration is the joining of supply chain management and building information model-modeling-management using the state-of-the-art communication technologies. Table 10.1 shows in detail how each technology can contribute to the digital supply chain and operational management of a port:

Another example of application was developed a few years ago at the Hamburg Port, Germany. This example shows the added value of designing a virtual port with the aim of supporting port infrastructure management.

Another example of digital transformation is Hamburg Port, the largest port in Germany, where a virtual reality model was developed to improve port infrastructure management. According to the Hamburg Port Authority, "*the use of innovative digital future technologies have been developed over several years in order to manage the port infrastructure more efficiently and intelligently.*" Moreover, they worked in collaboration with local universities in Hamburg and the State Enterprise of Geoinformation and Surveying (Landesbetrieb Geoinformation und Vermessung [LGV]) to form a virtual port. Based on their report, it is stated that all asset data (digital assets) of the port area should be integrated, visualized, and coordinated in real time.

In both of the examples, it is evident that the paradigm of integrating the virtual with physical reality becomes a norm (in the United Kingdom, there is an investigation on the impact of this norm known as BIM). As a summary, the use of BIM paradigm could add value to an improved supply chain management mechanism where *asset data of the entire port area should be integrated, visualized, and coordinated in real time. It is thus possible for real-time coordination of ship movements in the port area to superimpose the AIS ship location data with the current water depth model or to evaluate different planning variants of digital building models in the realistic context of the harbor environment.*

10.5 Future of port information modeling

One of the key aspects of the future of the digital supply chain and PIM is related to the use and development of immersive technologies such as virtual reality, augmented reality, mixed reality, and digital twins in which stakeholders can visualize certain information in a more realistic and useful way. High-quality models that can be navigated and edited in a virtual or semi-virtual world can contribute in a significant way to the improvement of port management. As a result, the maturity level of information, known as LOD, could contribute to port stakeholders to operate the asset in a more efficient and effective way. The data in the information model can be "accurate" and "informative" in a way to support the operational and asset management. However, computational techniques and AI could process that information. As a result, better information quality could be provided and more importantly stakeholders can be given access to this data in a timely manner. Consequently, the integration of advanced digital technologies with the provision of certain details (virtual objects—digital assets), including those provided by

Table 10.1 The digital supply chain and operational management

Category	Technology	Role in the digital supply chain	Potential impact in the operation of the port
Network	Digital architecture for data transfer process	How the data will be transferred from Point A to Point B	Data transformation from dock or ship or other areas in the port to the back-office management team
	Security	Use of those tools and techniques that can keep the level of security at high level during the transformation from Point A to Point B	Ensure safe data transfer from any point in the port to the back-office management team
Data	Data collection	Those technologies can be used to gather data in the digital port synchronously or asynchronously	Data can be collected using synchronous technologies: • IoT (objects, e.g., containers and cranes) • RFID (objects, e.g., containers and cranes) • GPS (location) Asynchronous technologies for the port are the following: • Laser scanner • Photogrammetry • Traditional surveying tools
	Data save	Where data will be saved, including relevant information modeling at later stage	common data environment or blockchain or an integrated environment that could allow access to certain stakeholders to certain data
	Data process	Processing those data is required in order to turn them to valuable information. This information has to be proceeded and be available to be visualized by stakeholders	On this basis, there is a need of a high performance computer (HPU) unit to ensure high-speed processor to process a large volume of data into information. Minimum specs for the volume is i7. Soon quantum processors will be available

(Continues)

Table 10.1 (Continued)

Category	Technology	Role in the digital supply chain	Potential impact in the operation of the port
	Data transfer	Data and later information have to be clear how they are going to be transferred through a secure channel	According to the latest communication technologies, the most secure networks are the following: • 5G (big data can be transferred faster) • Wi-Fi/Li-Fi (data can be transferred within the port by being controlled by digital security officer) • Ethernet (traditional data transfer through cable) • Optical network
Information	Information model and management	Model data to information is a journey that requires a systematic approach. This will enable developers and users to understand the process, technologies, and where end users need access and why. In the domain of information systems, the relevant tool is used. It is identified as a Unified Markup Language (UML)	The end users in the port will be able to tackle where a problem has occurred or even to enable new mechanisms; wherever, this is a requirement to be taken. UML shows exactly all these points in which further development can be taken to improve port operational process in the future in the case of further expansion
	Information modeling	The first stage of modeling is performed by converting the 2D drawings to 3D models using relevant software solutions. In the market, there are many different software solutions that support 3D Information Modeling such as • Autodesk Revit for 3D design • Bentley AECOsim for 3D design • Trimble for 3D design and scheduling with Microsoft HoloLens 2 • Autodesk Navisworks for clash detection	The design modeling process is reviewed by those stakeholders who have instant access and responsibilities to the project. Additional data can be added, e.g., data and costs that could visualize the project process

	Information visualization and simulation	Visualization technologies are • Virtual reality • Augmented reality • Mixed reality	With the support of gamification and its tools, it is possible to simulate the operational process of the port using relevant mechanisms such as HoloLens 2, HTC Vive, or other VR/AR/MR tools. All of the previously mentioned tools can be used not only during the design stage to review port information models but also during the construction phase through digital twins
Knowledge	Stakeholders' accessibility to the data and information of the port will give them the capacity to provide constant and constructive feedback, when the port is in operation, and make recommendations in real time	Collaborative culture is the digital twin place where people, processes, and technologies are integrated. This innovative interactive environment will allow stakeholders to share, exchange, and correct insufficient or inaccurate digital information model(s) of the port	By giving access to stakeholders of the port, in a collocated, mobile, and distributed way, the following will be achieved: • to make best use of the technologies • to develop and integrate a system that will allow data and information to be accessible to relevant stakeholders only • allow the right people to have access to the right information. As a result, the future of the port sector is to provide intelligent services in which people will follow integrated supply chain procedure by using the state-of-the-art digital technologies
Decision (wisdom)	At this point, stakeholders of the port need to take decisions	Decision Information Systems, with the use of artificial intelligence, will help people do a number of works, e.g., check and review certain scenarios in an effective manner	In this intelligent and interactive environment, information sharing can support people to feel more confident and to make effective decisions for the operation of the port. As a result, stakeholders will develop more proactive behavior and support reactions with more confidence

manufacturers' specifications, such as quantity, size, shape, location, and orientation, could support the optimum operation of the port as an asset and could also contribute toward improving supply chain management and thus reach Industry 4.0.

10.6 Conclusion

This chapter has shown the added value of investing in the digitalization of ports to aid in the improvement of supply chain operations. Digital transformation is divided into five levels sorted by the degree of impacts on business, which are generated from the transformation process. Therefore, there should be an opening to a more collaborative culture in which data could be collected and shared to external/internal stakeholders regardless of their location. Collocated, mobile, and distributed ways to collaborate are currently available but businesses are not mature enough in order to (1) make best use of the technologies, (2) develop an integrated system that will allow data and information to be accessible to relevant stakeholders only, and (3) allow the right people to have access to the most appropriate information.

The advanced use of different technologies in the integration of port operational processes and asset management will add value and support to the day-to-day operation of ports. In particular, from the data level, more timely input will be provided due to the utilization of IoT and its connectivity to the web, whereas the data will be transferred to a fast data network (i.e., 5G) allowing fast analysis and visualization of data. Spatial data collected using GNSS, laser scanners, and photographs can be used to collect and update spatially correct data that can be integrated with GIS to aid navigation keep track of assets. On this basis, the information can be visualized in a timely and spatially correct manner via AR/VR/MR to aid stakeholders in their understanding of problems and, therefore, help them react in a more responsive mode regardless of location to aid in better decision-making processes.

In addition, external business performance can be improved, as goods can arrive quicker and with added security and accuracy and, thus, can eliminate risks and inefficiency associated with uncertainty. The preliminary research at Ningbo Port has shown that the use of an integrated innovative solution to the port asset and operation management could add value that could aid the senior facilities/asset teams in the decision-making processes. Stakeholders can get benefited from up-to-date accurate information about their resources and can use this information in order to maximize the use of assets and thus move toward a smoother operation of the port. The example of Hamburg Port shows that there is added value in the effective visualization and integration of port data using spatial coordinates in real time.

The future of the port and supply chain should provide a high level of detailed information to help intelligent services and, thus, would attract more clients and offer high-quality services at both the national and the international levels because of the more integrated strategy that moves toward SCx and PIM.

References

[1] McArthur S., Sankur A., Shah K. and Singh V. 'Why building a digitally enabled supply chain is hard, and what to do about it', McKinsey Report, Feb. 2013. [Online]. Available: https://www.mckinsey.com/business-functions/operations/our-insights/digital-supply-chain-transformation-with-a-human-face. [Accessed on: January, 2020].

[2] Tabrizi B., Lam E., Girald K. and Irvin V. Digital transformation is not about technology, 2019. [Online]. Available: https://hbr.org/2019/03/digital-transformation-is-not-about-technology.

[3] Venkatraman N. '*IT-enabled business transformation: from automation to business scope redefinition*', Sloan Management Review, 1994, Vol. 35, No. 2, pp. 73–87.

[4] TWI. What is Industry 4.0?, 2019. [Online]. Available: https://www.twi-global.com/what-we-do/research-and-technology/technologies/industry-4-0.

[5] Reid H. *How the shipping container created today's global economy*, 2019. [Online]. Available from: https://moneyweek.com/511580/how-the-shipping-container-created-todays-global-economy [Accessed 24 January 2021].

[6] Ningbo Port. *Company's Profile* [Online]. Available from: http://www.nbport.com.cn/gfww/gsjs/gkgk2 [Accessed 24 January 2021].

[7] Jin R., Hancock C.M., Tang L. and Wanatowski D. '*BIM investment, returns, and risks in China's AEC industries*', Journal of Construction Engineering and Management, 2017, Vol. 143, No. 12, ISSN 0733-9364, https://doi.org/10.1061/(ASCE)CO.1943-7862.0001408.

[8] Linknnovate. Maersk, 2020. [Online]. Available: ttps://www.linknovate.com/affiliation/maersk-1340492/all/?query=&utm_source=blog.linknovate.com&utm_medium=referral&utm_campaign=industry%20trends&__hstc=255305623.1bb630f9cde2cb5f07430159d50a3c91.1531785600074.1531785600075.1531785600076.1&__hssc=255305623.1.1531785600077&__hsfp=1773666937.

[9] Heilig L., Schwarze S. and Voss S. *'An analysis of digital transformation in the history and future of modern ports'*, Proceedings of the 50th Hawaii International Conference on System Sciences, 2017, Waikoloa Village, Hawaii, USA. http://hdl.handle.net/10125/41313, ISBN: 978-0-9981331-0-2.

[10] Buss A., Freeborn L. and Silva J. The digital ready bank how ready are European banks for a digital world. [Online]. IDC Financial Insights. Available: http://www.mynewsdesk.com/material/document/57850/download?resource_type=resource_document. [Accessed on: April 2, 2020].

[11] Kapogiannis G. and Sherratt F. '*Impact of integrated collaborative technologies to form a collaborative culture in construction projects*', Journal of Built Environment Project and Asset Management, 2018, Vol. 8, No. 1, pp. 24–38. https://doi.org/10.1108/BEPAM-07-2017-0043.

[12] Sacks R., Eastman C., Teicholz P. and Liston K. *'BIM Handbook: A Guide to Building Information Modeling for Owners, Managers, Designers,*

Engineers and Contractors', 2018, London: Wiley Publishing ISBN: 978-1-119-28753-7.

[13] International Organization for Standardization. Organization and digitization of information about buildings and civil engineering works, including building information modelling (BIM)—Information management using building information modelling—Part 1: Concepts and principles (ISO/DIS Standard No. 19650), 2019. Retrieved from https://www.iso.org/standard/68078.html.

[14] Boehler W., Borda V.M. and Marbs A. *'Investigating laser scanner accuracy'*, Proceedings of CIPA XIXth International; ISPRS/CIPA, Vol 34, 2003, Antalya, Turkey, p 696–701.

[15] Chen C., Tang L., Hancock C.M. and Zhang P. '*Development of low-cost mobile laser scanning for 3D construction indoor mapping by using inertial measurement unit, ultra-wide band and 2D laser scanner*', Engineering, Construction and Architectural Management, 2019, Vol. 26, No. 7, pp. 1367–1386. https://doi.org/10.1108/ECAM-06-2018-0242.

[16] Kapogiannis G, Yang T., Jonathan R. and Hancock C. *'An innovative dynamic gamificative BIM environment'*, FIG Working Week 2020 Smart Surveyors for Land and Water Management, 2020, Amsterdam [Online], 2020. Available: https://www.fig.net/resources/proceedings/fig_proceedings/fig2020/papers/ts02c/TS02C_georgios_craig_et_al_10627.pdf. [Accessed: 28-Apr-2020].

[17] Margaret, R. 'Internet of Things (IoT)', *IOT Agenda*. Retrieved 14 August 2019.

[18] Schank R.C. 'Where's the AI?', *AI Magazine*, 1991, Vol. 12, No. 4, p. 38. https://doi.org/10.1609/aimag.v12i4.917.

[19] Yin M., Wu Z., Wen Y. and Tang L. '*Visualize GIS, BIM and IoT data for comprehensive green building parameters monitoring*', The 8th International Conference on Innovative Production and Construction, 2019, Hong Kong.

[20] Kapogiannis G., Gaterell M. and Oulasoglou E. '*Identifying uncertainties toward construction projects*', Journal of Procedia Engineering, 2015, Vol. 118, pp. 1077–1085, https://www.sciencedirect.com/science/article/pii/S1877705815022067.

[21] Deloitte the Netherlands. Smart ports point of view, 2017. [Online]. Available: https://www2.deloitte.com/content/dam/Deloitte/nl/Documents/energy-resources/deloitte-nl-er-port-services-smart-ports.pdf.

[22] Mussomeli A., Gish D. and Laaper S. The rise of the digital supply network Industry 4.0 enables the digital transformation of supply chains, 2016. [Online]. Available: https://www2.deloitte.com/us/en/insights/focus/industry-4-0/digital-transformation-in-supply-chain.html.

[23] HM Government. Construction 2025, industrial strategy: government and industry in partnership, 2013, London.

[24] Baldwin R., Kawai M., and Wignaraja G. *'A World Trade Organization for the 21st Century: The Asian Perspective'*, 2014, Massachusetts, USA. Edward Elgar Publishing.

[25] Muhamad A. and Thörnberg B. '*High precision laser scanning of metallic surfaces*', International Journal of Optics, 2017, Vol. 2017, Article ID 4134205, https://doi.org/10.1155/2017/4134205.
[26] Lee H. The triple-A supply chain, 2004. [Online]. Available: https://hbr.org/2004/10/the-triple-a-supply-chain#:~:text=Great%20companies%20take%20care%20to,maximize%20only%20its%20own%20interests.
[27] Raslan A., Kapogiannis G., Towey D., Cheshmehzangi A. and Tizani W. '*A framework for assembling asset information models (AIMs) through permissioned blockchain*', 2020 IEEE 44th Annual Computers, Software, and Applications Conference (COMPSAC), 2020, Spain.

Further reading

[1] https://www.scmr.com/article/do_you_have_the_dts_why_its_time_to_ bring_digital_transformation_down_to_ea
[2] https://www.scmr.com/article/do_you_have_the_dts_why_its_time_to_ bring_digital_transformation_down_to_ea
[3] https://www.forbes.com/sites/dinamedland/2016/06/19/collaboration-is-key-to-digital-transformation-so-is-changing-a-mindset/#79ea51ef5b73ccb ukycu
[4] https://sloanreview.mit.edu/article/how-digital-leaders-inspire-engagement/
[5] https://hbr.org/2004/10/the-triple-a-supply-chain
[6] https://sloanreview.mit.edu/article/how-digital-leaders-inspire-engagement/
[7] https://www.scmr.com/article/do_you_have_the_dts_why_its_time_to_ bring_digital_transformation_down_to_ea
[8] https://www.forbes.com/sites/dinamedland/2016/06/19/collaboration-is-key-to-digital-transformation-so-is-changing-a-mindset/#79ea51ef5b73
[9] https://sustainableworldports.org/project/port-of-hamburg-virtual-reality-for-model-based-port-infrastructure-management/
[10] https://www.mckinsey.com/business-functions/organization/our-insights/unlocking-success-in-digital-transformations
[11] https://www.aspiresys.com/sites/default/files/WhitePapers/Multidimensional-Framework-for-Digital-Transformation_0_0.pdf
[12] https://hbr.org/2019/03/digital-transformation-is-not-about-technology
[13] https://bohatala.com/the-history-of-supply-chain-management/

Chapter 11

ICT innovation in port–hinterland transport services: developing cost-effective horizontal integration tools

Valentin Carlan[1]

This chapter discusses the development path of information communication technology (ICT) innovation cases, points out the costs and benefits they generate to maritime supply chain (MarSC) stakeholders, and analyses how ICT providers reach cost-effective outcomes. The research is carried out from the perspective of the ICT providers that have the goal of achieving a high number of users for their applications. The results show that by applying a mixed development approach (bringing together a multidisciplinary team), structurally investing in market update and using a public–private funding source for faster operation, ICT innovation achieves better cost-effectiveness in the supply chain (SC).

11.1 Introduction

Hinterland transport operators act in a highly competitive market and therefore face low profit margins. To ensure profitability, these types of stakeholders leverage on ICT innovation to deliver their services. Yet, the cost of implementing these solutions is perceived as a barrier and benefits are not always proven. To break this vicious circle, ICT providers unite a critical mass of users around one solution to spread development costs and prove the benefits of their solutions to multiple users. This is more so the case with emerging technologies such as blockchain, Internet of Things or artificial intelligence that are expected to be implemented in the next decade and to reshape the operations in the SC [1]. This change (achieving a critical mass of users) will occur when ports and MarSC stakeholders (users) adopt ICT platforms developed at a global scale.

Although important steps have been made and lessons with regard to digital technologies have been previously drawn [2], to achieve the global integration performance, several integrative steps are still to be made. One area where this applies is the port–hinterland interface. Road carriers aim at optimising their

[1]Department of Transport and Regional Economics, University of Antwerp, Antwerp, Belgium

operations and decrease their costs by making the use of data that are commonly shared. More specifically, they use ICT application to list real-time information with regard to orders and communicate this to their dispatchers, who plan the optimal use of their fleet. To carry out this type of optimisation, data need to be integrated with regard to, but not limited to, order details, vehicles' location, structural delays or waiting times. This type of integration creates an ecosystem that brings benefits to every participating road carrier that contributes to this common listing of real-time orders data. Thus, the more users of the same applications exist, the more accurate optimisation actions can be carried out.

Evidence with regard to horizontal collaboration practices and their cost-effectiveness is scattered. This type of collaboration tends to be addressed as horizontal data-exchange practice and refers thus to the data exchange (or integration) between MarSC stakeholders with similar activity scope (e.g. road transport operators, terminal operators and shippers). While this integration is taking place, there is little evidence-based literature on the topic and therefore a lack of elaboration of the benefits brought by such an integration.

In light of the previous, this chapter addresses the cost-effectiveness of data-exchange platforms that change or enable new data communication practices between road transport carriers in the ports' hinterland. This is done by means of a comparative analysis of how different characteristics of the innovation process influence the cost-effectiveness of ICT introduction. This comparative analysis is carried forward by addressing three ICT use cases that enable horizontal integration in the MarSC. The research answers the following main research question: *Which external resource factors can bring more cost-effective outcomes for ICT tools introducing horizontal integration in hinterland road transport?*

To answer this question, this chapter follows a dedicated research approach. Following the introduction, the subsequent section highlights questions emerging from the existing literature, concerning the introduction of ICT-related innovations in the port–hinterland interface. This is then followed by a description of the methodologies used to assess innovation introduction in the transport and logistics sector. This section explains how a cost-effectiveness approach can be used as a method to analyse the outcome of innovation introduction. Equally, it presents the key elements that need to be defined when setting up the framework of a cost-effectiveness analysis (CEA). Following on, the next section presents the case studies and their key characteristics used in the later analysis. The later section looks more in details at the costs, benefits and cost-effectiveness of each case study, providing a discussion on how the characteristics of the innovation paths influenced these results. The final section concludes the chapter.

11.2 ICT innovations in the port–land interface

According to Van de Voorde [3], as the ports' infrastructure reaches its capacity limits, ICT platforms are introduced to assist the activity of stakeholders and help them in daily operations. ICT platforms therefore introduce cooperation practices.

In a later publication, Van de Voorde and Vanelslander [4] highlight that vertical cooperation is one of the emerging trends in the future development of the ports and MarSC. Yet, this observation contrasts the limited research carried out around horizontal cooperation practices. In the later context, no encountered research investigates the success of ICT platforms that enable horizontal data-exchange practices in the MarSC.

Looking further at publications that study port innovation from an economics perspective, the following observations are made. Aronietis *et al.* [5] assess the economic impact of innovative developments. They use the investment costs as the main input element for assessing the impact of innovation solutions. Furthermore, Acciaro [6] notes that the innovations' main objectives are not often linked to cost minimisation. His findings suggest that before reducing the cost of an activity, companies having their activity related to seaports also consider other objectives, such as optimising operations and reducing congestion, of similar or higher importance. Moreover, Vanelslander *et al.* [7] find that a unique 'recipe' for innovation success does not exist. Yet, they show that ICT innovation success can be done by looking at its declared goals and finding a consistent measurement unit to show whether these goals are achieved throughout its life cycle or not.

In many cases, an already developed ICT platform suffers multiple upgrades or generates new development before its foreseen life cycle ends. Sys *et al.* [8] present the example of the e-gate cases or the barge transport system (BTS) at the port of Antwerp. While the initiators of the former opted for a step-wise implementation (where the same fundamental idea is applied for the subsequent processes in the SC and lessons from the previous implementation are added, e.g. extended gateway up to the hinterland depots for import, export and customer warehouse), the initiators of the latter developed extra functionalities over subsequent versions (e.g., the Port of Antwerp BTS v1.0 up to BTS v4.0 [9]). This raises difficulties in indicating the shutdown or replacement time of an ICT system or data-exchange platform. Research shows that the life cycle of an ICT project in the transport sector depends either on its applicability (freight or passenger use) or on the transport mode that it serves (road, rail, water navigation, air or a mix) [10]. Andrae [11] demonstrated that the life cycle of ICT equipment (hardware) is on an average of 5 years, whereas ICT applications (software) are foreseen to be in use for up to 10 years. The previous studies show thus that ICT innovation can have multiple successful factors and is indicated to look over a longer period of time when accounting for its development efforts.

Achieving a critical mass of users within its early life cycle period is a critical factor (besides the more traditional return of an investment element) that points to the success of ICT innovation. For example, cities have developed smart solutions that are continuously improved to answer mobility problems for their inhabitants [12]. This solution's success is relative to the number of users it reaches. Similarly, ports have started to develop Port Community System platforms since the early 1980s to enhance their activity and to increase their users' competitive position [13]. Therefore, the number of users an ICT platform reaches is a relevant measurement of its effectiveness in achieving success. The next section develops further on this theory and shows how the number of achieved users can be used as indicator in cost-effectiveness analyses.

11.3 CEA and its presence in applicative studies

Academia has developed studies that apply several methods that look into the costs and the benefits of innovation. This section frames the CEA and documents it in relation with other types of analysis methods like cost-feasibility, cost-utility, cost–benefit and social cost–benefit. CEA [14–17] is used to analyze the impact of interventions that can be expressed with a key variable for which the evaluator has a quantifiable indicator.

Another method identified is the cost-feasibility analysis [16,18,19]. This method refers to estimating the cost of innovation and evaluates whether it exceeds predetermined boundaries. In contrast to the CEA, the cost-feasibility analysis does not account for the impact of an innovation.

Next, there is the cost-utility analysis that determines which alternative yields a given level of utility at the lowest cost [14,15]. It combines multiple measures of analysis into a single estimate of utility, and it can also weight each measure of outcome. The later use of weight can introduce subjectivity into the results. A popular method of analysis is the cost–benefit analysis that determines whether the benefits of a given alternative outweigh the costs [16,20–22]. This method requires that the benefits of innovation are expressed in monetary values. An extension of this method, being the social cost–benefit analysis, as an extra, introduces societal costs and benefits into the analysis [3,23]. Based on this overview, the CEA is further applied in the context of this research as it is a method that uses two quantitative elements: cost and achieved outcomes (regardless of the cost-feasibility analysis). Equally, it eliminates potential methodological conversion biases by keeping these quantitative elements in their original unit of measurement (regardless of the other cost-utility or cost–benefit analyses that quantify outcomes in monetary units). The next section further details the use of the CEA in transport and logistics-related studies.

A series of journals in transportation were used as the main sources to point out the usage of CEA. For this reason, research publications that had the aim of applying the CEA in practical case studies were analysed and a non-exhaustive literature review was conducted focusing on the 2000–2017 period. Serving the scope of the literature review, four studies conducted before the year 2000 were also considered in the current overview. The main aspects that were investigated are linked with the transport sector for which the CEA was used are as follows: the scope of the analysis, the elements that were considered for analysis, the quantification practice and the practice of how each study puts forward its findings. Observations for each category are detailed in the following subsections.

The literature review shows that the CEA has been applied also to decisions taken in the transport sector. Looking at the rail segment, Mulvey [24] creates a ranking of Amtrak routes that should be modernised by using a CEA. He puts in balance the cost of upgrading the Amtrak service and the number of passengers, who will be encouraged to use the train for long-distance travels. Rufolo [25] creates a CEA taking into account the cost of introduction of articulated buses in comparison with the time gained for boarding of the passengers. Wang [26] calculates the cost-effectiveness of mobile source emission control measures by dividing the total costs by the total emissions that are being reduced. Barkan [27] determines the cost-effectiveness of a new refuelling system by comparing the installation cost of the new equipment with the yearly volume of fuel spared from spilling, but without quantifying the pollution saving.

From reviewing publications that use transport as an applicative field, it is observed that the road transport sector has benefited from a lot of attention from researchers applying CEA. The scope of studies in this case was to prove the effectiveness of different measures either on social or environmental impact [26–28]. Furthermore, the rail sector has benefited from the attention of CEA applications. The review of studies shows that the scope of CEA in this case was mixed, being focused on either social, economic or environmental matters [27,29,30]. Hence, applications of CEA have also been developed to determine the effect of maritime policies on environmental matters. Here is to be mentioned the effect of speed limitations [31,32] or the consequences of policies regarding container repositioning [33] on maritime transport emissions. CEA was not preferred in analysing issues dedicated to air transport. Stewart and Mueller [34] use CEA (as an extension of a wider cost–benefit analysis) to point out the cost-effectiveness of measures taken by several airports against terrorist attacks. Their findings indicate that any additional measures against terrorist attacks would be too expensive to be justified by their effectiveness. They do that by referring to the number of extra attacks that could be avoided through extra safety measures.

The literature review shows finally that the scope of CEA in transport studies is to determine the effectiveness of measures for which a complete estimation of benefits is difficult to determine [17]. For this reason, most CEA are limiting their applicability to a specific topic such as assessing the environmental impact of an investment [26,30,35,36] or quantification of costs for measures addressing social benefits in short term [28]. Researchers applying CEA have quantified either the average cost of emission mitigation measures or actions aimed at decreasing the number of road fatalities to prove the cost-effectiveness of different investments. In some cases, CEA can also use monetary value as outcome indicators; these studies have a purely economic purpose [27,37,38]. The next section presents how the CEA method can be used in analyses of ICT introduction in the MarSC.

11.4 CEA for ICT innovation – an evaluative framework

The following section presents the applicative framework of the CEA for ICT innovation that aims at horizontal data integration in the MarSC. This research follows a four-step approach as shown in Figure 11.1. The analysis is based on information retrieved through desk research and interviews with industry representatives. As highlighted in the next figure, the initial step of this research is to identify, in literature, the use of the cost-effectiveness for innovation assessments in transport and logistics. The later research builds further on the literature and carries

1. Desk research
In-depth documentation of the use of cost-effectiveness analysis in transport and logistics

2. First round of interviews with industry representatives
Identification of general characteristics of ICT innovation cases that enable horizontal data-exchange practices in the MarSC

3. Second round of interviews with industry representatives
Inquiry for data with regard to the expenses and the achievements of ICT innovation

4. Analysis and results
In-depth cost-effectiveness analysis with regard to the expenses and outcomes of ICT innovation enabling horizontal integration in the MarSC

Figure 11.1 Research approach. Source: Own composition

an in-depth analysis on data disclosed through two interview rounds. The interviews follow a semi-structured approach and are carried out in the 2015–16 period. A total of six interviews are carried out with ICT innovation initiators (CEO or project managers of ICT applications with a horizontal data-exchange scope).

The approach aims to reveal the stated preference and the perception of the interviewees over the development process of the case studies. Although this approach might introduce subjectivity in the dataset, the later analysis using these quantitative elements and results is cross-validated with the interviewees for correctness. For each of the ICT innovation cases, a first interview was carried out to seek the general characteristics of each case, followed by a second interview where more in-depth data with regard to the expenses and the achievements (regarding the number of users) of each case were addressed and validated.

The structural design of each round of interviews is detailed in Table 11.1.

The final quantitative assessment of ICT innovation outcomes is done according to a CEA. The cost and outcome elements used next are presented in the following paragraph.

To answer the question addressed by this research, the CEA is used to quantify all necessary expenses done by the innovation developer to reach as many users as possible for its innovation. This quantitative assessment determines a ratio as an indicator of the cost-effectiveness. This ratio is obtained by dividing the cost of developing the innovation by a quantitative element that measures its outcome. A low value of this ratio indicates high success of innovation (naturally assuming that incurring low costs and accomplishing high outcomes are desired). This result is given thus from the perspective of the developing party (IT developing company). The ICT development cost and the number of users reached are the two key quantitative elements used in the analysis.

Table 11.1 *Items addressed through interviews carried with innovation representatives*

First round of interviews	Second round of interviews
• Details with regard to the problem addressed through the ICT innovation • ICT innovation scope, market addressed (type of MarSC stakeholders) • The use of external knowledge • Details with regard to the development approach • The funding strategy carried out • Revenue strategy	• The expenses incurred to develop the innovation (type, period, value, etc.) • The type of benefits brought to the MarSC stakeholders • Qualitative elements to measure the success of innovation • The targeted and the achieved outcomes

Source: Own composition.

The ICT development cost usually refers to the following categories of expenses: ICT equipment, communication equipment and software development. In detail, the first two cover the acquisition of equipment and its related software used in production. The latter includes the acquisition of pre-packaged software, customised software and/or algorithms developed in-house [39,40] as part of the innovation process.

The total costs that are taken into account in this assessment are defined by Carlan *et al.* [13] as ICT platform development costs. The types of costs referred to as C_j^{tot} are as follows:

- **Capital costs:** These are investment costs (initial costs) that are incurred on the development of a new ICT platform. In some cases, these elements can be assimilated with costs for the acquisition of developer rights and.
- **Service or operating costs:** These include personnel training cost, employment of new personnel, regular acquisition of new equipment, market uptake, hosting and cost of personnel operating the platform (where applicable).

Determining which outputs are representative for measuring the innovation outputs is an important step in applying CEA. For this reason, this research reports the financial effort of developing ICT innovation to the number ***of users that it reaches***. This second CEA element is referred to by this research as B_j.

Further, the cost-effectiveness ratio is calculated for each innovation j. This ratio is calculated by dividing the total costs C_j^{tot} by the innovation outcome B_j [41,42]. The output of the cost-effectiveness ratio is defined as the cost of achieving one unit of the innovation outcome.

$$E_j = \frac{C_j^{tot}}{B_j} \tag{11.1}$$

The elements used to calculate the cost-effectiveness ratio in the case of a company developing an ICT platform are thus defined as follows:

- C_j^{tot} – total cost of the j ICT platform development. If the development period is spread over a long period, the investment costs incurred each year are discounted and actualised to the assessment date.
- B_j – quantification of the innovation's outcome (for innovation j).

Following up on the previously exposed methodology, the next section presents the case studies addressed in the empirical analysis.

11.5 Horizontal integration in road transport: case studies

This section presents ICT innovation cases that enable horizontal data-exchange practices (or integration). From this category of innovation, the development process and the outcomes of three ICT innovation initiatives in the MarSC are described.

The selection of these three innovative concepts is determined by data availability. Moreover, due to confidentiality reasons, the three application cases will be referred to as ICT innovation cases 1, 2 and 3, respectively. These applications are similar in scope as they address the transport of cargo linked to the port of Antwerp, enabling a comparative analysis. Other cases can be used in same comparative analysis, but then one should be aware of the limitations set by the geographical use of these applications (the port of Antwerp), the time frame (the first 3.5 years of live activity of each case) and the period for which data have been in use (recall the period between 2009 and 2016).[*]

The **first platform** was initiated by a trucking company. This ICT application brings together road transport operators from the port area that have as scoped the haulage of cargo to and from the hinterland area. Their goal is to complete these transport tasks as effectively as possible. This first ICT tool provides the transport operators with a virtual space where they can communicate and agree on consolidating their tasks. Within this business framework, the selection and the decision regarding the consolidation of transport tasks still remains at the will of each platform user (road transport operators). Moreover, this platform enables companies to exchange information in order to reduce the transport of empty containers. It started as a tool that optimised the transport tasks of one road transport operator.

Undoubtedly, its development required the contribution of an ICT developing entity that has been involved in the process as internal knowledge capability. The development pattern of this application has followed a bottom-up approach. After gaining success at the level of the developer company, it attracted the interest of several other road transport operators and the access was then opened for third parties.

[*]There is potential of expanding the present analysis by carrying out one of the following approaches. On the one hand, one could bring extra cases in the same analysis framework by using to the same conditions on scope, time and period. On the other, the same method of analysis can be applied for cases that do not meet these criteria, yet a new analysis framework needs to be redefined.

Access to its services is based on a monthly subscription fee and leaves the users to pick and combine their own transport tasks. This brings the innovation in the category that refers to horizontal integration in the MarSC.

The second case study was initiated by a software developer company. This seaport ICT application matches full transport orders with planned empty trips of containers. Through this application, shipping line operators (or container owners) can communicate with land operators and integrate their transport necessities with the unused land transport capacities. By using this platform, the unproductive transport of empty containers is being capitalised on, so that it brings benefits to all the involved parties. An independent software developer saw that this application as an opportunity and developed it for the market. The developer started by answering the need of shipping line operators and container owners to use their equipment (containers) more efficiently. This application finds opportunities to reuse empty containers that need to be relocated. In this case, the user access is enabled through a pay-per-use fee. This way, the goals of the ICT application administrator remain directly linked with the user's interests.

The third case refers to an initiative started to help SC stakeholders manage their services. This innovation addresses the issue of information visibility along the SC. It integrates different SC stakeholders in one communication network. It starts from the premise that each participant in an SC has fragmented knowledge about the real-time status of a (transport) order. As the goods proceed along the SC, information is automatically delivered to interested stakeholders. This innovation addresses issues such as relatively long dwell time for data, lack of a standard communication tool among different stakeholders, the existence of dependencies in an SC and the absence of an infrastructure for real-time data communication. The success of this innovation depends on achieving a critical mass of users. Looking at an SC, for the system to be effective, ideally all the stakeholders need to adhere to it.

The outcome and usage of these platforms are protected by a sign-in interface making this innovation's outcome semi-open. Hence, the developer keeps full control of the innovation's outcome. Further, this research also shows that the development pattern, characteristics and applied business strategy of these innovation cases differ. These distinctions are brought into a comparison that investigates the link between the characteristics of the innovation path followed and its outcome.

Table 11.2 summarises further the main development characteristics of the three ICT platforms.[†]

As seen in Table 11.2, there is no common development recipe followed by the three ICT innovation cases. Two of them have used external knowledge, while one has used a completely closed process. The development approach was different for each case as well. The monthly subscription is used as a revenue strategy in two of the cases, and only innovation case 2 applies a pay-per-use fee. Private funding is preferred as a source by the first two innovation cases and the last one has used a

[†]The names of the ICT platforms are known by the author and kept anonymous for confidentiality reasons.

Table 11.2 Characteristics of ICT case studies

	Use of external knowledge (open/closed innovation process)	**Development approach**	**Revenue strategy**	**Funding source**
Innovation case 1	No (closed process)	Bottom-up	Monthly subscription	Private
Innovation case 2	Yes	Top-down	Pay-per-use fee	Private
Innovation case 3	Yes	Mix	Monthly subscription	Public–private

Source: Own compilation based on interviews.

mixed public–private funding capital. Hence, a further comparative analysis regarding the costs, benefits and cost-effectiveness generated by these cases is presented in the next section.

11.6 The costs, benefits and cost-effectiveness of horizontal integration

The costs and the benefits are the main factors that determine the successful implementation of ICT innovation [43]. First insights with regard to the type of cost and benefits are given. Later, the cost-effectiveness is calculated from the perspective of the ICT developing companies that target the achievement of a wide mass of users.

To further detail the context of this analysis, the achieved outcomes are shown through applying the third step of the research approach, carrying a second round of interviews with SC stakeholders. This section makes the use of the framework developed by Carlan *et al.* [13]. It provides an aggregated list of the cost and benefit elements generated by the three studied cases. Moreover, the stakeholder types involved in each of the innovation cases as users are given as well in Table 11.3 as follows: shipping line operators, shipping agents, road transport operators and ICT developing companies.

Table 11.3 identifies the cost and benefit elements generated by the three innovation cases used in the framework of this analysis. The type of cost or benefit elements generated by the first case are marked with '1', '2' indicates the stakeholders that incurred those cost and benefit elements for the second case and '3' for the third.

From this table, it is clear that the ICT developing entities accounted for the development and maintenance costs of these innovation initiatives. From a user perspective, for each case, the road transport operators made investments in hardware or software acquisition. Cost elements for training and setting up a connection were also incurred by a shipping line operator or a shipping agent, respectively.

Table 11.3 The costs and benefits generated by three ICT innovation cases introducing horizontal integration

		ICT users			ICT developing company
		Shipping line operators	Shipping agents	Road transport operators	
Cost elements					
Operational cost					
Co1	Platform development	2	3	1	**1,2,3**
Co2	Market uptake				**1,2,3**
Co3	Operating costs and maintenance				**1,2,3**
User cost					
Cu1	Connection cost		3	1	
Cu2	Hardware/software investment costs	2	3	**1,2,3**	
Cu3	Training cost	2		1,3	
Cu4	Transactions fees (license fees)			2	
Benefit elements					
Digital transformation benefit					
Bu1	Reduce cost of information access/receive	2	3	**1,2,3**	
Bu2	Reduce cost of information send	2	3	**1,2,3**	
Bu3	Avoid cost of illicit transactions and/or their prejudice				
Bu4	Avoid cost for error correction		3		
Bu5	Avoid cost of information matching	2	3	2	
Bu6	Avoid cost of information search	2	3	**1,2,3**	
Bu7	Gain from efficient use of resources	2	3	**1,2,3**	
Community attendance benefit					
Bu8	Reduce cost to access broad (external) information		3	**1,2,3**	
Bu9	Gain from added-value services	2			
Bu10	Avoid cost for compliance with standard and regulation		3		

Source: Own composition.

From the overview of benefit elements brought by these innovation initiatives, Table 11.3 shows that all cases not only support receiving, sending or searching information but also improve the resource use for road transport operators. While one case offers similar benefits to shipping line operators alone, the other brings these benefits to shipping agents. The further framework of analysis is given from the perspective of the ICT developing company.

As the earlier results show, ICT developing companies have incurred the development costs of the three cases analysed. Moreover, the qualitative information obtained during the first round of interviews shows that user benefits are gained only if a wide mass of users is reached. The latter is confirmed as well by each ICT developing company and the achievement of users is unanimously pointed out as a key preliminary goal of each innovation case. The following section presents thus a CEA in this framework, where numerical data are further collected, first for the development costs and second for the reached number of users (where the latter is used as a proxy for preliminary benefits of ICT introduction).

11.7 ICT innovation breakthrough costs

This section presents a comparative analysis of the three cases and examines the costs associated with each. In order to enable a reliable comparison, the analysis looks into the expenses accounted by each case within 3.5 years of activity. This period is the longest duration in-service for one of the ICT innovations in this analysis and it is taken as reference (i.e. ICT case 1 – 8 years in operation; ICT case 2 – 4.5 years in operation and ICT case 3 – 3.5 years in operation). Therefore, data are centralised for this period with regard to expenses on software development, market uptake and operation and maintenance. These values are then discounted to year 1 using a discount rate of 5% [44]. Further, these values are indexed to the value of the lowest expense identified in the dataset. The latter is attributed with the value 100 and the others are related to it. This practise ensures that the confidentiality of this type of data is respected.

Figure 11.2 presents the indexed cost carried in the account of each case. The comparison of the cost leads to the following remarks. First, with regard to the platform development cost, ICT case 3 has the lowest platform development cost which is incurred in the first year of development, while ICT case 2 has the highest development cost (50 times more). This difference shows that ICT case 3 entered quickly in operation, while case 2 carried more development before the market uptake. This type of cost shows no other significant differences for neither of the cases in the following years. Second, ICT cases 2 and 3 dedicate equal amounts each year to market uptake expenses. Third, Figure 11.2 shows that the highest operation and maintenance costs are generated by the third ICT case. Lastly, ICT case 3 makes the most expenses from the three analysed cases. The ICT innovation case that is most successful is determined in relation with the outcome it achieves.

A further comparison of these three ICT innovation cases is made from the perspective of a CEA. This analysis applies the formula presented in Section 11.3.

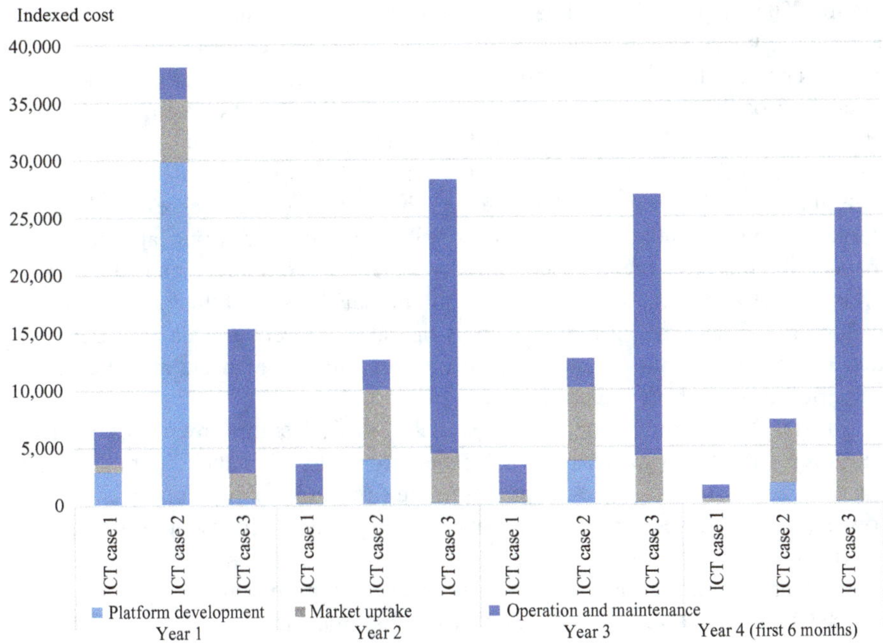

Figure 11.2 Indexed cost of each ICT innovation case. Source: Own composition

Table 11.4 CEA outcome for innovation access

	ICT platform		
ICT innovation outcome	**ICT case 1**	**ICT case 2**	**ICT case 3**
Total costs/active users [Euro/user]	2,043	491	197

Source: Own calculation.

It uses the nominal values of cost and reports them to the number of active users reached in the first period that the innovation is active (e.g. first 3.5 years of activity). The number of active users is put forward and agreed by the interviewees as being the proxy measuring the success of the ICT innovation cases. The result is referred to as the cost-effectiveness ratio. The lower the ratio the more a cost-effective innovation outcome is obtained. The final outcome of the CEA is given in Table 11.4.

Table 11.4 shows that ICT innovation case 3 has the most cost-effective outcome. This result is approximately 10 and 2.5 times better than the results of cases 1 and 2, respectively.

The data presented in the previous sections are used to formulate a comparative analysis between the three presented innovation cases and interpret the CEA outcome. The characteristics of each innovation development path and the achieved outcome are visualised in Figure 11.3.

The characteristics shown with regard to the development path are the development approach, the use of knowledge, the type of user fee chosen, the development stakeholder and the market uptake expenses relative to the total expenses. Equally, the CEA outcome shows which case has the most cost-effective outcome.

Figure 11.3 shows that the initiator of the ICT innovation case 1 did not use external knowledge but followed exclusively a private bottom-up approach. The expenses in this case are focused initially on the development and later on the market update and operation. The stakeholders that developed this case are an ICT developing company and a road transport operator. Moreover, this case generated much less costs than the other cases; hence, the number of users achieved in the first period of operation is also low.

The IT developing company of case 2 followed a top-down approach and brought up a relatively wider range of benefit types to SC stakeholders. The initiating stakeholder invests in development in the first active period, involves external knowledge and choses for a pay-per-use pricing model. It also keeps the same investment rate for operation in the following period. Furthermore, it relies exclusively on private capital and spends a relatively high amount on market uptake. This case manages to reach much more users than the first case and delivers better success with regard to its cost-effectiveness.

Similar to innovation case 1, case 3 applies a monthly subscription fee for its revenue strategy. Moreover, it follows a similar development pattern as innovation case 2

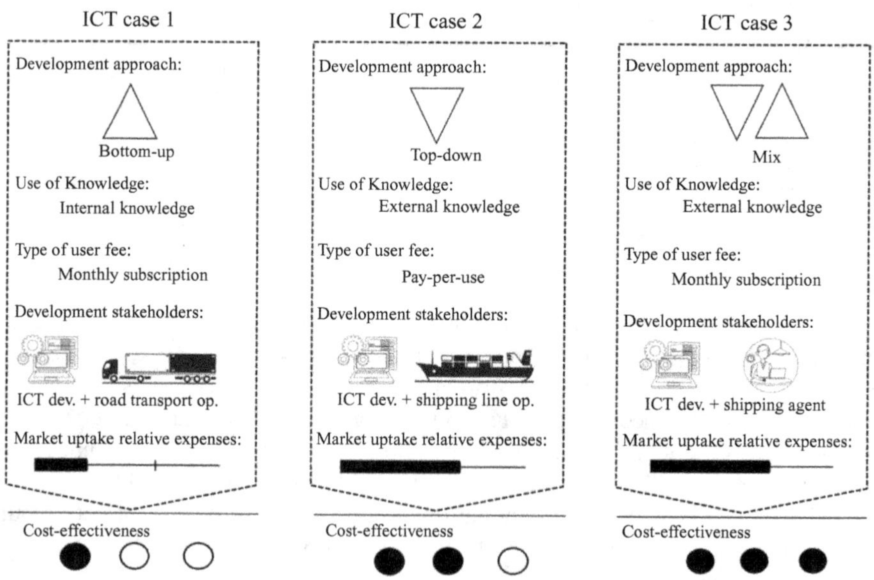

Figure 11.3 The characteristics and relative cost-effectiveness of the three ICT innovation cases

where external knowledge sources have been used and equal market uptake costs are structurally made. As main differences, however, it is noticed that innovation case 3 has followed a mixed development approach, was initiated through private capital and that it later benefits from further public subsidies (ending in a public–private funding structure). This development path mixes ICT expertise with the knowledge gained from the SC stakeholders during early development. The developed outcome is thus flexible enough to be further commercialising towards other actors. By using these two approaches, the developers of the third ICT innovation case, although make higher market uptake costs, achieve higher success with a better cost-effectiveness. The next section wraps up this research with concluding remarks and suggestions for further research.

11.8 Conclusions

Companies from all business sectors innovate in order to stay competitive. According to the existing literature, investing in innovative ICT concepts is a trend that is encouraged by contemporary business managers. This chapter presents ICT innovation as it is currently developed by IT developing companies for MarSC stakeholders active in the ports' hinterland. The analysis of ICT innovation cases similar in scope was pursued to underline the influence of the external resource factors on innovation outcome. With this as a target, a comparative analysis was carried out. This analysis used the characteristics of three ICT case development paths and the outcome from a CEA to determine how a more successful ICT innovation that aims at horizontal integration in the SC can be achieved.

An initial step of this research was to examine the characteristics of the development paths of each of the three studied cases. The research looked into the type of costs and benefits generated by MarSC stakeholders. From this overview, it was observed that the benefits of SC stakeholders are scattered, and while one application might generate a limited range of benefits, the other brings wider advantages. This type of outcome was linked to the expertise of the innovation initiators and their capacity to deliver flexible innovation. Equally, it is observed that IT developing companies are carrying the same types of costs for developing innovation.

The analysis then presented the differences in expenses made for development, market uptake and operation and maintenance. While the development and operation and maintenance costs point to different expenses paths, the market update cost shows that the three cases have followed a structured approach to gain users.

The results of the analysis show that innovation case 3 has achieved a better cost-effectiveness. This result was explained by the characteristics of the development path. Hence, it shows that, by using sector knowledge and setting public–private financial resources on enabling early operation, ICT developing companies achieve a more flexible product. Likewise, the structured market uptake expenses help innovation developers reach their users. By applying a mixed development approach, using a public–private funding source and following a structured market uptake expense path helps ICT innovation in the SC to achieve higher success in terms of cost-effectiveness.

A limitation of this study is the number of case studies used in the quantitative assessment. So, caution with generalisation is needed. This limitation was imposed by data availability. Therefore, further research is suggested in extending the number of ICT cases and in collecting data with respect to innovation's costs. In parallel, additional information over the innovation outcome is also necessary. Another follow up research direction that can be perused is the identification of aggregated benefits or full cost–benefit analysis of ICT innovation in SCs. Nonetheless, the results of this research are equally relevant for policy makers and innovation enablers to stimulate collaborative future developments. Equally, this method is transferable to bring up insights of other innovation and ICT cases focused on data-exchange practices between other types of MarSC stakeholders.

Acknowledgements

This research was developed with the financial support of the BNP Paribas Fortis Chair on Transport, Logistics and Ports at the University of Antwerp, to which I am grateful. I also would like to show my appreciation, for sharing their professional experience and knowledge, to Prof. Dr Christa Sys and Prof. Dr Thierry Vanelslander.

References

[1] R. Franklin and W. Hofman, "Blockchain or Distributed Ledgers: Limitations and Realities," Working paper of the ETP ALICE group (Alliance for Logistics Innovation through Collaboration in Europe), 2018.

[2] A. Calatayud, V. Carlan, C. Sys, and T. Vanelslander, "Digital Innovation in Maritime Supply Chains: Experiences from Northwestern Europe," Inter-American Development Bank, IDB, 2018, doi: 10.18235/0001070.

[3] E. V. D. Voorde and W. Winkelmans, "A general introduction to port competition and management," De Boeck Ltd., Antwerp, 2002.

[4] E. Van de Voorde and T. Vanelslander, "Trends in the Maritime Logistics Chain : Vertical Port Co-operation: Strategies and Relationships," Brussels: University Press Antwerp, 2014, pp. 121–140.

[5] R. Aronietis, E. Van de Voorde, and T. Vanelslander, "Competitiveness determinants of some European ports in the containerized cargo market," *Proceedings of the BIVEC-GIBET Transport Research Day*, 2011.

[6] M. Acciaro, "Innovation in the maritime sector: Aligning strategy with outcomes," *Research outcomes within the BNP Paribas Fortis chair of Transport, Logistics and Ports*. University of Antwerp, 2015.

[7] T. Vanelslander, V. Carlan, and C. Sys, "Innovation among seaport operators: A QCA approach for determining success conditions," International journal of transport economics, vol. 43, no. 3, pp. 291–314, 2016.

[8] C. Sys, T. Vanelslander, A. Roumboutsos, G. Giuliano, and M. Acciaro, "Port Related Innovation: The Answer to Today's Constraints and Challenges in Seaports Related Operations," Antwerp: University of Antwerp, 2016.

[9] F. Feng, Y. Pang, and G. Lodewijks, "Integrate multi-agent planning in hinterland transport: Design, implementation and evaluation," *Advanced Engineering Informatics*, vol. 29, no. 4, pp. 1055–1071, 2015, doi: 10.1016/j.aei.2015.08.001.

[10] Y. Tyrinopoulos, A. Kortsari, and M. Chatziathanasiou, "11. A critical overview of ICT deployment in transport in Europe," *ICT for Transport: Opportunities and Threats*, Cheltenham, UK: Edward Elgar Publishing, 2015, p. 249.

[11] A. S. G. Andrae, "European LCA standardisation of ICT: Equipment, networks, and services," in Towards Life Cycle Sustainability Management, M. Finkbeiner, Ed. Netherlands: Springer, 2011, pp. 483–493.

[12] M.-P. Pelletier, M. Trépanier, and C. Morency, "Smart card data use in public transit: A literature review," *Transportation Research Part C: Emerging Technologies*, vol. 19, no. 4, pp. 557–568, 2011.

[13] V. Carlan, C. Sys, and T. Vanelslander, "How port community systems can contribute to port competitiveness: Developing a cost–benefit framework," *Research in Transportation Business & Management*, 2016, doi: 10.1016/j.rtbm.2016.03.009.

[14] R. Sugden and A. Williams, "The Principles of Practical Cost-Benefit Analysis," Oxford University Press, OUP Catalogue, 1978. Accessed: Dec. 07, 2015. [Online]. Available: https://ideas.repec.org/b/oxp/obooks/9780198770411.html.

[15] R. Robinson, "Cost-benefit analysis," *BMJ*, vol. 307, no. 6909, pp. 924–926, 1993, doi: 10.1136/bmj.307.6909.924.

[16] H. M. Levin and P. J. McEwan, "Cost-effectiveness analysis: Methods and applications," vol. 4, Sage Publications, Inc., Thousand Oaks, CA, 2001.

[17] M. Išoraite, "Evaluating efficiency and effectiveness in transport organizations," *Transport*, vol. 20, no. 6, pp. 240–247, 2005, doi: 10.1080/16484142.2005.9638027.

[18] S. J. Murdoch and G. Danezis, "Low-cost traffic analysis of Tor," *2005 IEEE Symposium on Security and Privacy*, May 2005, pp. 183–195, doi: 10.1109/SP.2005.12.

[19] Q. Chen, Z. Tang, Y. Lei, Y. Sun, and M. Jiang, "Feasibility analysis of nuclear–coal hybrid energy systems from the perspective of low-carbon development," *Applied Energy*, vol. 158, pp. 619–630, 2015, doi: 10.1016/j.apenergy.2015.08.068.

[20] R. Layard and S. Glaister, "Cost-Benefit Analysis," Cambridge University Press, UK, 1994.

[21] H. A. Nijland, E. E. M. M. Van Kempen, G. P. Van Wee, and J. Jabben, "Costs and benefits of noise abatement measures," *Transport Policy*, vol. 10, no. 2, pp. 131–140, 2003, doi: 10.1016/S0967-070X(02)00064-1.

[22] A. E. Boardman, D. H. Greenberg, A. R. Vining, and D. L. Weimer, "Cost-Benefit Analysis: Concepts and Practice, 3rd Edition," 2006, Accessed: Mar. 10, 2016. [Online]. Available: http://works.bepress.com/david_weimer/4/.

[23] K. De Langhe, "The importance of external costs for assessing the potential of trams and trains for urban freight distribution," *Research in Transportation*

Business & Management, vol. 24, pp. 114–122, 2017, doi: 10.1016/j.rtbm.2017.07.002.

[24] F. P. Mulvey, "Amtrak: A cost-effectiveness analysis," *Transportation Research Part A: General*, vol. 13, no. 5, pp. 329–344, 1979, doi: 10.1016/0191-2607(79)90031-1.

[25] A. M. Rufolo, "Cost effectiveness of articulated buses when passenger time is treated as a cost," *Transportation Research Part A: General*, vol. 20, no. 5, pp. 337–343, 1986, doi: 10.1016/0191-2607(86)90001-4.

[26] M. Q. Wang, "Examining cost effectiveness of mobile source emission control measures," *Transport Policy*, vol. 11, no. 2, pp. 155–169, 2004, doi: 10.1016/j.tranpol.2003.10.003.

[27] C. P. L. Barkan, "Cost effectiveness of railroad fuel spill prevention using a new locomotive refueling system," *Transportation Research Part D: Transport and Environment*, vol. 9, no. 4, pp. 251–262, 2004, doi: 10.1016/j.trd.2004.02.002.

[28] L. I. Rizzi, S. Cumsille, F. Fresard, P. Gazmuri, and J. C. Muñoz, "Cost-effective measures for reducing road fatalities in the short term," *Transport Reviews*, vol. 31, no. 1, pp. 1–24, 2011, doi: 10.1080/01441641003736572.

[29] D. M. Z. Islam, K. Laparidou, and A. Burgess, "Cost effective future derailment mitigation techniques for rail freight traffic management in Europe," *Transportation Research Part C: Emerging Technologies*, 2015, doi: 10.1016/j.trc.2015.06.017.

[30] J. M. Matute and M. V. Chester, "Cost-effectiveness of reductions in greenhouse gas emissions from High-Speed Rail and urban transportation projects in California," *Transportation Research Part D: Transport and Environment*, vol. 40, pp. 104–113, 2015, doi: 10.1016/j.trd.2015.08.008.

[31] P. Cariou and A. Cheaitou, "The effectiveness of a European speed limit versus an international bunker-levy to reduce CO_2 emissions from container shipping," *Transportation Research Part D: Transport and Environment*, vol. 17, no. 2, pp. 116–123, 2012, doi: 10.1016/j.trd.2011.10.003.

[32] J. J. Corbett, H. Wang, and J. J. Winebrake, "The effectiveness and costs of speed reductions on emissions from international shipping," *Transportation Research Part D: Transport and Environment*, vol. 14, no. 8, pp. 593–598, 2009, doi: 10.1016/j.trd.2009.08.005.

[33] D.-P. Song and J.-X. Dong, "Effectiveness of an empty container repositioning policy with flexible destination ports," *Transport Policy*, vol. 18, no. 1, pp. 92–101, 2011, doi: 10.1016/j.tranpol.2010.06.004.

[34] M. G. Stewart and J. Mueller, "Cost-benefit analysis of airport security: Are airports too safe?" *Journal of Air Transport Management*, vol. 35, pp. 19–28, 2014, doi: 10.1016/j.jairtraman.2013.11.003.

[35] K. A. Small and S. J. Frederick, "Cost-effectiveness of emissions control strategies for transit buses: The role of photochemical pollutants," *Transportation Research Part A: General*, vol. 23, no. 3, pp. 217–227, 1989, doi: 10.1016/0191-2607(89)90004-6.

[36] D. Zhu, H. D. Kuhns, J. A. Gillies, V. Etyemezian, S. Brown, and A. W. Gertler, "Analysis of the effectiveness of control measures to mitigate road dust emissions in a regional network," *Transportation Research Part D: Transport and Environment*, vol. 17, no. 4, pp. 332–340, 2012, doi: 10.1016/j.trd.2012.02.003.

[37] H. Klemick, E. Kopits, A. Wolverton, and K. Sargent, "Heavy-duty trucking and the energy efficiency paradox: Evidence from focus groups and interviews," *Transportation Research Part A: Policy and Practice*, vol. 77, pp. 154–166, 2015, doi: 10.1016/j.tra.2015.04.004.

[38] W. Liu, Y. Yin, and H. Yang, "Effectiveness of variable speed limits considering commuters' long-term response," *Transportation Research Part B: Methodological*, vol. 81, no. Part 2, pp. 498–519, 2015, doi: 10.1016/j.trb.2014.12.001.

[39] OECD, "Information and Communication Technology (ICT) – ICT Investment – OECD Data," OECD, 2011. http://data.oecd.org/ict/ict-investment.htm (accessed Mar. 28, 2017).

[40] V. Spiezia, "ICT investments and productivity: Measuring the contribution of ICTs to growth," *OECD Journal. Economic Studies*, vol. 2012, no. 1, p. 199, 2012.

[41] European Commission, "Guide to cost-benefit analysis of investment projects," *Policy*, vol. 2014, p. 2020, 2014.

[42] European Commission, "Cost-effectiveness analysis. Evaluation methods," *Policy*, vol. 2005, 2005, Accessed: Nov. 27, 2015. [Online]. Available: http://ec.europa.eu/europeaid/evaluation/methodology/.

[43] A. Roumboutsos, "Comparative analysis of cases: Systems' innovation approach." *BNP Paribas Fortis Innovation Event*, Antwerp, 2015.

[44] R. Bojanc and B. Jerman-Blažič, "Towards a standard approach for quantifying an ICT security investment," *Computer Standards & Interfaces*, vol. 30, no. 4, pp. 216–222, 2008, doi: 10.1016/j.csi.2007.10.013.

Chapter 12

Decarbonisation technologies in shipping and the question of transition fuels

Natalie Gupta[1]

12.1 Introduction

This chapter examines different decarbonisation scenarios in shipping with a view of discussing some of the core technologies enabling the transition towards zero-emission propulsion for vessels. This is significant partly as a result of the lack of consensus, concerning the role played by transition fuels, in the transition to zero-carbon fuels by 2050 and the heavy focus on regulation to achieve this. A review of the different perspectives concerning the role of transition fuels in the changeover to a zero-carbon operation for commercial vessels shows that in addition to the question of the role of regulation (e.g. mandatory decommissioning of old vessels, marine fuel tax) to achieving the transition to zero-carbon fuel types, the timing of the commercialisation of operational efficiency measures and technologies are fundamental as well, and there is still uncertainty over this. Unlike regulation, which is mainly understood in terms of economic impact and challenges concerning implementation, so-called transition technologies also have the challenges of demonstrating the ability to be safe and to be effective in terms of operationalisation. Financial returns on the investments are also a fundamental factor particularly for vessel owners and ports. An example of this is the focus on the role to be played by transition fuels which are considered already tested and fit for use, and while serving a core role in the pathways towards decarbonisation of the industry, at the same time are not yet seen to fulfil the full requirements set out by a complete decarbonisation of the industry.

This raises the broader point about the speed of the transition pathway more generally and the fact that the type of fuel used needs to be considered in the context of a much broader ecosystem of requirements that the industry has. The fundamental challenge that the industry faces in respect to transition fuels is not being able to answer the question of how fast these will be replaced by zero-carbon fuels, either by legislation or market forces or both. This, in turn, weighs heavily on the feasibility of making the required investments in such technologies today, and

[1]Port Processes Ltd, London, UK

of infrastructure such as bunker hub requirements. In fact, it also weighs heavily on the question of whether or not existing targets will be met. In view of these challenges, the purpose of the chapter is to review the key areas of disagreements in the debate and the role of some key greening technologies discussed in this context in enabling the transition to zero-carbon shipping.

The chapter discusses some of these challenges in relation to transition fuels and the extent to which the existing and future technologies being developed are able to address the regulatory and commercial challenges of the industry under conditions of a decarbonisation path. To achieve this, the chapter is structured as follows: the first part of the chapter provides a brief commentary on the historical context of the evolution of transition fuels. This provides a historical perspective of the development of the technologies around alternative fuels over time and therefore how existing pathways to decarbonisation relate to past thinking about the issue.

The following section provides an overview of the current emission targets set by the International Maritime Organisation (IMO) and other international organisations for the maritime sector. This is important to frame the discussion and also understand how significant and ambitious the targets currently set actually are, particularly for an industry that has in the past been very slow to change compared to what are arguably more dynamic service sectors such as aviation. This is in fact changing very rapidly but is a recent phenomenon.

The third part of the chapter translates the previous internationally set targets into scenarios developed by a number of reports on the topic. The scenarios recently developed by DNV-GL in their 2020 forecast for the maritime sector will be discussed as an example of the lack of certainty over the role played by transition fuels.

The fourth part of the chapter moves on to discuss three areas of disagreement or debate around the role played by transition fuels to meet these targets. The first is the role of scrubbers. The second is the role of liquefied natural gas (LNG) as a transition fuel. The third is the viability of other types of fuels in the future. The final part of the chapter concludes.

12.2 Historical pathways to alternative fuels

For many decades, the base fuel for commercial shipping has been fuel oil. Fuel oil replaced coal between First World War and Second World War. This transition was a fundamental one for the industry and was not immediate but took place over a number of years.

A paper by the authors in [1][*] discusses this in the context of emissions calculations. Based on the information drawn from historical accounts and United Nations (UN) and International Energy Agency data, the authors record that coal consumption for the commercial fleet peaked in 1913 and dominated up until 1920,

[*]https://agupubs.onlinelibrary.wiley.com/doi/epdf/10.1029/2006JD007630

which is the time when steamers replaced sail ships. Gradually, newbuilds, expanding the global fleet, were equipped with shift to diesel engines and oil-fired steam boilers, and therefore also the move towards the use of residual oil for propulsion. This explains the gradual reduction of coal as a means of propulsion and a shift towards fuel oil, diesel oil and distillates, which is what we currently have still today.

That said, the full replacement took many decades and was driven by a number of factors. One includes the broader trends in world trade, which clearly influence the speed of the renewal of the global fleet; second, the nature of the installed engine systems in newbuilds was a fundamental factor in the transition. The advent of modern engines and other technologies was extremely important in cutting global emissions stemming from international commercial shipping.

Moving on to more recent times, while technology remains a powerful enabler of change, it is also important to note that it is regulation that is increasingly seen to be the driver of change. This is particularly the case with respect to mitigation measures to achieve the reduction of sulphur oxides (SO_x), nitrogen oxides (NO_x) and carbon dioxide (CO_2) from international shipping (discussed further in the next section). The majority of the world fleet is currently running on residual fuel, now of the low sulphur type since January 2020 when the so-called global sulphur cap was implemented.

Yet regulations concerning emission make it very important for vessel owners to clearly understand what engines to invest in for their fleet and given the fast-paced nature of the technologies around alternative marine fuels, it is important to record the challenges that vessel owners face in making these decisions. This is particularly the case when it comes to the main engines of large vessels, which according to the International Council on Clean Transportation (ICCT) accounts for the majority of emissions in the sector.[†] This is in large part independent of the service provided by the vessel itself (e.g. dry bulk, containers), leaving the decision-making complex.

12.3 Decarbonisation targets for the sector

The IMO is the leading intergovernmental organisation setting climate-change-related targets for the maritime sector as a whole. It is leading in that its members are nation states, including both developed and developing countries, and it is global in scope. For this reason, the IMO targets can be seen as the key targets that the maritime sector, including ports and shipping, make reference to. These targets, in turn, refer to more global targets such as those set by broader inter-sectoral agreements.

For the purpose of this chapter, we focus on the goals set by the IMO as well as note the role played by the European Union (EU) and multilateral development banks (e.g. the European Bank for Reconstruction and Development – EBRD, the

[†]ICCT [2].

International Finance Corporation), which play an important role in infrastructure financing in developing countries. These development banks are not only a large source of financing for the sector but they also play a very big role in pushing through the greening agenda in the sector.

12.3.1 IMO targets

The relevant activities of the IMO, concerning combatting emissions in the sector, include pushing the principles of Article 2 (2) of the Kyoto Protocol, which states that the parties included in Annex 1 will pursue limitations or reduction of emissions of greenhouse gases (GHGs) not controlled by the Montreal Protocol from aviation and marine bunker fuels. In addition to this, the scope of the IMO framework includes the United Nations Convention on the Law of the Sea Convention and The International Convention for the Prevention of Pollution from Ships (MARPOL). We specifically focus on MARPOL.

The subject of MARPOL Annex 4 is the prevention of air pollution from ships. CO_2 is considered harmful under MARPOL. The scope of the IMO framework also extends to market-based mechanisms, namely economic mechanisms for mitigation. The key target that is commonly quoted and provides the framework for everything else is the commitment to reducing GHG emissions by 50% from shipping by 2050 compared to 2008 emissions with a strong focus on reaching zero emissions soon after that.

The IMO's work on the reduction of GHG emission is ongoing and it is often difficult to keep pace. As of the writing of this chapter, the international working group on the reduction of the GHG from international shipping met in October 2020 to develop short-term measures to reduce the carbon intensity of existing fleet by 40% by 2030 compared to 2008. The result of the work is proposed amendments to MARPOL Annex 6, to be presented to the International Panel on Climate Change for adoption in 2021. These measures combine a new Energy Efficiency Existing Ship Index addressing the technical performance of the ship and measures to address the operational carbon intensity of ships based on indicators. These are short-term measures but the key thing to note is that they are binding. This comes alongside the IMO emission targets which, if understood in the context of decarbonisation pathways, tend to fall between what tend to be modelled as 'do-nothing scenarios' on one hand, and very ambitious scenarios on the other. There are to achieve 50% emission reductions by 2050 (from 2008 levels), and to cut these further by 2070 (Figure 12.1).

The strategy laid out to achieve the IMO pathway includes a number of measures. Efficiency gains by the global fleet are a very important area for this and the transition from existing fuels to zero-emission fuels and electricity (or others, for instance wind) as means of vessel propulsion. Some key ones are listed as follows:

- Different construction materials of vessels particularly refer to using lighter weight materials that would, in turn, make vessels less heavy and therefore enable the use of less fuel consumption. According to the OECD, this could contribute up to 10% of fuel consumption [3].

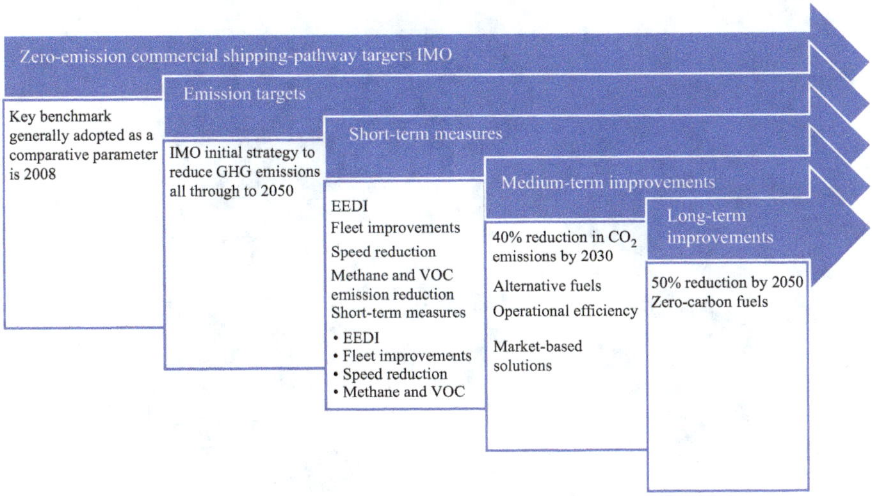

Figure 12.1 IMO targets for the maritime sector. Source: Author

- Propulsion devices to improve efficiency (e.g. different types of propellers).
- Waste-heat recovery systems aimed to recuperate thermal energy from the engine or the exhaust gas and convert it into electrical energy.
- Renewable power technologies are an important area of research, and it needs to be said that based on the literature, we are still in early days. So far, using ports as potential locations where renewable energy sources such as solar panels can be located is an increasingly interesting area of work. That said, renewable power technologies for ports and shipping are still very much in need of advancement. The International Renewable Energy Agency has sought to strengthen this line of work by producing technology briefs on the issue. Projects outlines include harnessing wind energy via different types of sails [4]. This also includes harnessing hybrid solar photovoltaic solutions for smaller vessels. Finally, biofuels as a source of fuel [5].
- Vessel design optimisation, such as changes in hull form optimisation and bow structures, as well as ballast water system improvements.
- Auxiliary systems and main engine improvements.
- Cold ironing (i.e. shore-based power) for ships, which involves investments by ports as well as by vessel owners.
- Electrical hybridisation, which is a very important dynamic for ports, in particular, which in many cases are still using diesel-fuelled cranes and quay and yard equipment. While in many countries, the shift to electricity-powered equipment is market driven, in others it is increasingly being fostered by regulation as well.
- Speed reduction, namely slow steaming, has had a big impact on shipping in recent years and has been estimated to have a significant impact on emission reductions. That said, service reliability tends to decrease in some cases and this,

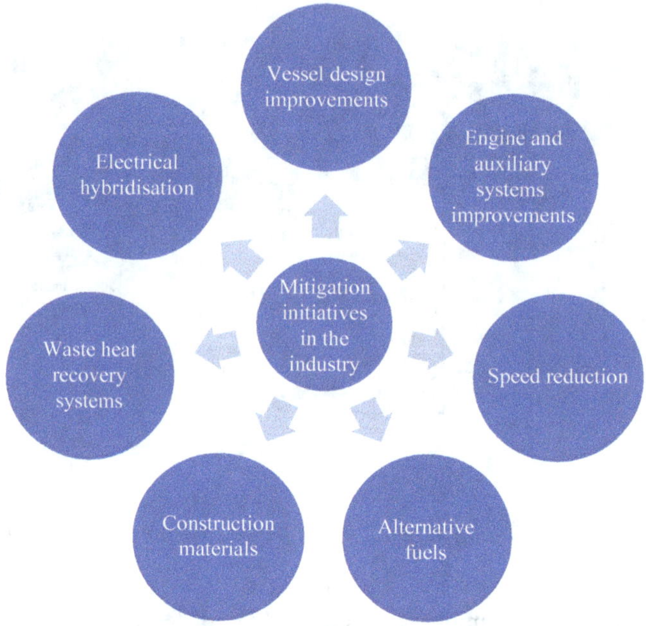

Figure 12.2 Examples of mitigation initiatives in the industry

in turn, may have consequences in ports linked to increased waiting times. It is therefore a matter of understanding how to increase efficiency at the marine–land side interface together with recognition that slow steaming may continue to offer a solution for mitigation at least in the short-to-medium term (Figure 12.2).

Zero-carbon fuels and alternative fuels are a key component of the IMO strategy. One of the areas gaining interest is how to analyse the emissions along the life cycle of each fuel, to essentially start converging on what fuels can indeed enable the zero-carbon transition and when. Neither of the questions have been determined at this point of time, as the chapter discusses in the next section following the review of the EU level regulation and the role of International Financial Institutions (IFIs).

12.3.2 European Union Emission Trading Scheme (ETS)

In addition to the IMO, the EU is to be credited as one of the key drivers of emission-related regulation when it comes to the shipping sector. This, in turn, very clearly reflects in the way in which transition fuels and its related investments are being implemented in Europe. Information provided by DNV-GL clearly shows the dominance of onshore energy for vessel investments in European ports, as well as LNG-propelled vessels.

Concerning the first, for example, of the zero-emitting fuel sources, ship batteries are predicted to be one of the core solutions to zero-carbon propulsion for vessels in the future and are already commercially available. Currently, this is considered as part of a wider set of options for vessel propulsion and serves to operate without causing emissions particularly in harbour areas. According to the DNV database, there are just under 450 ships confirmed with battery installations globally. Of these, about half are operational and the remainder are under order. These will need shore power sources, and investments in this respect are also reported to be taking place in Europe in key ports. The Port of Antwerp is a good example of this as they also publish their energy transition plan which shows increasing onshore power as one of their targets.

At the core of the European strategy so far is the EU Emissions Trading System. The legal basis of the EU ETC is the EU Directive 2003/87/EC of the European Parliament and of the Council of 13 October 2003 establishing a scheme for GHG emissions allowance trading within the community and amending Council Directive 96/61/EC. The regulation is based on a 'cap and trade' principle which recommends a maximum cap on the total amount of GHG that can be emitted by participating installations. Allowances for emissions are auctioned off. No free allocation is allowed anymore. These can be subsequently traded.

That said, as the EU current commissioner for transport in a recent speech[‡] explains, operational efficiency measures will not be sufficient to decarbonise international shipping in line with the IMO targets. This also applies to market-based systems. Therefore, the EU is seeking to promote alternative fuels, through standard development and concrete requirements for their use. To achieve this, the EU is developing a number of instruments. One is a so-called 2030 Climate Target Plan that seeks to bring shipping into the European Climate Law. Another is the Strategy for Sustainable and Smart Mobility, concerning green and digital transition of the transport sector, including maritime.

Concerning alternative fuel, the EU is seeking to develop new rules for decreasing the carbon intensity of fuels used in shipping, and ensuring an appropriate infrastructure for these fuels is available across Europe. This strategy will include strict emissions standards for ships in ports. It will also include tighter rules in the existing Alternative Fuels Infrastructure Directive and the Renewable Energy Directive, ensuring that the provision and distribution of alternative fuels are met with sufficient levels of demand.

Finally, there is an increased number of emerging literatures addressing the question of bunker fuel taxation, which will involve adding a levy on bunker fuel sold in Europe that will then be pooled into a fund that will, in turn, be used to finance a number of mitigation initiatives. The key aspect to note about this tax is that the EU has expressed the possibility of implementing this unilaterally therefore not waiting until such a tax is adopted at the global level. This would lead to a potential loss of competitiveness for EU-based bunker hubs and ports, since bunkering in particular is a market that is very much determined by price.

[‡]https://ec.europa.eu/commission/commissioners/2019-2024/valean_en

12.3.3 International multilateral banks

There is another important force at play which is often not fully recognised in the literature but actually increasingly plays an important role in the so-called greenification of the industry, and that refers to the banking sector. We consider the role of the EBRD as an example.

The EBRD has placed the greenification of infrastructure project it finances, including maritime, at the centre of its activities in several ways. This is the case with several other IFIs but that said, the EBRD has arguably pushed more strongly than its counterparts at least as of very recently. This has been and is being pursued in various ways. One example is the EBRD's Green Economy Transition approach for 2021–25. The approach has taken EBRD green investments to over 40% of its total investment in 2019, from an average of 24% in the decade to 2015. New goals target for the green share of EBRD investment to reach more than 50% by 2025. It also commits the EBRD to focus on aligning with the Paris Agreement goals. Another focus is enhancing policy engagement to develop long-term decarbonisation and resilience pathways.

The challenge to achieve this of course is ensuring that the private sector follows through, and enabling interventions to achieve this are in the form of environmental and social advisory around projects seeking financing from the EBRD. This trend is further penetrating other segments of the financial sectors such as private banks and crucially central banks.

Furthermore, trade flows are changing rapidly in part not only as a result of COVID-19 and its scarring impact but also as a result of the trade and technology disputes between China on one hand and Western countries, including the United States, on the other, which impact the rest of the world in more than one way. This has led to a tightening of credit for many organisations and therefore increased difficulty for ports and maritime sector entities to finance greenfield or brownfield projects. This would increase the leverage that IFIs like the EBRD and others may have in pushing this green agenda forward.

12.4 Decarbonisation scenarios

As already noted earlier, alternative fuels play a major role not only in the zero-carbon pathways set out for the industry both by regulators but also in the private sector. The following subsection of the chapter draws on the DNV-GL report published in September 2020 as well as other reports to highlight the areas of uncertainties around key decarbonisation scenarios for the industry.[§] The DNV-GL report is a good example of the difficulties of reaching consensus in the industry regarding the role of transition fuels in the IMO pathway to zero-carbon shipping and furthermore, the large questions emerging regarding the nature and timing of the technologies required to enable zero-emission fuels to be commercialised. The Maritime Forecast to 2050 report Energy Transition Outlook 2020 report published

[§]Reference DNV GL report 2020.

in September 2020 is one of the hundreds of papers and reports on the topic. DNV-GL stands out for its long-standing work as a classification society for the industry. The report presents the findings from the modelling of 30 different scenarios under three umbrella frameworks: a do-nothing scenario, an IMO scenario and a radical decarbonisation scenario.

For the purpose of this discussion, a 'do-nothing' scenario is ignored in light of the increased prioritisation of the greening of the sector. There is an increasing amount of evidence to show that a 'do-nothing' scenario is unlikely to apply. This is particularly a case in the context of the EU jurisdictions but the not only. Most recently China and Japan pledged to become carbon neutral by 2060 and 2050, respectively.[¶] The China pledge is particularly significant and important, as well as ambitious. It is very important for the sector as well given the fact that China has the busiest ports in the world as well as an increasing portion of the global commercial fleet.

The decarbonisation scenarios vary but generally discuss the IMO-led pathway based on the above-mentioned targets of 2050 as well as more ambitious scenarios partly driven by the EU targets. Generally, the scenarios focusing on alternative fuels predict a dual transition. Each is treated in turn.

The first is that between currently used fuels and so-called transition fuels. This first includes heavy fuel oil (HFO), namely residual fuel oil, which has been the standard for many decades. HFO is sold at different viscosity grades. Global fuel consumption is estimated to be around 330 million tonnes per annum of fuel oil and MGO, with the remainder made up of LPG and LNG. HFO at different grades have been the dominant type of fuel sold for the global commercial fleet so far.

As of 1 January 2020, all ships without exhaust gas cleaning systems (or scrubbers) were required to burn compliant fuel, very low sulphur fuel oil (VLSFO) of no more than 0.5% sulphur content. This has meant a redistribution of the proportion of type of fuels demanded by the market. If ship owners determine that scrubbers are not the option to go for, vessel owners can decide to use LSFO, including VLSFO and ultra LSFO (ULSFO), or distillates (e.g. MGO and MDO).

The following diagram displays the distinction between LSFO and options such as scrubbers that allow vessels to continue to use Heavy Fuel oil (HFO) (Figure 12.3).

Transition fuels mainly include LNG, as well as LSFO, distillates, potentially synthetic fuels and biofuels. These fuel types are called 'transition fuels' in so far as they are not carbon neutral. This means that while improving on existing fuel types when it comes to emissions, they are not the long-term solution required to achieve carbon neutrality for the industry by 2070 or later.

The second transition is the one between transition fuels and so-called zero-carbon fuels. The following diagram distinguishes between the two. The key difference between the two is the fact that zero-carbon fuels are as their name suggest, fully green. That said, the source of these fuels and how they are produced matters

[¶]See, for example, China's pledge expressed at the UN General Assembly in September 2020 to become carbon neutral by 2060.

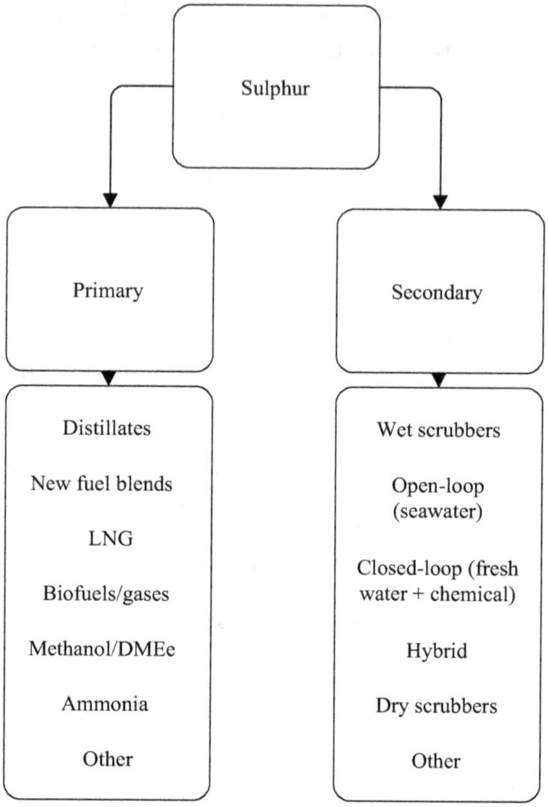

Figure 12.3 Types of fuels, including zero-emission fuels

Table 12.1 Carbon neutral and zero-carbon marine fuels

Carbon neutral	Zero-carbon
• Biofuels (biogas, bioethanol, biodiesel) • Synthetic methane • Synthetic methanol • Synthetic gasoil	• Liquid hydrogen • Liquid ammonia • Battery electric

in such respect, since there is not only the issue of the emissions generated when burning the fuel for energy, but also the emissions generated in the process of producing such fuels (the so-called well-to-wake question) (Table 12.1).

As highlighted by a number of reports, there are a number of key areas of disagreement between experts on the role of transition fuels in the decarbonisation pathway. Areas of disagreement and key questions include the following:

1. What new regulation will be implemented to increase the pace of the transition to carbon neutral and zero-carbon fuel and when?
2. What will the uptake of scrubbers be in the short- and medium term? This, in turn, determines whether HFO demand will increase, stay the same, or fall, in the short term.
3. The length of time during which demand for LSFO and LNG as a propulsion option for vessels will increase, or in other words, when will LSFO/distillates, LNG and biofuels be taken over by hydrogen and ammonia.
4. When new alternative fuels will become commercially viable and come to market.
5. Whether the IMO targets will change and what fuel will be the most successful before 2050 and beyond.

Each is treated, in turn, next with particular emphasis on the role played by the timing and nature of the technologies enabling the transition.

12.4.1 Decarbonisation pathways by regulation

Regulation is an important dynamic generating uncertainty in the market. The question being asked is how stringent the regulation will, in fact, be and how fast will this regulation be implemented. This weighs heavily on the feasibility of investments going forward.

One of the issues being debates is that of the decommissioning of old vessels making up the global fleet. The rate at which the alternative fuel types will gain space in the market will depend on how quickly existing vessels using conventional fuel oil will be decommissioned. All newbuilds going forward will be developed to operate using alternative/new fuel types. While the timing of this is uncertain, what is certain is that the number of types of fuels and propulsion methods used by the global fleet is set to increase going forward.

One of the issues that has dominated the current transition is sulphur content. The question being considered in respect to mitigation measures and abatement, of course, is not only sulphur (local emissions) but also GHG emissions (global emissions). While the continued use of fuel oils is considered incompatible with achieving the IMO targets, carbon neutral fuels and particularly zero-carbon fuels are not. These, therefore, are predicted to dominate in the longer term.

The transition period however remains uncertain. The current global fleet is relatively new furthermore. Assuming that decommissioning does not become mandatory, this would be that the market for retrofits will be a fundamental driver of what becomes possible in the transition phase. If on the other hand mandatory decommissioning does become mandatory, the shipbuilding sector will clearly be flooded and that, in turn, will raise questions concerning capacity and the financial sustainability of several segments of the shipping sector, revenues of which are driven by the older segment of the global fleet.

12.4.2 Scrubbers and retrofitting

Part of the earlier discussion stems from the solutions concerning retrofitting and the returns on investments associated with this. The first segment of retrofitting that has dominated the debate for the shorter term solutions includes scrubbers. The economics of scrubber technologies have been analysed extensively in this sector.

Currently, DNV-GL estimates that just over 4,400 vessels either have or already ordered scrubbers. This is equivalent to approximately 11% of the applicable fleet of some 40,000 heavier vessels worldwide that would normally use HSFO.[||] Scrubber economics has been analysed in detail by several consultancies. Overall, there is an increasing level of doubt as to whether scrubbers are to be considered a medium-term solution, or in other words, a solution to be included in the series of options for the transition. The 'key challenge' is the continued use of HFO as a source of energy for the vessel, which, in turn, is fossil-fuel-based and therefore not part of the long-term solution. However, the other aspect that emerges as important is the wastewater dimension of using scrubbers. Some ports have banned the use of open-loop scrubbers, including Singapore, many in China, some in Europe, the United States and Fujairah. Very recently, the Suez Canal Authority has also banned the discharge of open-loop scrubber wastewater while transiting the canal (see appendix). Vessels may comply with this regulation by simply changing fuel type when entering these waters and instead using the open-loop scrubbers white in open sea. That said, there are those who disagree, suggesting the demand for HFO fuel oil will not only continue but actually increase in the next 10 years, thus placing the continued use of HFO directly at the centre of the transition.

Another example of the retrofitting debate concerns LNG. The uptake of LNG-fuelled vessels is forecast to remain a relatively small proportion of the main shipping sectors. This is mainly to do with bunkering challenges. Trucks can only supply a relatively small amount, and LNG installations may be difficult to integrate in already existing port complexes. That said, Europe is leading the way on this front.

On the one hand, several marine service providers and stakeholders have supported the use of LNG as a transition fuel. LNG is recognised to be both being a low sulphur and a low oxide (SO_x) and NO_x emissions options. Second and more significantly for vessel owners, LNG is less expensive than distillate fuels such as MGO. That said, there are strong views against this argument.

A report published by the ICCT (2017) argued that ships fuelled by LNG do not deliver the emissions reductions required by the IMO targets, and that using LNG could actually worsen shipping's climate impacts. This is the case, according to the report, regardless of the engine technology. The direct implication of this is that LNG is not suitable as a transition fuel and therefore would either not generate the returns required in the case of regulation or markets-pushing vessel owners away from LNG as a source of fuel, or, alternatively, would not support reaching the decarbonisation pathways thereby hinder progress.

[||]https://theicct.org/sites/default/files/publications/LNG%20as%20marine%20fuel%2C%20working%20paper-02_FINAL_20200416.pdf

12.4.3 *Low-sulphur fuel/very low sulphur fuel oil/ultralow sulphur fuel*

There has been a sharp increase in demand for LSFO in major bunkering hubs since 1 January 2020. Price differentials between HFO and LSFO have furthermore been decreasing, making it less significant to invest in scrubbers for the purpose of compliance. Generally, reports, concerning fuel quality, have also been positive or better than expectations. LSFO/VLSFO/ULSFO availability not only will increase as the proportion of global energy is sourced by gas but may also decrease with the increase in cooking. Supply, therefore, is not given since there are factors that may push supply in either of the two directions. Again, the investment decisions of refineries are very important in this determination.

There are debates around the lifespan of LSFO over time. First, LSFO in its variants continues to be sources from fuel oil and therefore by no means can be considered a clean break from past sources. It is therefore the case that in the DNV-GL report as an example, in the decarbonisation scenarios, fossil VLSFO/MGO is in rapid decline by mid-century or even phased out in the most ambitious scenarios. Second, it should be noted that bunkering competes with alternative sectors for both fuel and blending components. Available can be an issue particularly given that returns may indeed be better in alternative markets for the supply. Availability of coarse influences price, since demand and supply is largely the main driver of price for bunkers. That said, it also influences reliability, which for container vessels, in particular but not only, is an important requirement to ensure schedule integrity during service.

Overall, the role of LSFOs as transition fuels vary, and one of the real reasons is that there is a more fundamental debate concerning whether or not transition fuels are actually required or desirable. Clearly, if transition fuels are the requirement, LSFO being the dominant type of fuel today will continue to play a role on the short- to medium term. That said, some argue that there is no significant argument for transition fuels, and that the industry needs to go straight to zero from 2030 or similar to in fact truly achieve the decarbonisation pathway that is truly required.

12.4.3.1 Biofuels and alternative fuels

The longer term solutions that are increasingly becoming more relevant as technology develops are biofuels and alternative options such as hydrogen, methane and ammonia. Major shipping lines are pulling their weight on the discussion.

Maersk has been vocal on the role played by ammonia and methanol. MSC on the other hand has backed hydrogen and biofuels. Each seems to have their own advantages and disadvantages. Part of the reasoning behind decision-making and weighing on one versus the other are the constraints on the ship side. These include having available engines and fuel systems, as well as the capacity to installing them and training crews to use them [6].

On the bunkering side (land side), there are also severe potential constraints concerning the availability of a necessary primary-energy production and supply infrastructure, such as dedicated berths, tankages, pipelines and delivery.

That said, the technology and investments are moving quickly, also greatly supported by governments who benefit greatly from the sector, and that therefore do not want to lose competitiveness as shipping or marine hubs. These debates are already reflected in the development plans of global bunker Hubs. In the complex comprised of Antwerp, Rotterdam, Hamburg and other ports in the region published their plans for the transition to alternative fuels and placed alternative fuels such as at the centre of their longer term transition,[**] noting the objective of including methanol, hydrogen gas and electricity in the bunkering market.

For transition fuels, supply side factors will be considered in relation to the cost of production and supply of different fuels in the same horizon. Different technologies are assumed to reach a level of maturity sufficient for commercial application at different times. Hydrogen fuel cells, for example, may be ready by 2030. In addition, vessels that will be transiting to alternative fuels will likely be capable of running on more than one type of vessel.

Finally, the argument for wind and renewables remains very strong. Technologies such as battery-hybrid solutions and drawing directly from the grid are considered as possible energy-efficiency measures in the longer term. These solutions are likely to apply starting from smaller vessels. The real challenge for the industry is tackling the energy requirements for larger vessels transiting the Asia–Europe route and similar.

12.5 Conclusions

The challenge of finding alternative sources of propulsion for the global fleet is one of the most important segments of research and development currently in the maritime sector. The number of reports and information on this issue is testament to how important and significant the topic is today.

This chapter has sought to highlight some of the areas of debate and disagreement, concerning the role of transition fuels in the decarbonisation pathway, set out for the industry by the IMO and other international organisation. So-called future fuels provide a very good opportunity for countries producing the relevant sources of energy to become sources of supply for the global fleet going forward. It also is an opportunity for service providers in the sector to offer solutions for the challenges faced by the sector in this time of transition.

That said, there are fundamental uncertainties still characterising the debate that make it very difficult for vessel owners and investors, to know and assess the exact trajectory of the pathway that regulators and the market will take. Some of this uncertainty is directly related to the lack of understanding of when the technologies required to shift to zero-emission fuels will be made commercial, and when therefore, to make the required investments to support the meeting of the global targets set for the sector.

[**] https://safety4sea.com/wp-content/uploads/2019/03/Port-of-Antwerp-Transition-to-a-Multi-Fuel-Port-2019_03.pdf

References

[1] Endresen, O, E. Sorgard, H.L. Behrens, P.O. Brett, and I.S. Isaksen (2008) "A historical reconstruction of ships fuel consumption and emissions", Journal of Geophysical Research, Volume 112.

[2] ICCT (2017) Greenhouse gas emissions from global shipping, 2013–2015, International Council on Clean Transportation, Washington, DC. Available at: theicct.org/publications/GHG-emissions-global-shipping-2013-2015 (accessed November 2020).

[3] OECD (2018) Decarbonising maritime transport: pathways to zero-carbon shipping by 2035 (case specific policy analysis). Available at: https://www.itf-oecd.org/sites/default/files/docs/decarbonising-maritime-transport.pdf (accessed November 2020).

[4] IRENA (2015) Renewable energy options for shipping: technology briefs. Available at: https://www.irena.org/-/media/Files/IRENA/Agency/Publication/2015/IRENA_Tech_Brief_RE_for-Shipping_2015.pdf?la=en=hash=D1A6873AAFFD31689CE2B23C63E4EC0545A511C0 (accessed November 2020).

[5] IEA Bioenergy (2017) Biofuels for the maritime shipping sector. Available at: http://task39.sites.olt.ubc.ca/files/2013/05/Marine-biofuel-report-final-Oct-2017.pdf (accessed November 2020).

[6] DNV GL (2020) Maritime 2020 energy transition outlook 2020. Available at DNV-GL website: https://eto.dnvgl.com/2018/maritime (accessed November 2020).

Chapter 13

The advent of shore-based navigation: from vessel traffic services (VTSs) to e-navigation maritime service portfolio (MSP)

Michele Fiorini[1]

Fundamental requirements for all transportation systems are safety, efficiency and sustainable movability of any goods from shipper to clients with the entire delivery on time. Nowadays, this is ensured by contributions of different components and subsystems of the transportation systems.

The basic components and subsystems are:

- transport means (in maritime transportation: vessels of various types and sizes);
- drivers (ship crews incl. captain, officers, engineers, A.B.s, Motor men, etc.);
- transport paths (open sea, coastal waters and inland fairways);
- traffic management (e.g. ship reporting systems, vessel traffic service (VTS) systems); and
- organizational components (including international and national bodies and institutions like International Maritime Organization (IMO), International Hydrographic Organization (IHO) as well as classification societies, national water and shipping administrations that provide, implement and enforce standards, rules and regulations).

This chapter introduces structure and operations of two shore-based components of the maritime transport system (MTS), VTS and fleet operations centres (FOC), as well as coastal surveillance systems (CSS) and discusses selected aspects of the integration of these systems. The potentials and future roles of VTS/FOC/CSS in the MTS under the e-navigation framework are discussed also. The chapter includes a case study described in some technical details on appendixes.

[1]Leonardo S.p.A., Rome, Italy

Disclaimer: The opinion expressed in this work reflects the professional experience of the author alone and does not necessarily represent the position of the company.

13.1 VTS at present

Until today, according to the established rules and regulations, crews on-board operate the merchant ships. Additionally, especially in areas with high traffic density, ships are monitored, and their crews are supported from shore-based surveillance and control centres.

VTS systems are well known and recognized in this respect. There are IMO Resolutions that provide definitions, standards and guidance on specification of the services and how to implement and to maintain them. A VTS, as defined by IMO Resolution A.857(20), Guidelines for VTS, is:

> A service implemented by a competent authority, designed to improve the safety and efficiency of vessel traffic and to protect the environment. The service should have the capability to interact with the traffic and respond to traffic situations developing in the VTS area.

The purpose of a VTS system is to improve the safety and efficiency of navigation, safety of life at sea (SOLAS) and the protection of the marine environment and/or the adjacent shore area, worksites and offshore installations from possible adverse effects of maritime traffic.

A clear distinction may need to be made between a port or harbour VTS and a coastal VTS. A port VTS is mainly concerned with vessel traffic to and from a port or harbour, while a coastal VTS is mainly concerned with vessel traffic passing through the area. A VTS could also be a combination of both types. The type and level of service or services rendered could differ between both types of VTS; in a port or harbour VTS, a navigational assistance service (NAS) and/or a traffic organization service (TOS) is usually provided for, while in a coastal VTS usually only an information service is rendered. The benefits of implementing a VTS are that it allows identification and monitoring of vessels, strategic planning of vessel movements and provision of navigational information and assistance. It can also assist in the prevention of pollution and coordination of pollution response. The efficiency of a VTS will depend on the reliability and continuity of communications and on the ability to provide good and unambiguous information. The quality of accident prevention measures will depend on the system's capability of detecting a developing dangerous situation and on the ability to give timely warning of such dangers. The precise objective of any VTS will depend upon the particular circumstances in the VTS area and the volume and character of maritime traffic.

13.2 Types of services in VTS

The IMO Resolution A.857(20) states that a VTS should comprise at least an information service (INS) and may also include others, such as an NAS or a TOS or both.

13.2.1 Information service (INS)

An information service (INS) provides essential and timely information to assist in the on-board decision-making process. It involves maintaining a traffic image and allows interaction with traffic and response to developing traffic situations.

13.2.2 Navigational assistance service

An NAS may be provided in addition to an information service and/or TOS. It is a service to assist in the on-board navigational decision-making process and to monitor its effects and is provided at the request of a vessel or when deemed necessary by the VTS. It may also involve the provision of information, warning, navigational advice and/or instruction. The NAS is especially important in difficult navigational or meteorological circumstances or in the case of defects or deficiencies. An NAS is an important supplement to the provision of other navigational services, such as pilotage. NAS may be provided at the request of a vessel, irrespective of whether a pilot is on-board, or when a navigational situation is observed and intervention by the VTS is deemed necessary.

13.2.3 Traffic organization service

A TOS is a service to prevent the development of dangerous maritime traffic situations and to provide for the safe and efficient movement of vessel traffic within the VTS area. A TOS provides essential and timely information to assist in the on-board decision-making process and may advise instructor exercise authority to direct movements. It concerns the operational management of traffic and the planning of vessel movements and is particularly relevant in times of high traffic density or when vessel movements may affect the traffic flow. A TOS should be responsible for separating traffic in the interest of safety. This separation could be defined in space, time and/or distance.

13.2.4 Other services (pilotage)

Among other services a VTS centre is offering, pilotage is probably the most common one used to get in and out of ports.

Pilotage is a specialized (and usually licensed) service to navigation, specifically in restricted and/or navigationally challenging waters. The skill of the pilot draws on local knowledge related to geographic and hydrographic points of interest, marine aids to navigation, submerged features, hazards to navigation, the navigable depth of water and any limitations associated with the waterway, in which pilotage is being conducted. Pilots must also possess a high degree of ship handling skills, be cognisant of the local tides, currents, and climatic conditions, as well as the handling characteristics for the specific ship receiving the pilotage service.

When a pilot boards a vessel, he/she is given 'conduct of the vessel', but not 'command'.

The role of the pilot often includes:

- giving necessary instructions to ship's personnel who operates equipment essential to the safe navigation and manoeuvring of the vessel;
- assisting local communication with a VTS centre, port control and other vessels;
- communicating instructions to tugs and linesmen if berthing or sailing;

- providing current and specialist knowledge of:
  - local conditions and traffic;
  - operational status of marine aids to navigation;
  - sailing directions;
  - restrictions applicable to the piloted vessel; and
  - being able to quickly adapt to:
    - operational culture aboard the vessel;
    - the vessel's handling characteristics; and
    - the state of the navigation equipment aboard.

From an engineering point of view, VTS can be simplistically considered as a system that collects environmental and traffic data in order to create a comprehensive traffic image continuously analysed to assess if there are any developing or already existing risks that require interaction with the traffic. If a VTS operator becomes aware of a vessel for instance not complying with the rules, violating regulations, or in case she/he detects a situation with a risk of grounding or collision, then VTS can intervene by sending out information, warning/advice or instructions to all or individual vessels using VHF voice communication at first.

13.3 Equipment and function requirements for VTS

IALA Recommendation V-128 and IALA Guideline G1111 provide guidance on the preparation of operational and technical performance for VTS systems.

VTS equipment might be defined as the individual items of hardware and software which make up the VTS system. The VTS system is considered to be the hardware, software and their behaviour as a coherent entity. This excludes personnel and procedures.

Traffic density and structure, navigation hazards, local climate, topography, environmental requirements, commercial aspects and the extent of a VTS area set the requirements for VTS equipment, and these factors will have substantial impact on life cycle costs of a VTS and the acquisition of VTS equipment. Equipment examples include:

- communications (VHF, telephone, satellite telephone, mobile telephone, e-mail, automatic identification system (AIS) messaging);
- radar system;
- AIS;
- electro-optic systems;
- radio direction finders;
- hydro-meteorological equipment; and
- external information gateway equipment.

The composition of the system is not fixed and may vary according to the mandate (VTS/FOC/CSS), the geographic context, the needs of final user and so on. Many other different configurations are possible. An example of a possible national CSS system is described later in this chapter.

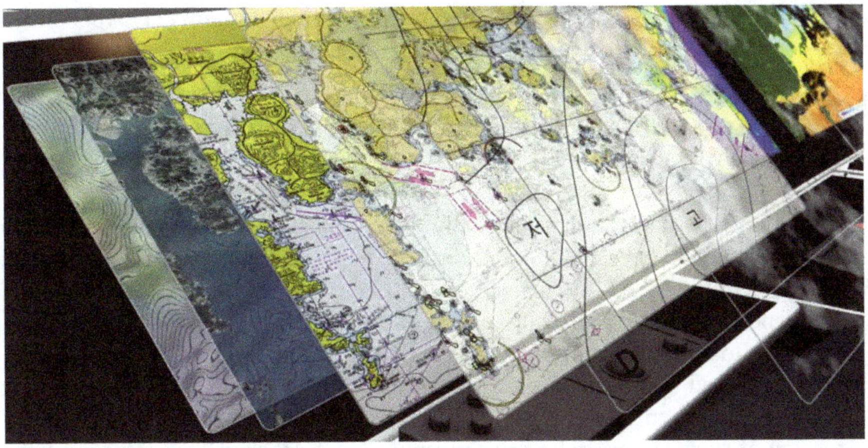

Figure 13.1 COP layers view. Source: Author collection

The data coming from the various sources of information (sensors, external systems or manually inputted by operators) contribute to a common operational picture (COP) built up with different levels of details and with different periods of refresh (i.e. update of data) (Figure 13.1).

An example of a data layer could be AIS information which could have a global coverage making the use of satellite connections.

13.4 Fleet operation centres

Besides the recognized monitoring of vessel traffic in sea areas of the national territories by VTS, there are an increasing number of company-based FOC. They monitor the safe and efficient progress of company-owned ships as well. In the same way as VTS, an FOC operates 24 h a day, 7 days a week. On the other hand, contrary to VTS, FOCs even monitor ships on a worldwide scale. So far, the relation between the navigators on-board and the operators ashore is characterized by the status quo that the shore operator provides additional information in order to support the decision-making processes of the bridge team on-board. However, until now, there is no guidance about any potential relations between VTS and an FOC.

IMO's e-navigation initiative makes an effort to introduce state-of-the-art technologies to improve and harmonize the collection, processing and presentation of data and information to support human operators on-board and ashore to ensure safety of navigation from berth to berth [1]. Even though e-navigation clearly addresses the human operators, the captain and the whole bridge team as well as the VTS supervisor and his watch standing operators in shore-based centres, e-navigation is obviously supporting even more enhanced services [2]. Together with the increasing level of digitalization and automation, it is supporting the introduction of autonomous navigating and even unmanned ships. Making them a reality will fundamentally change the existing transportation system [3]. Appendix C reports some projects of unmanned ships currently undergoing.

13.5 E-navigation – a keystone of future maritime developments

The definition of e-navigation is formally described as the harmonized collection, integration, exchange, presentation and analysis of marine information on-board and ashore by electronic means to enhance berth-to-berth navigation and related services for safety and security at sea and protection of the marine environment [4].

Referring to this IMO definition, it has to be highlighted that e-navigation originally addresses ships and operation of ships falling under IMO's convention, especially the international convention on the SOLAS. However, shipping on a worldwide scale is a composition of ships falling under this convention 'SOLAS ships' and ships not falling under this convention 'non-SOLAS ships'. On the other hand, e-navigation is a concept that not only addresses the ship's perspective but also covers all aspects of processes linked to shipping, including shore-based monitoring and support for the purposes of ensuring a safe and efficient vessel traffic flow along the coasts from and to ports.

E-navigation inputs could be summarized as follows:

- real-time (or near real-time) update information:
  - AtoN (e.g. position/status);
  - Maritime Safety Information (MSI);
  - radar;
  - position fixing system;
  - ships sensors (e.g. heading);
  - echo sounder;
  - sonar (optional);
  - inertial navigation system;
  - ship–ship, ship–shore and shore–shore communications;
  - AIS data;
  - meteorological data/chats; and
  - updates to charts and publications.
- long-lead (reference) information:
  - digital charts and publications;
  - AtoN infrastructure and relevant information; and
  - predicted meteorological, oceanographic and hydrographic information (e.g. seasonal weather patterns, tides).
- organizational:
  - training and procedures;
  - quality management;
  - data access and security protocols;
  - conventions, registrations and guidelines (IMO, IALA, IHO);
  - communication protocols (ITU);

- international standards (ISO, IEC, etc.); and
- legal instruments.

The e-navigation core is deployed upon three main pillars (also known as 'the three side of the coin'):

- shore-side e-navigation components, including human–machine interfaces to operators;
- integrated communications, including AIS, LRIT data, U/VHF and satellite communications; and
- ship-side e-navigation components, including human–machine interfaces to on-board operators.

E-navigation outputs could be summarized as follows:

- safe navigation by enhanced:
  - anti-collision and anti-grounding processes;
  - route planning and monitoring;
  - pilotage and berthing;
  - under keel and air draft clearance management; and
  - alert management.
- efficiencies and other benefits:
  - standardized and automated maritime reporting;
  - logistics efficiencies, including port operations;
  - potentially reduced insurance costs;
  - improved security;
  - search and rescue (SAR) and pollution response;
  - strategic analysis for infrastructure refinement, investigations and incident analysis;
  - reduced human inefficiencies and errors;
  - improved on-board efficiencies;
  - decision support mechanisms;
  - improved ship–shore cooperation;
  - sharing risk analysis between ship and shore; and
  - others.

13.5.1 The development of e-navigation

Since the 81st session of IMO's Maritime Safety Committee (MSC 81) in 2006, which is commonly referred to as the official birth of the e-navigation initiative and the beginning of the development of that concept, until now, the involved stakeholders and all maritime users are continuously further developing the concept for smooth and comprehensive implementation. The materialization of the concept itself has been completed in 2020. Some countries in the European and Asian region have already started to implement this concept, by establishing test bed areas using developed prototypes under real conditions.

13.5.2 E-navigation maritime service portfolios (MSPs)

One component of the e-navigation concept, presently of highest relevance, is the established strategic implementation plan (SIP). This plan, among others, calls for comprehensive and profound gap analysis through formal safety assessment in order to identify and form the core of e-navigation maritime service portfolios (MSPs).

13.6 Maritime service portfolios

The following text describes the already identified e-navigation MSPs by IMO-navigation SIP NCSR 1/28.

13.6.1 MSP 1 – information service (INS)

The INS is defined as 'a service to ensure that essential information becomes available in time for on board navigational decision making'. Relevant information is broadcasted at fixed times and intervals or provided when deemed necessary by the VTS or at the request of a vessel.

A INS involves maintaining a traffic image and allows interaction with traffic and response to developing traffic situations. An information service should provide essential and timely information to assist in the on-board decision-making process.

The INS is designed to improve the safety and efficiency of vessel traffic and to protect the environment. Among others, such services include catalogue such as Routing, Channel info, Security level, Berthing, Anchorage, Time slot, Traffic monitoring and assessment, Waterway conditions, Weather, Navigational hazards, any other factors that may influence the vessel's transit, Reports on the position, Identity and intentions of other traffic.

13.6.2 MSP 2 – navigational assistance service (NAS)

The NAS is defined as 'a service to assist on board navigational decision-making and to monitor its effects'. NAS may be provided on request by a vessel in circumstances such as equipment failure or navigational unfamiliarity.

13.6.3 MSP 3 – traffic organization service (TOS)

The TOS is defined as 'a service to prevent the development of dangerous maritime traffic situations and to provide for the safe and efficient movement of vessel traffic within the VTS area'. The purpose of the TOS is to prevent hazardous situations from developing and to ensure safe and efficient navigation through the VTS area.

13.6.4 MSP 4 – local port service (LPS)

The local port service (LPS) is applicable to those ports where it has been assessed that a VTS, as described earlier, is excessive or inappropriate. The main difference arising from the provision of LPS is that it does not interact with traffic, nor is it required to have the ability and/or the resources to respond to developing traffic situations and there is no requirement for a vessel traffic image to be maintained.

Examples of LPS may include:

- berthing information;
- availability of port services;
- shipping schedules; and
- meteorological and hydrological data.

A number of web-based LPS services are being developed such as AVANTI (Access to Validated, Nautical Information) and later PRONTO (Port Rendezvous of Nautical and Terminal Operations), an initiative of the International Harbour Masters' Association.

13.6.5 MSP 5 – maritime safety information (MSI) service

The Global Maritime Distress and Safety System (GMDSS) as described in SOLAS chapter IV defines the seventh functional requirement as 'Every ship, while at sea, shall be capable of transmitting and receiving maritime safety information'. The MSI service is an internationally coordinated network of broadcasts of MSI from official information providers, such as:

- National Hydrographic Offices, for navigational warnings and chart correction data;
- National Meteorological Offices, for weather warnings and forecasts;
- Rescue Coordination Centres, for shore-to-ship distress alerts; and
- the International Ice Patrol, for Oceanic ice hazards.

Specific information on aids to navigation and restrictions on safe navigation are part of MSI services provided by national authorities.

13.6.6 MSP 6 – pilotage service

The aim of the pilotage service is to safeguard traffic at sea and protect the environment by ensuring that vessels operating in pilotage areas have navigators with adequate qualifications for safe navigation. Each pilotage area needs highly specialized experience and local knowledge on the part of the pilot.

13.6.7 MSP 7 – tugs service

Efficient tug operations depend on, among other things, the effectiveness of the communications and information exchanges between relevant stakeholders. The aim of the tugs services is to safeguard traffic at sea and protect the environment by conducting operations such as:

- transportation (personnel and staff from port to anchorage) operations;
- ship assistance (e.g. mooring) operations;
- salvage (grounded ships or structures) operations;
- shore operations;
- towage (harbour/ocean) operations;
- escort operations; and
- oil spill response operations.

13.6.8 MSP 8 – vessel shore reporting

The aim of vessel shore reporting is to safeguard traffic at sea, ensure personnel safety and security, ensure environmental protection and increase the efficiency of maritime operations. Single Window is one of the most important solutions to reduce the mariners' workload (amount of time spent on preparing and submitting reports to shore-based authorities). To achieve this, reports should be automatically generated as much as possible from on-board systems.

13.6.9 MSP 9 – telemedical assistance service (TMAS)

The TMAS centres should provide medical advice for seafarers 24 h/day, 365 days/year. TMAS should be permanently staffed by physicians qualified in conducting remote consultations and who are well versed in the particular nature of treatment on-board ship.

The COVID-19 pandemic in 2020 has shown the prevailing view that a standardization of the TMAS services is both necessary and wanted.

13.6.10 MSP 10 – maritime assistance service (MAS)

The primary mission of MAS is to handle communication between the coastal states, ship's officers requiring assistance and other players in the maritime community. The MAS is on 24 h/day alert to deploy rapid assistance and professional support for ships in connection with combating pollution, fire and explosions on-board, collision, grounding, maritime security, terror mitigation, etc., including pirate attacks. The MAS is responsible only for receiving and transmitting communications and monitoring the situation. It serves as a point of contact between the master and the coastal state concerned if the ship's situation requires exchanges of information between the ship and the coastal state.

13.6.11 MSP 11 – nautical chart service

The aim of the nautical chart service is to safeguard navigation at sea by providing information such as nature and form of the coast, water depth, tides table, obstructions and other dangers to navigation, location and type of aids to navigation. The nautical chart service also ensures the distribution, update and licensing of electronic charts to vessels and other maritime parties.

13.6.12 MSP 12 – nautical publications service

The aim of the nautical publication service is to promote navigation awareness and safe navigation of ships.

Nautical publications include:

- tidal currents, aids to navigation system, buoys and fog signals; radio aids to marine navigation, chart symbols, terms and abbreviations, sailing directions; and
- a Chart and Publication Correction Record Card system can be used to ensure that every publication is properly corrected prior use by mariners.

13.6.13 MSP 13 – ice navigation service

The ice navigation service is critical to safeguard the ship navigation in ice-infested waters, given how quickly the ice maps become outdated in the rapid changing conditions of the ice-covered navigational regions. Such services include:

- ice condition information and operational recommendations/advice;
- ice condition around a vessel;
- vessel routing;
- vessel escort and ice breaking;
- ice drift load and momentum; and
- ice patrol.

13.6.14 MSP 14 – meteorological information service

The meteorological service is essential to safeguard the traffic at sea by providing weather, climate digital forecasts and related information to mariners who will use these types of information to support their decision-making. Such information includes:

- weather routing, solar radiation and precipitation;
- cold/hot durations and warnings;
- air temperature, wind speed and direction; and
- cloud cover and barometric pressure.

13.6.15 MSP 15 – real-time hydrographic and environmental information service

The real-time hydrographic and environmental information service is essential to safeguard navigation at sea and protect the environment. The provided services are as follows:

- current speed and direction;
- wave height;
- marine habitat and bathymetry;
- sailing directions (or pilots): detailed descriptions of areas of the sea, shipping routes;
- harbours, aids to navigation, regulations, etc.;
- lists of lights: descriptions of lighthouses and light buoys;
- tide surge prediction tables and tidal stream atlases;
- ephemeris and nautical almanacs for celestial navigation; and
- notice to mariners: periodical (often weekly) updates and corrections for nautical charts and publications.

13.6.16 MSP 16 – search and rescue (SAR) service

The SAR service is responsible for assisting, coordinating SAR operations at sea. In maintaining a state of full readiness, the Maritime Rescue and Coordination Centre (MRCC) may perform the following rescue functions:

- survivors of any aircraft (not in an act of war) crashes or forced landings at sea;
- the crew and passengers of vessels in distress; and
- survivors of maritime accidents or incidents.

The SAR services must also coordinate the evacuation of seriously injured or ill persons from a vessel at sea when the person requires medical treatment sooner than the vessel would be able to get him or her to a suitable medical facility.

E-navigation can provide additional information such as number of persons on-board, type of ship, port of destination, etc. and enable provision of additional information such as available SAR resources on-board ships.

Information on other vessels in the area can be crucial for an effective rescue.

Communication solutions used for e-navigation will be able to exchange information about SAR areas and allocate search patterns and provide facilities for MRCCs to set up a common information sharing log or chat room for MRCCs, on scene coordinator and other resources to share and update information during an SAR incident.

13.7 Is e-navigation disruptive technology?

We live in the beginning of the era of disruption, where the desire of 'instant and fast' communication started to grow up. The change in our daily communication in terms of its speed and convenience can be seen as a result of a worldwide development and competition in information and communications technology (ICT). Each organization or company races to become an innovator. Meanwhile, users of all kinds of domains are constantly challenged to use this technology effectively. The maritime sector has introduced e-navigation as a new way of communication between the numerous different involved parties that offer a seamless way. But is this concept benefitting the industry or vice versa? It seems, however, the vibrant e-navigation concept is to become a problem solver for the complex tasks related to marine transportation and sea navigation.

Christensen [5] stated that the technology can be divided into the two types 'sustaining' and 'disruptive'. As indicated by Rouse [6], sustaining technology relies on incremental improvements to an already established technology, while disruptive technology lacks refinement, often has performance problems because it is new, appeals to a limited audience and may not yet have a proven practical application. In this sense, e-navigation can be seen as an attempt to provide disruptive solutions by applying latest ICT solutions for instance in system integration but facing the problem of a widely inconsistent existing technical infrastructure. A good sample might be the introduction of the Electronic Chart Display and Information System (ECDIS). The electronic 'sea chart' is to replace the traditionally used paper chart. Navigators will no longer need to take care of indicating the actual position in the chart but can check in the display the progress of the voyage and if the ship keeps the planned route at all times.

There is clear potential for many other functions and services. However, for several reasons ECDIS, although being made mandatory for SOLAS ships, nowadays is still 'coexisting' with paper charts. ECDIS is considered to be one core element integrating a huge amount of the data and information flows needed for the implementation of the e-navigation concept on-board ships as well as ashore.

However, in relation to the currently ongoing developments in respect to autonomous navigation and even unmanned shipping, e-navigation as a whole is a concept that focuses on the human operators involved in processes on-board or ashore providing assistance for improved situational awareness and decision-making. Systems and services under development or tested, e.g. 'route exchange', 'augmented reality and head-up displays' or 'dynamic manoeuvre prediction', let the human in the loop and consider the change of skills and knowledge that is needed to make e-navigation smoothly functioning [7]. In this respect, e-navigation clearly is 'sustaining technology'.

13.7.1 Issue of mixed environment

Current research on e-navigation [8] demonstrated the implementation of the e-navigation concept on a global and even on a regional level, strongly depending on the provision of a harmonized infrastructure and environment.

With regard to the link between shore-based services and their link to on-board process, the shore-based e-navigation part is defined as it stated on MSC 85/26/Add.1 Annex 20 as the management of vessel traffic information and related services from ashore enhanced through better provision, coordination and exchange of comprehensive data formats that will be more easily understood and utilized by shore-based operators in support of vessel safety and efficiency.

In this way, e-navigation will allow for a much more enhanced management of the traffic flow; however, the prerequisites need to be very well studied in advance. Such investigations need to take into consideration the three sides of the 'e-navigation coin', when looking at Figure 13.2: the shipborne and the shore-side systems as well as the connecting integrated communication. On the one hand it is expected to have a wide spectrum of technology-driven technical solutions for new functions and systems on the ship-end and on the shore-side end. Integrated communication will depend on existing technical infrastructure (availability and abilities of communication networks, broadband, satellite, etc.) and geographical conditions in the respected areas (dimensions, topographic, hydrographic, climate and other characteristics). It is obvious that all these parameters are varying significantly for each specific region.

13.7.1.1 E-navigation-based applications

Besides the e-navigation initiative, the current technological developments and research mention not only 'unmanned' but also 'autonomous' ships often synonymously. IMO at the recent sessions of its MSC has introduced the abbreviation term MASS for 'maritime autonomous surface ship' and recognized the need for a clear definition of unmanned ships and shall also address different levels of automation, including semi-autonomous and unmanned ships as well [8].

However, per definition, unmanned ships have no crew, nor even any human operator on-board. An unmanned ship can be remote controlled from any

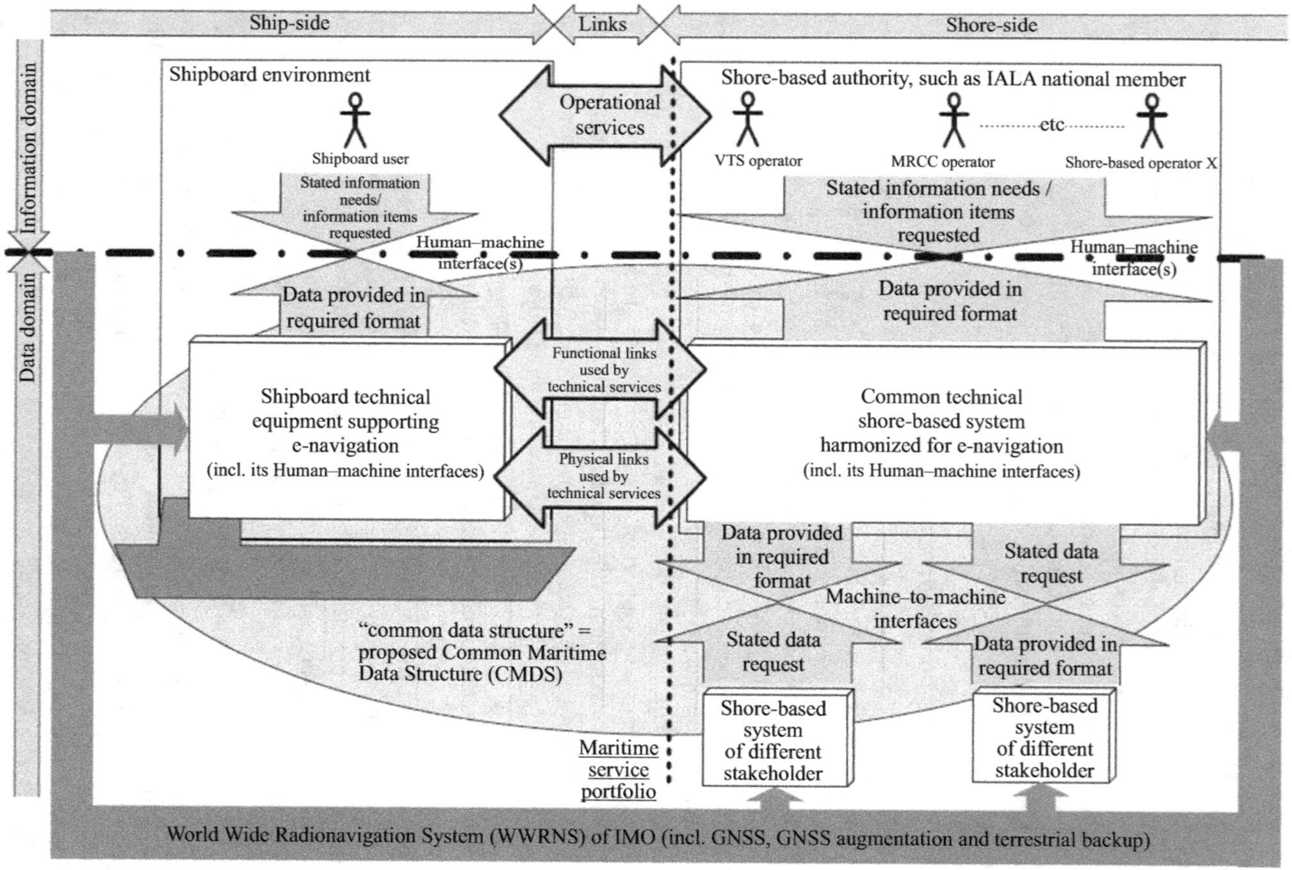

Figure 13.2 Overarching e-navigation architecture. Source: IMO e-navigation Strategy Implementation Plan (SIP) NCSR 1/28

monitoring and control centre either ashore or on another mobile station or it is in autonomous mode. A ship in autonomous mode has systems, which steer the ship and 'decide' about any change of the control settings by her own. Autonomous operating ships do not necessarily have to be unmanned [9].

As in all other transport modes also in the maritime sector, a range of autonomy and its potential to manning levels are suggested. Typically, the level ranges from fully manned to unmanned vessels, and definitions have been suggested by several institutions, e.g. by Lloyd's Register or by Scandinavian Maritime Administrations. However, a working group of the global standard setting maritime organization IMO has defined the following four levels of operation for MASS:

1. ship with automated processes and decision support,
2. remotely controlled ship with seafarers on-board,
3. remotely controlled ship without seafarers on-board, and
4. fully autonomous ship.

Moreover, IMO provided guidelines for MASS trials (2018) and intends to deliver draft-specific regulations for MASS by 2025. It is expected that regulations take into account all aspects of operation. Training and education issues might be addressed as well, but it is not yet on the agenda. Experiences and related studies are obviously lacking. Also, while remote control of ships is addressed, the interaction with VTS is not.

Present concepts proposing and researching the introduction and operation of MASS usually contain a kind of shore-based control centre that monitors the status of such ships and the navigational and technical processes. In the maritime unmanned navigation through intelligence in networks project [7], such a centre was named ship control centre (SCC) and provided for direct remote-control options in the case of its need [8].

However, the introduction of such transportation systems has not only various technical–technological but also administrative–organizational [9] and human-factor-related challenges. While the technical–technological aspects are very well addressed by numerous research and technical development projects (see e.g. [10]), the consideration of the operational aspects, e.g. about the interaction with other conventional ships and with a VTS, is still just beginning and in a rather initial phase of studying. First efforts have been made and provided a good basis for further research into the topic [11–14].

So far, it has not been considered in very detail and is comparably seldom subject to ongoing studies and research activities. Operational and human-factor-related aspects are, on the other hand, on the research agenda.

One of the major challenges is seen in the field of addressing the interaction of SCC and VTS. If there will be no spatial separation of unmanned and remotely controlled ships from conventionally manned and navigating ships, certain communication needs will arise. Ship-shore (including shore-ship and ship-ship) communication needs are to be addressed in detail at least on the basis of series of case studies for specified scenarios or by systematic consideration of potential situation and environmental parameters of traffic scenarios. Those challenges include among others technical ways of and procedures for communication between VTS and SCC operators or even a MASS; not to raise the question who will be the responsible party in the case of any accident that might happen. Will there be a need for overriding rights for any of the involved parties? Shall such questions given to coastal states be governed by national rules and regulations exclusively? Is there a need for an overall harmonized approach and provision of a framework of operational procedures to just mention a few samples?

However, operational integration quite obviously requires addressing legal and operational aspects as well as technic–technological issues. Experimental studies are to approach such challenges in a systematic way.

13.8 Summary and conclusions

Once shipping moves into the digital world, e-navigation has started providing the infrastructure framework for data and (digital) information exchange for the benefit of maritime safety, security and protection of the marine environment, reducing the administrative burden and increasing the efficiency of maritime trade and transport.

At the same time, the global community is slowly moving from systems to services enhancing the initial offer of services provided by VTS to the e-navigation. The MSP includes and extends far more the offer of services to address actually all needs from shore to ship communications and vice versa. This sets the scene for the introduction and further development from robotics to full autonomous ships.

There is also degree of communalities among VTS, FOC and even CSS on mandate, functions and user needs to address.

Acknowledgements

The author express his gratitude for many thought-provoking conversations, exchange of ideas, contributes and long friendship cooperation with Dr.-Ing (PhD) Michael Baldauf, Hochschule (HS) Wismar, University of Applied Sciences, Technology, Business and Design, Department of Maritime Studies, Wismar, Germany.

References

[1] Burmeister HC, Bruhn WC, Rodseth ØJ, and Porathe T (2014). Can unmanned ships improve maritime safety? In: Proceedings of TRA, Paris/France, 14–16 April 2014.

[2] Kitada M, Baldauf M, Mannov A, *et al.* (2019). Command of vessels in the era of digitalization. In: Kantola J, Nazir S, and Barath T (eds.) Advances in Human Factors, Business Management and Society. AHFE 2018, Advances in Intelligent Systems and Computing, vol. 783. Springer, Cham. DOI: 10.1007/978-3-319-94709-9_32.

[3] Lloyd's Register (2017). Shipright design and construction, additional design procedures: Design code for unmanned marine systems. Lloyd's Register, London.

[4] IMO (2021). *Vessel Traffic Services*. [Online]. Available from: https://www.imo.org/en/OurWork/Safety/Pages/VesselTrafficServices.aspx [Accessed 20 March 2021].

[5] Christensen CM (1997). The Innovator's Dilemma: When New Technologies Cause Great Firms to Fail. Boston, MA: Harvard Business School Press.

[6] Rouse M. Disruptive Technology. TechTarget. [Online]. Available from: https://whatis.techtarget.com/definition/disruptive-technology. [Accessed 20 March 2021].

[7] Baldauf M, Kitada M, Mehdi R, and Dalaklis D (2018). e-Navigation, digitalization and unmanned ships: challenges for future maritime education and training. In: 12th Annual International Technology, Education and Development Conference (INTED), Barcelona, Spain.

[8] IMO (2018). Regulatory scoping exercise for the use of maritime autonomous surface ships (MASS). MSC 99/5/12, 27 March 2018.

[9] Quinisio BS (2018). Development of a strategy for management of autonomous ships by coastal states. World Maritime University Dissertations, Malmö.

[10] Kongsberg Maritime (2018). Autonomous ship project, key facts about YARA Birkeland. [Online]. Available from: https://www.kongsberg.com/maritime/support/themes/autonomous-ship-project-key-facts-about-yara-birkeland/. [Accessed 20 March 2021].

[11] Porathe T, Prison J, and Man Y (2014). Situation awareness in remote control centres for unmanned ships. In: Proceedings of Human Factors in Ship Design & Operation, London, UK 26–27 February 2014.

[12] Ramosa MA, Utne IB, and Mosleh A (2019). Collision avoidance on maritime autonomous surface ships: Operators' tasks and human failure events. Safety Science, Vol. 116, pp. 33–44.

[13] Wrobel K, Montewka J, and Kujala P (2018). Towards the development of a system-theoretic model for safety assessment of autonomous merchant vessels. Reliability Engineering & System Safety, Vol. 178, pp. 209–224.

[14] Thieme CA and Utne IB (2017). Safety performance monitoring of autonomous marine systems. Reliability Engineering & System Safety, Vol. 159, pp. 264–275.

Appendix A

VTS/FOC/CSS systems engineering approach to system's architecture

Michele Fiorini[1,*]

The following appendix describes a decomposition of a system into hardware component items and computer software component items based on author's on-field experience in the concept, realization and delivery of vessel traffic service (VTS) and coastal surveillance systems (CSS) in different countries.

The system deployed at national level is supposed to be structured on a multilevel architecture based on:

- central/national observation level (with possible back-up/redundant centre) (COs);
- regional observation level (ROs);
- local observation level (LOs);
- observation posts (POs), usually composed by unmanned tower with sensors;
- patrol boats (PBs); and
- mobile radar stations (MRS).

Any modern surveillance system (VTS/CSS/FOC) is based on a modular architecture: the system is subdivided into autonomous building blocks, so that each block can be singularly maintained, upgraded and/or replaced. In this way, the system has been designed to guarantee a high degree of modularity and expandability.

The system ensures the highest degree of interoperability towards internal components and external systems. This will be achieved thanks to the implementation of an open architecture, based on standard TCP/IP over Ethernet protocol stack for all the networking and data communication among the different levels of control of the system (local, regional and national). External systems as existing radar systems, port management systems as well as existing telecommunication systems, can be integrated at the only condition of establishing logical (software) and physical (hardware) interface and setting the adequate administrative authorizations for resource access.

[1]Leonardo S.p.A., Rome, Italy

[*]The opinion expressed in this work reflects the professional experience of the author alone and does not necessarily represent the position of the company.

An example of decomposition of CSS system is the following split into eight subsystems as follows:

1. command and control (C2) subsystem,
2. radar subsystem (RAD),
3. electro-optic subsystem (EOS),
4. radio subsystem (RDS),
5. simulation and training subsystem (STS),
6. site security access subsystem (ACS),
7. perimetric subsystem (PRS), and
8. power supply subsystem (PWS).

A.1 Command and control (C2) subsystem

The C2 subsystem is composed of processing servers, database servers, operators and supervisors consoles, located at the local/regional/national control centres. The distributed structure of this subsystem allows users to manage the whole system from the different centres in an integrated approach.

The C2 receives data from different type of sensors, both internal and external (e.g. radars, cameras, automatic identification system (AIS) and radio), integrates them, shares the information through the distributed structure and allows data management from control centres. Integrated data are elaborated in order to yield the traffic picture, consisting in tracks and plots, measurements, identifications, predictions, to permit the navigation management, including communications with vessels, anti-collision controls, port approaching, transits, anchorages, and to provide decision aids functions. All data are organized and handled in databases.

This subsystem represents the main human interface, performing the graphical data presentation (e.g. maps, tracks, video alarms) on the user consoles, as well as on video wall screen and projectors. The C2 performs also recording and playback functions, by means of recording servers, including the capabilities to record and playback traffic data, radar tracks, AIS tracks and voice communications. Also supervises the physical and data security of the system and the data safety, the user's accesses and communications.

The C2 is composed of the following configuration items (CIs), linked as shown in Figure A.1:

- POINT (operator console),
- CMS: central main server,
- DBS: database server,
- CAS: Central Archive Server,
- GPS: Global Positioning System synchronization kit, and
- ODV: output devices (video walls, projectors and printers).

Figure A.1 reports subsystem (in this case C2) items and the internal and external interfaces among C2 and with external subsystems. The interfaces identify the flow of data. More detailed documents describing the type of data are foreseen in project specification but are out of the scope of the present description and often contain proprietary (company confidential) information.

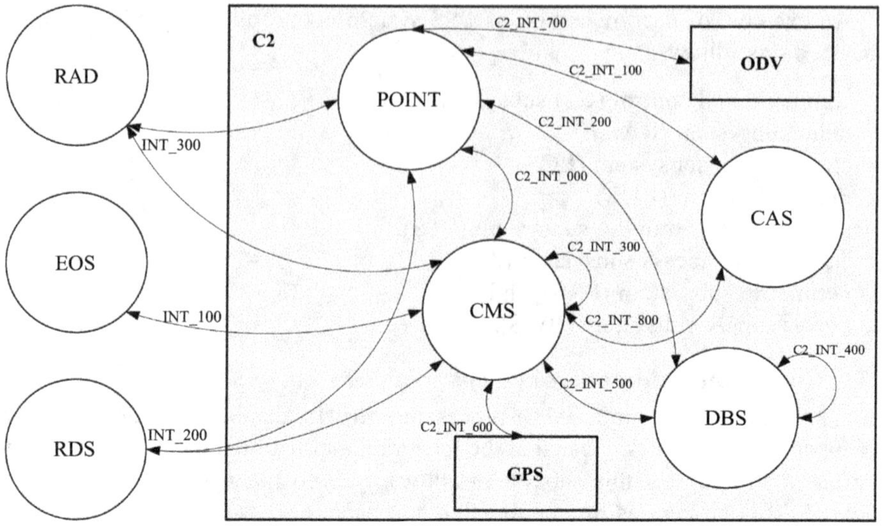

Figure A.1 C2 configuration items and internal and external interface identification

Table A.1 C2 configurations

	Element				
CI	**COs**	**ROs**	**LOs**	**PBs**	**MRS**
POINT	X	X	X	X	X
CMS	X	X	X	X	X
DBS	X				
CAS	X				
GPS	X			X	
ODV	X	X	X		

A service console is also provided for the installation and setting up of the server's software. Servers and POINT workstation are hosted in a rack server.

The C2 is distributed on different elements, namely COs, ROs, LOs, PBs and MRS. Each element has a different C2 configuration, as shown in Table A.1.

A.1.1 The POINT

The POINT CI is the console operator and represents the main human–computer interface of the system. It is installed at COs, ROs, LOs, PBs and MRS (Table A.2). Depending on the role of the default operator, the POINT CI has different configurations, namely

- CSU: Console dedicated to the system user (CSU) in charge of the processing and analysis of the border situation. CSU provides the display of the real-time situation, the monitoring of system status and the control of subsystems (RAD, EOS) parameters. In case that more than one CSU console is present at a single station, a hierarchy is created;

Table A.2 POINT configurations

		Element				
No.	**Configuration**	**COs**	**ROs**	**LOs**	**PBs**	**MRS**
1	CSU	2	2	2	1	2
2	CEI	1				
3	CSO	1				
4	CSA	1				

- CEI/CSO: The console/station of external information (CEI), dedicated to external users with functionalities of common operational picture (COP) visualization, and console dedicated to the assistant of the shift operative (CSO) officer, with functionalities of COP visualization;
- CSA: Console of System Administration (CSA), provided with the functionalities of system software and databases configuration, management and maintenance, and role management.

A.1.1.1 The authorization procedures and start-up of the POINT console presentation

The authorization procedure links the user, just authenticated by user name and password or any other form/type of secure authentication, to the appropriate operative role. An operative role defines the functions that a user could perform on the system. A user could have more than one operative role; however, only one can be performed at a given time.

The association between user, operative role and functions follows the logic schema shown in Figure A.2.

The operative functions (cf. Appendix B) available to a user depend on the operative role (e.g. operator, supervisor system administrator) selected among those authorized for a specific user.

A.1.2 Central main server (CMS) subsystem

The CMS CI is the server in charge of system data processing and performing the navigational control on the target shipping in the relevant maritime area. CMS CI functionalities are:

- multisensor fusion (elaboration of real-time surface and – eventually – aerial picture based on data coming from different sources);
- navigation controls (e.g. controls on tracks in order to grant the safety and security of navigation: position and collision avoidance, checkpoint crossing, speed limits, restricted and no-traffic areas, navigation assistance surveillance);
- alarms generations, related to traffic control;
- decision support (including artificial intelligence algorithms related to safety and security of the monitored area);
- accounting management (access granting of system users); and
- web DBS functions (services request, querying or data updates/additions).

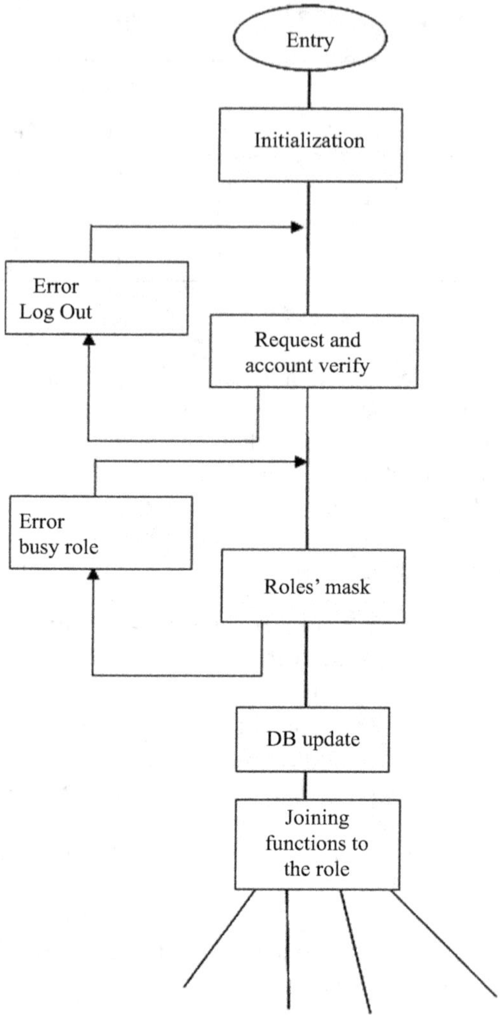

Figure A.2 Schematic representation of log-in initialization process on a multifunction operator console

A.1.3 Database subsystem (DBS)

The DBS CI is the server in charge of static and dynamic data storage related to the system targets. The database is accessed by real-time applications running both on POINT and on CMS.

DBS CI could be centralized at COs or distributed on each site and is based on the following commercial hardware/software aggregations:

- operating system (such as SUSE Linux Enterprise for instance);
- server (any brand); and
- database management system (such as Oracle Server Enterprise Edition for instance).

A.1.4 Central Archive Server

The CAS is the server in charge of the storage of system COP data to be used during playback session requested by an authorized user. Recording is performed by capturing all messages exchanged on the system network by means of proper software application. During playback sessions, recorded messages are copied on the system network with the same time frame.

Following the centralized database approach, the CAS CI is installed at COs and is based on the following hardware/software aggregations:

- operating system (such as SUSE Linux Enterprise for instance); and
- server (any brand).

The CAS at COs will be used to accumulate synthetic data of traffic image, related to last 12 (configurable parameter) months for instance. These data will be archived on its hard disk or on removable media.

A.1.5 Global Positioning System

The GPS CI is the kit performing the time synchronization of digital systems. Other GNSS (Global Navigation Satellite System) are also possible and equivalent.

It is composed of a GPS smart antenna and a compact synchronization interface module. It represents the time reference.

A.1.6 Output devices (ODVs) (video walls, projectors, printers)

ODVs are directly connected to POINT CI and used to display or print data. There are three kind of ODVs, installed at COs, ROs and LOs, as shown in Table A.3.

The VDW control system allows the management and composition of visualized windows, the redefinition of windows groups and the switching of different sets of windows groups up to four windows coming from four different image sources.

A.2 Radar subsystem

RAD is composed for instance of 25-kW frequency diversity transceiver and 21 ft slotted waveguide antenna installed on high towers of POs and piloted in remote way from the LOs.

Table A.3 Output devices

	Element		
CI	**COs**	**ROs**	**LOs**
Videowall (VDW)	X		
Projector (PRJ)	X	X	X
Printer (PRT)	X	X	X

Radars are provided with local radar head processors (RHP) capable of automatically performing target detection and tracking, and feeding compressed video, plots and tracks to the pertaining LOs processor for multisensor fusion. The radar tracks from all the radar systems are also fed to the ROs and to the CO for regional and national-level multisensor correlation and traffic picture compilation.

The system utilizes mainly radar systems in order to monitor and control targets moving and flying over the sea of the area of interest.

Radar system outputs are mainly digitalized video (strings), plots and tracks relative to detected targets.

The following data are transferred from RAD to C2 (up to LOs level):

- video, plots messages and tracks messages, in order to be locally fused, after that, will be forwarded to C2 higher level for the correlation process; and
- information about radar status.

On the other side, users working in the LO can set the radar configuration via commands (e.g. antenna rotation, radiation, operational mode selection, blind sector settings) that can be activated from their console interface.

The RAD is composed of the following CI, linked as shown in Figure A.3:

- radar sensor; and
- RHP.

A.2.1 Radar sensor

Radar sensor irradiates electromagnetic energy in the space and picks up echo from observed targets, producing analogue (radar) video to RHP (radar surveillance function). Radar sensor performances are assured by opportune choices of radar pulse, pulse repetition frequency and radar antenna speed values. Using X band and short pulse length allows, respectively, high performances in azimuth and range resolution. The antenna operates in circular polarization so that effects of rain clutters are reduced.

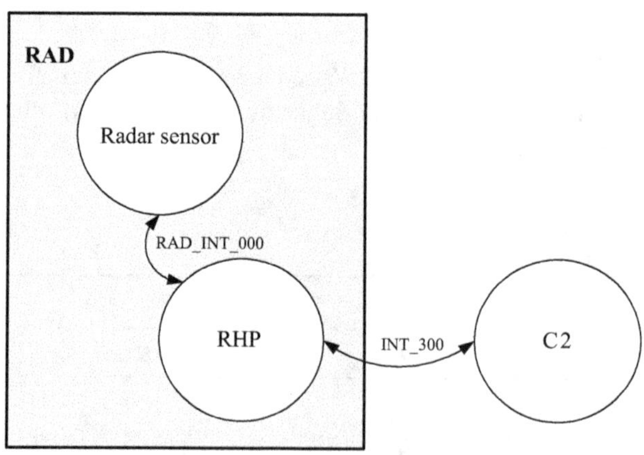

Figure A.3 RAD configuration items

The main components of radar sensors are (example):

- a slotted waveguide antenna,
- antenna safety switch,
- two frequency diversity transceivers,
- a service display,
- remote transceiver control and monitoring, and
- transceiver controller.

A.2.2 Radar head processor

The RHP performs the following functions:

- analogue-to-digital conversion of the radar signals,
- processing of the received signal for each transmitted pulse,
- processing in azimuth and string formation (video digitized),
- plot extraction and tracking functions,
- sending of the results of the reception to C2,
- local presentation of strings, plots and tracks,
- execution of the commands sent by the C2,
- reception of the status signals by the radar sensor, and
- radar sensor operative modes and status management.

The RHP system hardware architecture consists in a standard computer in which are installed on the peripheral component interconnect bus additional boards of calculation, acquisition data and interface. This architecture is modular and flexible as other cards could be added in order to increase the calculation power, carry out other functions and increase performances.

A.2.2.1 Generation of digital video

The radar video signal is converted into digital by a special compression technique, based on generation of strings, data structures representing digitized video by range, extension in azimuth and average amplitude (Figure A.4).

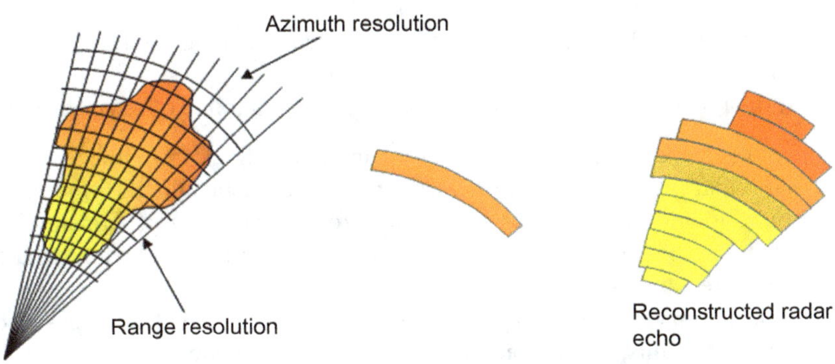

Figure A.4 Generation of digital video (video digitized)

Parameters are automatically or manually set by RHP. In the automatic mode, they depends on the radar operative mode and bandwidth of transmission line.

The 'string' element is characterized by:

- range,
- initial azimuth,
- final azimuth, and
- average amplitude on 8 bits for instance.

A.2.2.2 Plot extraction

The plot is the centre of gravity of the radar echo. It is characterized by a two-dimensional position (*x*, *y*) and by an average amplitude.

The extractor receives the radar video filtered by the video processor and performs the extraction of plots.

A.3 Electro-optic subsystem

The EOS is composed – for instance – of infrared (IR) camera and low light level TV (LLLTV) camera with automatic pan–tilt and zoom capabilities, installed at critical sites for ship identification and classification purposes. In the POs, the video from the cameras is encoded by the video and remote-control server, using MPEG format, and sent to the C2s over TCP/IP. In the relevant LO, the MPEG video is continuously stored by the digital video recorder. EOS is controlled and configured from the relative LO level.

In order to fulfil the requests coming from the C2, the main functions that the EOS supplies are as follows:

- Images acquired from sensors placed along the Polish coastline. The EOS provides mainly two types of videos (IR and LLLTV) according to the choice performed by the operator on its console. Each operator on its console is capable of:
  - selecting the monitor to display the video of cameras,
  - selecting at least two cameras of competence,
  - sizing the window cameras images,
  - modifying the image quality (by setting the contrast and the brightness), and
  - disabling or activating the representation of bearing line of selected cameras.
- Automatic target tracking. The operator can select a track from image of traffic scenario that must be tracked by a camera. The C2 sends the position of track (latitude, longitude) to server of the EOS for the pointing of cameras in the selected direction. On the console is possible to visualize an object, tracked by one or more cameras.
- Automated surveillance tour. The operator can draw a path, composed of a sequence of points, to be monitored by EOS in automatic way.
- Compression and forwarding of the images from PO to the relative control centre.
- Recording and playback of digital video.

The EOS is composed of the following CIs (see Figure A.5):

- video remote-control server,
- pan/tilt head,
- IR camera,
- LLLTV camera, and
- digital video recorder.

A.3.1 Video remote-control server

The video remote-control server is an electronic device that integrates a central processing unit, a power supply and an MPEG video frame grabber. It is a computer-based network appliance that allows remote control and communication with all the devices managed under its sensor configuration (pan/tilt head, LLLTV and IR cameras).

The server can be remotely configured using two possible alternatives, by serial port (via RS-232 at 9,600 Bd), or a web interface that can be accessed from a standard web browser on any authorized IP address within its network.

The server offers full remote operation and supervision of all the different functions and variables of the different sensors and devices connected to the server's serial port. These devices are interfaced by the server using the different drivers, abstracting the remote consoles or systems from their specific protocol details.

A.3.2 Pan/tilt head

The pan/tilt head is a two-axis assembly which provides accurate speed and position movement in azimuth and elevation. It is designed for heavy duty service in extremely demanding conditions; it consists in mechanic and electronic components which are ruggedized for reliable performance in an outdoor environment. The pan/tilt is provided with a slip ring connection which allows one to continue horizontal positioning. High

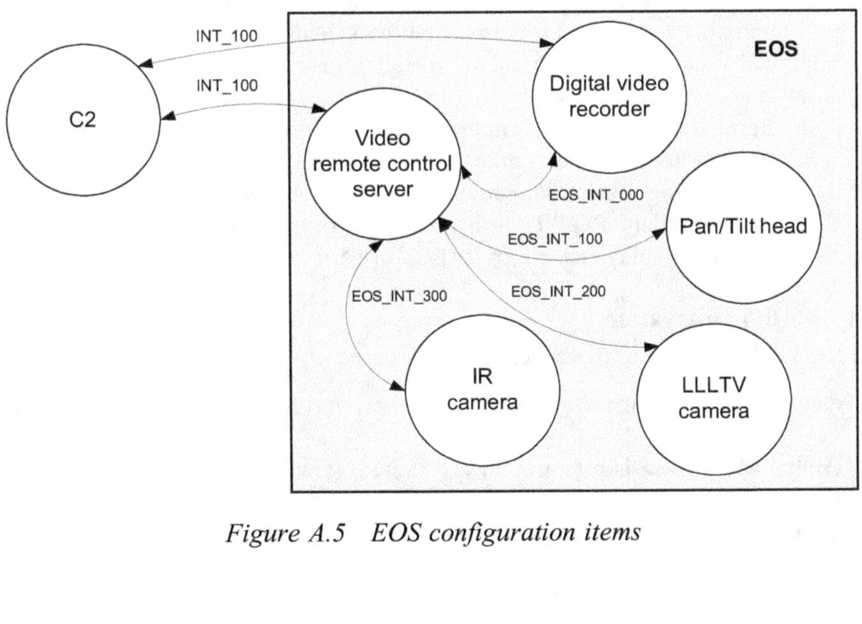

Figure A.5 EOS configuration items

speed is combined with a precise read-out of the vertical and horizontal angles allowing for accurate georeferencing of targets. All functions of the pan/tilt, including the sensors that are mounted onto the platform, are controlled over one single Ethernet cable. Serial control functions are converted to TCP/IP in the head itself.

A.3.3 IR camera

The IR camera could, for instance, be based on a 640×480 pixel quantum well detector with an excellent high-resolution image quality in all climate conditions and an operating range of more than 20 km. Three field of view optical systems can be configured, from a detection mode in the middle field of view to a six times magnified image that gives excellent identification at very long range. The 25° wide field of view provides instant situation awareness with overall thermal information.

IR camera is IP addressable which means it can easily connect to C2 for remote diagnostics and software upgrades.

A.3.4 LLLTV camera

The long-range surveillance camera is designed to fulfil a requirement for a low-maintenance, powerful, yet relatively light-weight camera system which provides colour imagery by day.

A.3.5 Digital video recorder

The digital video recorder offers the following functionalities:

- real-time recording of MPEG-4 ISO live streams to disk;
- remote-control commands for start and stop recording and create clips on the MPEG video stream associated to a given camera;
- remote-control status monitoring functions (e.g. health, BIT and specific functions like recording status, hard disk space, scheduler);
- programmable recording tasks managed by scheduler process;
- relational database for archival of metadata associated with recorded video assets and scheduler sessions;
- web interface for administration and configuration of services and settings;
- web interface to access and query stored video by camera, timestamp, user-defined recording session id, camera orientation and video settings; and
- support for standard Real Time Streaming Protocol Video Clients (e.g. VLC VideoLAN) with playback controls (fast forward, rewind, stop, pause and play).

A.4 Radio subsystem

The RDS could be part of the system network based on VHF radio links and used for

- data transfer from moving elements (vessels and MRS) to control centres and vice versa;
- Global Maritime Distress and Safety System (GMDSS) communications with vessels; and
- radio voice communications.

MRS consists of the following equipment:

- integrated multifunctional communication system (MCS), integrating radio communication with wired communication, including the necessary equipment of the surveillance centres;
- GMDSS system consisting of GMDSS servers with appropriate applications for the administrator and applications for operators;
- VHF voice, VHF data, VHF GMDSS voice, VHF GMDSS digital selective calling (DCS) base radio stations mounted in system cabinets installed in POs; and
- VHF data transmission base station on PBs.

The RDS is composed of the following CIs:

- MCS,
- operator console (SO),
- voice recording server,
- GMDSS server,
- Administrator Terminal (admin),
- GPS synchronization kit (GPS),
- radio control panel,
- VHF PBs Base Station,
- VHF Database Station (VHF data),
- VHF GMDSS Voice Base Station (VHF GMDSS Voice),
- VHF GMDSS Data Base Station (VHF GMDSS DCS),
- VHF Radio Telephone, and
- VHF Data Transmission Base Station.

These main functionalities are schematically depicted in Figure A.6.
The STS is composed of the following CIs (see Figure A.7):

- instructor station;
- radar simulator;
- POINT (integrated operator console); and
- training CMS.

A.4.1 Instructor station

The instructor station CI allows the creation of the exercises, assembling the components of a scenario, as geographical location, ships and weather condition. It has two main functionalities:

- Database objects (DBO) management that allows the definition of the elementary objects needed for the creation of the exercise. The following types of object are available:
  - geographical location (definition of the geographical localization of the exercise and to link the relevant geographical map), and
  - platform (definition of the features that will be created and animated during an exercise, such as object name, the kinematics parameters (as a range of values) and its radar cross section).

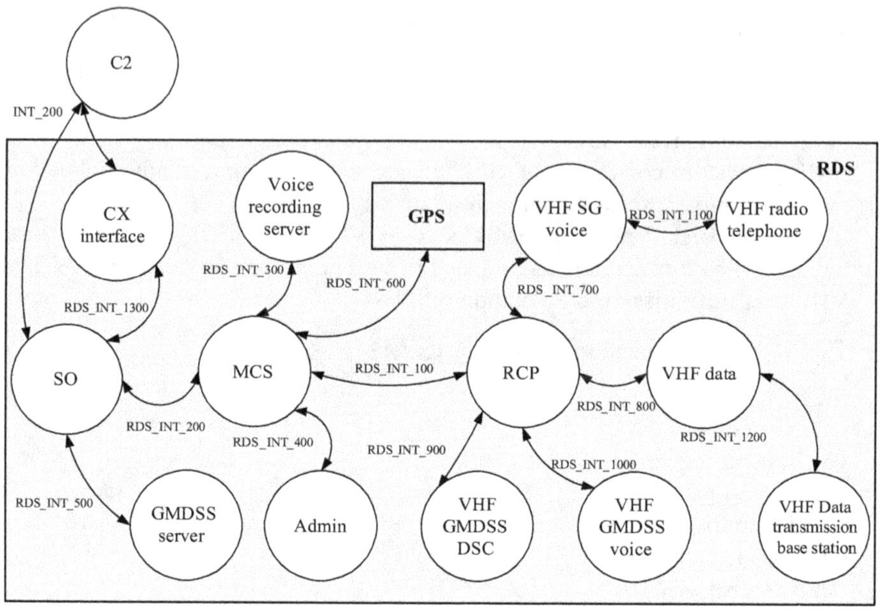

Figure A.6 RDS configuration items

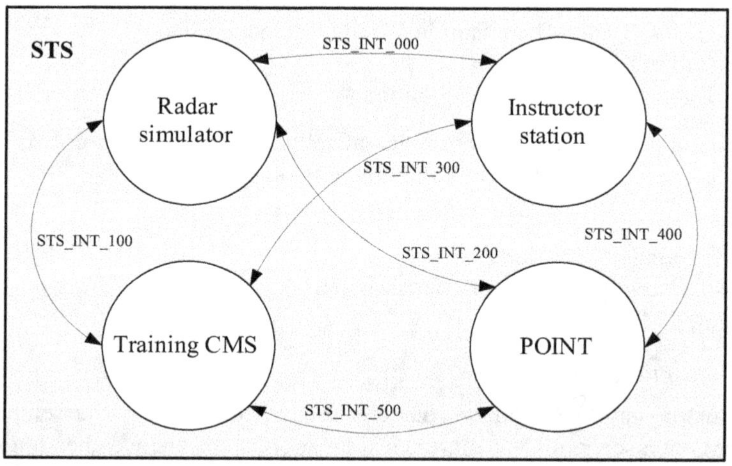

Figure A.7 STS configuration items

- Exercise creation management that allows the creation of the single exercise through the objects stored in the DBO. The compilation of the exercises will consist in some fundamental steps:
  - definition of the geographic locality of exercise,
  - creation of the platforms, defining initial cinematic parameters and manoeuvres, and
  - definition of the applicable weather condition for the exercise.

Through the instructor station is also possible to control and evaluate the training session. This functionality allows the trainer to start the execution of a predefined exercise and to interact with both simulator and trainees. From the instructor station, it is therefore possible to:

- select the exercises database of the exercise creation subsystem;
- send in execution the animation of a predefined exercise;
- manage the necessary simulators for the execution of the exercise;
- interact with the execution of the exercise (changing the route and the speed of the targets, inserting and deleting of targets);
- configure remotely the simulators; and
- display one of the selected trainee's consoles.

Finally, the instructor station performs the functionalities of:

- scenario generator: simulator of the coast environment, through the generation of the following features:
  - supplying of data referred to the coast line in order to be elaborated by the radar simulator,
  - supplying of 'simulated data' in order to guarantee the functionality of DBS, and
  - supplying of 'simulated data' in order to guarantee the functionality of C2 server.
- COs, ROs and LOs simulator: to supply all data for the right working of the real software of C2 server of COs, ROs and LOs.

The instructor station has two working modes: off line or run time. The off line mode consists of the composition and the exploiting of the DBO with an interactive environment, with scenarios of different complexity exploiting the exercises. The run time mode consists of selection and running the selected scenarios monitoring the course of the simulation.

A.4.2 Radar simulator

The radar simulator provides the simulated videos, plots and tracks which have to be sent to the system control centres in order to be fused and then to be analysed on trainees' consoles.

The radar simulator can simulate the contemporaneous activities of maximum four radars, allocated on coast line with the same performance of radars provided in the RAD, and a maximum of 2,000 objects/tracks (currently systems handle up to 10,000 objects approx.) for each training session.

A.4.3 Training operator console

The trainees' console has the same functionalities and configuration as the CSU (defined earlier).

A.4.4 Training CMS

The Training CMS has the same functionalities as the CMS subsystem described earlier.

The Training CMS is based on the following hardware/software aggregations:

- operating system (such as SUSE Linux Enterprise for instance), and
- server (any brand).

A.5 Simulation and training subsystem

The STS is composed of software and hardware components with the scope of providing a realistic environment to train operators by means of a real system reproduction. The STS is therefore composed of a reproduction of the real system, operator consoles and data processing servers, stimulated by a synthetic environment able to generate external inputs, as radar sensors data and communications.

The training centre represents the centre where the required facilities (hardware and software) for the arrangement of the training room are provided. An example of room consists in an area reserved to an instructor and the same consoles for trainees. The equipment are connected through an isolated local area network (LAN), separated from the rest of the system network. The trainer area is composed of a scenario generator, including radar simulators, and a monitoring station where the instructor can perform the necessary actions to configure, monitor, explain and analyse the exercises and the trainees' behaviours.

ROs and LOs training sessions shall be carried out remotely using predefined scenarios, without trainers but using training capabilities available at the training centre.

The STS can be split in three different levels in accordance to their different role:

- Training manager, consisting of those elements devoted to the performing of instructor actions. The instructor is the responsible for a training session not only with respect to the generation of the environment of exercises proposed but also with respect to the control of the actions performed by the trainees during the development of exercises. In particular, one of the most important roles of instructors is the explanation of an exercise in terms of relevant actions that shall be performed by an operator in case real situations similar to those simulated should occur.
- Integrated maritime picture (IMP) simulator, consisting of those elements necessary to simulate the complex surface image. In the training environment, the simulation of the IMP is performed in order to represent the maritime scenario, including the over imposing of radar data (videos, tracks and plots).
- User console simulator, consisting of those elements necessary to reproduce the functionalities of the users' consoles. The user console simulator role is the execution of the same actions that can be performed on consoles installed on system sites.

In the training centre, some functions are reproduced and some are simulated. The reproduced functions are realized by software or hardware with the same features of the real system, while the simulated functions are realized by special software and/or hardware.

A.6 Site security subsystem (ACS)

The site security subsystem is composed of remote site security cameras, performing a perimetric video surveillance of each PO, of alarm sensors installed at POs and of a site security console at LOs, in order to guarantee the remotes' site protection against unauthorized accesses.

The ACS is composed of the following CIs:

- alarm sensor,
- CCTV camera,
- ACS server, and
- ACS console.

Sensors and cameras send data to servers located at POs. Real-time situation and images are sent to ACS consoles located at LOs.

A.7 Perimetric subsystem

The PRS is composed of a surveillance system monitoring a specific area usually a national border or a such. The subsystem's main task is to detect persons and vehicles illegally trying to cross the country's border along the border. It could be up to km longer.

The PRS is composed of the following CIs:

- alarm sensor, i.e. sensor cable buried along the entire border section;
- TV cameras;
- recording system, for the sequential image recording on a digital video recorder, the recording date/time-stamped video, the search and playback of recordings, the recording events as continuous video with a 40-s pre-alarm and post-alarm;
- PRS console, providing manual display of all controlled cameras, visual and acoustic alarm of trespassing, and automatic display of images coming from the camera that generated the alarm.

A.8 Power supply subsystem

The PWS is composed of diesel generators and uninterruptible power supply (UPS) devices to guarantee the continuity of power supplies to all components. Diesel generators are located at POs, UPS are located at all centres (COs, ROs, LOs, POs, PBs and MRS).

Appendix B
VTS/FOC/CSS main functions

Michele Fiorini[1,*]

Main operational functionalities for a VTS/CSS/FOC system are:

- sensors and communications interfaces;
- traffic picture handling;
- navigational control and traffic management;
- decision aids;
- presentation management;
- database management;
- recording and playback;
- simulation and training management;
- security management; and
- system management.

B.1 Sensors and communications interfaces

- Main interfaces are as follows:
  - from/to RAD;
  - from AIS system;
  - from/to imaging sensors (LLLTV and IR cameras);
  - from/to radio communications subsystem (voice and GMDSS communication, VHF and microwave radio link network);
  - from anti-intrusion and security sensors (alarm sensors, closed circuit TV, fire sensors, etc.);
  - from hydrological and meteorological sensors;
  - from/to other systems VTMS, VMS, SWIBZ (notice to marines, system tracks and DB static data); and
  - from/to pilot section.

B.2 Traffic picture handling

- Management of radar video (radar echo processing);
- Management of radar plots (extractor settings);

[1]Leonardo S.p.A., Rome, Italy

[*]The opinion expressed in this work reflects the professional experience of the author alone and does not necessarily represent the position of the company.

- Management of tracks (traffic-integrated images formation):
  - local radar tracks management
  - AIS tracks management
  - system track management
- Management of LLLTV/IR video (camera settings);
- Identification of targets;
- Prediction of vessel movements;
- Monitoring of special 'tagged' vessels; and
- Drawing and editing control maps in the operational area.

B.3 Navigational control and traffic management

- **Information service**:
  - speed limit control,
  - collision avoidance,
  - anchorage and buoyage watching,
  - transit and anchorage in prohibited area;
- **Navigational assistance**:
  - data communications with patrol units transiting in the area;
- **Traffic organization service**:
  - monitoring and control of ships approaching and leaving competent areas/ports.

B.4 Decision aids

- Threats or emergencies management through the exploitation of the following capabilities:
  - **Mission planning**: capability to select and manage patrol units in the event of threats or emergency situations;
    - anti-pollution mission
    - search and rescue mission
    - target interception
    - target hunting
  - **Threats evaluation**: automatic identification of threats in accordance to target anomalous behaviours (suspected or hostile);
  - **Traffic statistics and reports**: alarms, target detection, statistical analysis.

B.5 Presentation management

- Cartography presentation:
  - S63 sea chart
  - V-Map Land Chart
- Traffic picture presentation:
  - radar video
  - tracks
    - radar plots
    - local tracks
    - system tracks

    - system tracks (hooked, manual, handover, cost)
    - hook window
  - AIS messages
  - recording and play back
  - IR camera
  - LLLTV
  - alarm
  - system status

- Display management:
  - graphical local functions
    - click and drag
    - overlapping
    - scale selection
    - rotation of presentation
    - panning
    - off centre function
    - full display function
    - cursor management
    - brightness symbols setting
    - unit selection
    - coordinates selection
    - forced position
    - track follow
    - zoom/unzoom
    - quick zoom
    - language selection
    - hard copy
    - status and dialog windows
    - colour selection
  - graphical external functions
    - video wall
    - projector

B.6 Database management

- Ships
  - static data (length, width, height, full and net loading height, displacement (net of moving water), IMO number, call-sign, name, country code, max speed, number of bridge, number crew members)
  - ship class
  - ship fleet/group
  - license
  - permission
- Users
  - password
  - user profile

  - user functions
  - command centre configuration

- External system interface
  - services that system exposes
  - messages received from external systems, in streaming TCP/UDP mode

- Navigation plan
  - date and hour of navigation plan creation;
  - command centre code that inserted the navigation plan;
  - estimated date and hour (of system) of arrival in check zone;
  - date and hour of identification;
  - invoice date related to last navigation;
  - passengers;
  - danger code;
  - number of occupied dock or to be occupied;
  - anchorage (YES, NO);
  - status monitoring;
  - pilotage status (on-board=ship on-board piloting, remote=distance piloting) and
  - shipping report (date of system of report insertion, latitude or azimuth of ship position, longitude or distance of ship position, speed, route).
    - Qualification of person
    - list of reports
    - Shipping Report (contact reports sent from ships)
    - On-board_staff

- Emergencies
  - harbour master number
  - emergency description
  - emergency date and hour
  - description of meteorological conditions
  - place of reporting organization
  - number of alarmed authority
  - pollution characteristics (in the case of polluting happening leak)
  - emergency status: 0=closed, 1=open, 2=in course
  - involved units (YES/NO, passengers number, crew members' number, possible number of people deceased during the accident)
    - Reporting organization
    - Mission
    - Means (description, availability, equipment description)
    - Intervention_type
    - Pollution type
    - Involved unities
    - Placet
    - Infringement (post noticing the infringement, date and hour of the infringement, text associated with infringement, status of infringement: open, closed).

B.7 Recording and playback

- Recording, archiving and reproduction of system real-time data.
- Each subsystem records and playback its own data.
- The following streams are handled:
  - Audio
  - Synthetic data
  - Video (E/O)
  - Playback
  - Security CCTV

B.8 Simulation and training management

The training environment is composed by:

- training-dedicated console (replicating operational environment),
- training central server (replicating operational server functions),
- training position:
  - scenario generator
  - radar simulator
- training environment functions,
- scenario configuration,
- execution of training session,
  - scenery data generation
  - CMS
  - simulation and training presentation (HMI)
- exercise playback,
- instructor monitoring position

B.9 Security management

- Based on the AAA principles:
  - Authentication
  - Authorization
  - Accounting
- Hierarchically managed (for all users – internal and external)
- Roles composed of user name, password and supported capabilities. Default roles are the following:
  - **System administrator**
    - database configuration, management and maintenance (e.g. security copies, loading and downloading elements keeping the data integrity, access rights, management);
    - recorded information management and maintenance;
    - accounting management;
    - import of new maps in the system;

- configure and control the information exchange with external systems;
- view and edit log files;
- change of display format and configuration of data fields (object characteristics);
- manual switching between national control centre and backup national control centre;
- remote control and monitoring of system IT equipment (e.g. server, workstation, switch and PC);
- act as supervisor or operator.

– **Supervisor**

- monitoring of system status;
- controlling the activities of the system's users working on its control centre (e.g. active sessions, entry logs);
- starting the record of scenarios and reproducing them;
- export the recorded data;
- accessing to log book, with the scope of check the error/fault/alarm occurred during the system operation (consults, printing and inform elaboration);
- hiding tracks and defining blind areas, where tracks shall not be presented to users;
- act as operator.

– **Operator**

- visualizing the integrated traffic picture;
- modifying the integrated traffic picture to its needs by means the performing of graphical local functions;
- remotely controlling radar sensors: modifying values of main parameter of radar sensors (STC, FTC) applied in the video extraction process, modifying the areas of inhibition and auto-acquisition in use in radar sensors;
- remotely controlling of EOS: modifying values of EOS applied in the video processing, modifying the areas of inhibition and auto-acquisition in use in EOS;
- generating reports;
- consulting and entering data in the system database;
- acquisition and identification of tracks and
- edit of local maps.

– **Guest**

A guest shall be able to see all the information gathered by the system. However, it shall not be able to interfere with operational functions.

B.10 System management

- System management defines all those aspects that are necessary for the correct functionality of the whole system.
- This section takes care of the following subjects:
  - system/subsystem status collection and presentation;
  - system/subsystem control capability;
  - system access procedure;
  - priority management on system resources accesses (E/O, VHF);
  - system log books and
  - radio communications management.

Appendix C
Autonomous vessel projects (MASS overview)

Aristotelis Komianos[1]

C.1 Introduction

The following appendix reviews the different types of maritime autonomous surface ship (MASS) projects currently taking place around the world and draws out some of the implications of the deployment of autonomous vessels for port services. Autonomous vessels can generally be divided into two distinct categories; remote ships, which are operated by remote-control means, and automated ships, structured with built-in cutting edge command and control (C2) systems on-board, which operate the ship with minimal or no human intervention. These vessels, which are equipped with self-steering systems and sensors for surface object detection and which are thus able to avoid collisions, will start a new era in maritime transportation.

C.2 MASS projects addressing the different segments of the global fleet

There are numerous autonomous shipping projects worth mentioning, including both relatively smaller automated barges, self-directed harbour tugboats, 120 TEU (Twenty-foot Equivalent Unit) crewless container ships, all the way to autonomous balk carriers. Others include Rosmorport's unmanned barge hopper [1], the ASKO AutoBarge [2], the Japanese remote-controlled tug [3], and Koreans' smart autonomous ship [4]. Following are the main projects currently playing a significant role in the advance of MASS which are greater than 24 m in length.

C.2.1 The maritime unmanned navigation through intelligence in networks (MUNIN)

The maritime unmanned navigation through intelligence in networks (MUNIN 2012–15) was sponsored by the European Commission, to examine the technical, economic and legal feasibility of unmanned ships. The project's vessel, which was either controlled by an automated system on-board or by land-based operators, was able to utilize safety and operational sensors to search, detect and evade floating objects in order to avoid collisions [5].

[1]Consultant, Fellow of the Nautical Institute, London, UK

C.2.2 The ReVolt

The project was developed by DNV GL. It is a 60-m long, crewless, autonomous short sea vessel, which is powered by batteries and is emission free. Based on an evaluation of current requirements along short sea routes, the ReVolt could save up to USD 34 million during its estimated 30-year lifetime if compared to a diesel-run ship, i.e. a saving of approximately one million USD annually. The extension of the project to include land-based charging facilities is one of the DNV GL priorities [6].

C.2.3 The Advanced Autonomous Waterborne Applications Initiative (AAWA)

Launched by Rolls-Royce in 2015, the objective of the Advanced Autonomous Waterborne Applications Initiative is to bring together academia, shipbuilders, manufacturers and classification societies in order to explore the economic, social, legal, regulatory and technological factors necessary to make autonomous ships a reality. The marine division of Rolls-Royce, comparing the smartphone revolution to the impact of the smart ship on ship design and operations, stated that shipping autonomy is the future [7].

C.2.4 The YARA Birkeland

A fully electric open top container feeder vessel, the Yara Birkeland, is the result of a YARA and KONGSBERG cooperation, which started in 2017 as a conventional crewed vessel but is now scheduled to be autonomous by 2020. Its use will reduce the number of truck journeys in populated urban areas by up to 40,000 rides, thus reducing NO_x and CO_2 emissions, improving road safety and relieving traffic congestion. It can transport 120 TEU. All cargo loading and unloading operations will be executed automatically using electric cranes, the battery packs will be used as permanent ballast and she will be capable of automatic berthing or go underway without the need of any special docking structures [8].

C.2.5 The cyber-enabled ship project

Developed by Lloyd's Register, the cyber-enabled ship project examines the processes and control of autonomous vessels, seen as a system of systems, including the system itself, the human–system, the software, the network and communications, the data assurance and the cybersecurity. Seven levels of autonomy (AL) have been studied: from entirely manual with no autonomous function to a fully autonomous operational mode, in which all the decisions during operations are made and all the actions are carried out by the system, totally unsupervised [9].

C.2.6 The autonomous marine operations and systems (AMOS)

The project was started by the Department of Marine Technology and the Department of Engineering Cybernetics at the Norwegian University of Science and Technology in cooperation with international and national partners. It focuses on autonomous systems, coordinated operation of a sensors' network to monitor unmanned vehicles using floating nodes, integrated underwater navigation, mapping, autonomous object detection and tracking in marine environments using IR sensors, sensor-based guidance and path optimization, coordinated and cooperative control architectures for intelligent task execution and collision avoidance in uncertain maritime environments [10].

The International Maritime Organization (IMO) has agreed on a four-AL description for MASS, ranging from vessels with limited crew to no one on board, and is currently working on establishing and analysing the influence that MASS vessels will have on-shipping regulations, e.g. collision regulations and safety of life at sea.

Currently, all regulations are produced for manned ships only. In June 2019, the 101 session of the Maritime Safety Committee approved interim guidelines for MASS trials ensuring that they demand the same level of safety, security and environmental protection required for conventional manned vessels. Only properly trained, competent personnel should participate in MASS trials, while all necessary precautions will be taken to ensure the effectiveness of cybersecurity [11].

C.3 Constraints related to the deployment and commercialization of MASS

Although numerous autonomous shipping projects have been initiated, some key questions and concerns still need to be addressed. From a strictly technical point of view, there is a need for dedicated testing areas during the initial phases of trials and integration of such systems in the port facilities structure. This will need to include special docks, designated channels or corridors, operation rooms manned with trained personnel, particular communication schemes, adaptation of the existing regulatory and statutory norms for port safety and security to the new reality, particular vessel traffic service (VTS) interaction with unmanned vessels, meteorological platforms to determine most suitable weather conditions and robust cybersecurity initiatives.

Additionally, ports will need to consider the deployment and use of measures such as:

- automatic berthing;
- digital platforms to control MASS arrivals and departures;
- adjusted infrastructure to load and unload MASS;
- automation to transfer cargo to trains;
- lorries and other secondary means of short distance transportation and electric recharging stations for battery-driven MASS;
- cybersecurity;
- communications;
- testing areas during the initial phase of using MASS;
- VTS interaction;
- separation channels;
- shore control centres (SCC);
- port safety and security;
- statutory adjustments;
- dedicated meteorological and digital platforms;
- automated cranes and MASS loading and unloading facilities and infrastructure; and
- charging stations for electric MASS.

The most important area treated next, in turn.

C.3.1 Cybersecurity

Whenever any system interacts with another by exchanging data in any form and by any means, the cyber (and digital) security is of paramount importance. Various methods of intervening, hacking, distorting and destroying information technology systems already exist on the dark Internet, not to mention the relevant capabilities of rogue states and criminal or terrorist organizations. Anyone with the knowledge of digital penetration, a laptop computer or even a smart phone and a desire to do harm, ask for a ransom or even play, could potentially affect these systems negatively.

It is the concern of the cybersecurity to assess risk and to deny or at least restrict access to anyone intent on such malicious, illegal actions. Software management procedures, identification and cyberthreat mitigation guidance, updated instructions on cybersecurity from industry and classification society, procedures for password management and a cybersecurity awareness programme accessible to all staff involved in MASS control are, among others, some of the required preventive actions needed. In recent years, the Protection and Indemnity clubs and insurance companies have even started to demand high-quality cybersecurity procedures and extensive digital protection for conventional manned ships.

C.3.2 Communications

Uninterrupted, short-distance, jam-free communications with high volume data capacity must be available at all times, while operations of MASS are in progress. In the event of a failure or jam in communications, the consequences could be dire, especially during the last phases of entering or exiting the port, when heavy traffic within extremely short distances is common. Actually, in the case of MASS, the communications are the 'nerves' of the system, issuing all commands and receiving the relevant responses. If, for whatever reason, this transfer of information is interrupted, disrupted or stopped, the whole system will produce errors and abnormal behaviour.

C.3.3 Testing premises

The initial phase of autonomous shipping integration in any port of call will require designated testing areas and piers with specialized equipment and infrastructure. In order to avoid any accidents and provide freedom for uninterrupted evaluation, these areas must be properly prepared and certified for autonomous ship trials.

It is therefore essential that they are selected very carefully in advance. These prerequisites may produce congestion and unwanted delays to ports; thus, very detailed forward planning is required sufficiently ahead of time in order to mitigate these phenomena. The most crucial technical and infrastructural needs are analysed in the following paragraphs. They include, among others, specialized short-distance communications, remote-control rooms, operation cabinets, automatic berthing facilities, automated cranes, VTS interaction procedures, charging stations for electrically powered MASS and state-of-the-art safety and security processes for the new MASS reality and needs.

C.3.4 Shore control centres

The safe and efficient C2 of MASS, during the final stage of entering the port and mooring at a dock or staying at anchor, require an advanced remote-control system and a fully functional operation room, even in the case of a fully automated unmanned vessel for safety reasons. Properly trained, competent personnel must be present in such rooms to control MASS in the vicinity of port limits and when entering the port. These rooms, separate or combined, may be named SCC as MUNIN project findings suggest in MUNIN-FP7 GA-No 314286.

C.3.5 Port safety and security

The International Safety Management Code and the International Ship and Port Facility Security Code have not taken into consideration the existence of MASS and the operation of such vessels at ports of call. A proper review of the relevant clauses and the full adoption and implementation of required actions and procedures by all ports that will facilitate MASS are of paramount importance. In this way, not only will the safety and security of the ports be improved and safeguarded, but the areas in the vicinity of such facilities will also be protected from accidents.

C.3.6 Statutory instruments adaptation

Statutory instruments should be used to change existing legislation so that it covers MASS entering or leaving ports of call in accordance with the growth of such technology.

These revisions will have an impact on various technical, commercial and operational standards, including chartering, ship management, insurance and best management practices. As previously mentioned, the IMO has already initiated the processes of defining MASS and including this new technology on the agenda.

C.3.7 VTS interaction

VTS are systems for exchanging information, advice and instructions for safe and secure passage between the crew of a vessel inside a VTS-established area and the VTS operators (VTSOs), who are the traffic controllers of that area. Since its establishment in the late 1940s, the system has proved very effective in controlling shipping-congested areas and approaches to ports while mitigating the risk of collision and grounding. Other functions of a VTS include the exchange of information regarding, for example, tug services, pilotage, berthing and a vessel's estimated time of arrival and departure, thus enhancing port operations.

Although all the above-mentioned information and advice are extremely useful for the safety and security of a ship, the overriding authority of the master of the vessel makes him responsible for taking the final decision, in accordance with best management practices, nautical science, seamanship and the peculiarities of the particular situation in space and time.

IMO Resolution A.578(14) states that '...VTS operations do not encroach upon the master's responsibility for safe navigation...', while IMO Resolution A.857(20) states that '...When the VTS is authorized to issue instructions to vessels, these instructions should be result-oriented only, leaving the details of execution, ..., to the master or pilot on board the vessel...'.

These instructions create a regulatory gap in the case of a MASS, where no master or pilot will be present on-board an unmanned vessel. Even when the regulatory and statutory issues are resolved by, for example, the delegation of the master's authority of a crewless vessel to a competent remote-control 'master' during the last phases of a MASS approach or berthing, the VTSOs must have access to relevant information.

C.3.8 Dedicated separation channels

During the initial phase of MASS operations and until further regulatory and statutory guidance is given, separate lanes must be designated exclusively for the entry and exit of MASS in ports of call. This practice will improve the safety of navigation in such areas and give the appropriate time for the personnel of remote control and operation rooms to react, in the event of any malfunction or human error. Additionally, vessels close to such lanes will be aware of the MASS presence, and in addition to the automated identification system and other navigational aids, they will properly adjust their course and speed to steer safely.

C.3.9 Meteorological platform

In order to determine the most suitable weather and sea state conditions for entering and exiting the ports of call, a meteorological platform specifically for MASS operations should be present in the remote control and operations centre (or SCC). The information provided will help the operators to better control the safe operation of MASS and assess any risks due to unfavourable conditions in advance. Integrated with the rest of the MASS systems, it will also help VTS, port state control (PSC), pilots, the shipping company operators and other vessels in the vicinity to recognize potential difficulties with MASS steering and navigation in almost real time.

C.3.10 Digital platform

In order to properly handle MASS arrivals and departures at ports of call, a digital platform should be established, capable of and exclusively for interacting with SCC, PSC, VTS, pilot services, charters and shipping companies alike. This platform could be a determining factor behind better traffic and berthing oversights mitigation, faster or even automatic chartering procedures, sufficient information regarding a vessel's required reports concerning such matters as cargo, dangerous shipments on-board, the draft, condition, maintenance requirements. All these data and information could be saved and analysed further, for the benefit of the MASS operations and the fiscal growth and availability of ports of call.

C.3.11 Automatic berthing

Automatic berthing technology is already in use and is gaining ground, due to the safety that it provides, facilitating humans and machines alike. Various automated berthing methods exist, like the magnetic, the vacuum pad and the autonomous mooring floating system. It is anticipated that they will, in most cases, cater for MASS needs. That said, an entire infrastructure must be designed and implemented for the ports to be ready to facilitate such technology. Proper electric power systems, magnetic shields for data storage protection, avoidance of any electromagnetic interference between the various systems and port facilities, extensive studies and forward planning would all ensure the proper use of such systems. Initially, the method of self-regulation or that of best practices would definitely be a guidance for such issues, until more specific procedures are adopted.

C.3.12 Automated cranes

Any autonomous vessel can easily be integrated with an automatic crane or a loading/unloading system. This kind of technology already exits and would further improve the total automation and efficiency of the port facilities. Supported by relevant artificial intelligence algorithms and designed with the best possible safety and security requirements, the overall system would minimize the time necessary for cargo operations and mitigate or even eliminate any accidents resulting from human error due to fatigue, poor judgment or other reasons.

C.3.13 Loading/unloading infrastructure

Current loading and unloading infrastructure will remain in use for the time being. A new approach to the design of future port facilities and even the reshaping of existing ones to better satisfy the MASS needs will give a competitive advantage to any port which, sooner or later, chooses to accommodate such vessels.

C.3.14 Cargo transfer automation

Automatic cargo transfer to port trucks, trains or any other mean of relevant transportation will minimize the required time usually needed for such operations and will maximize the profit from dock and pier usage while being in the best interests of the vessels waiting for their turn to load or unload. As with the automatic crane, such an initiative will eliminate potential accidents and will be beneficial for the overall logistic chain.

C.3.15 Recharging stations

There is a tendency for MASS to be powered by renewable energy in order to more efficiently tackle issues regarding environmental protection, social acceptance and increased independence from fossil fuel. This trend will force ports to re-evaluate their refuelling/refilling infrastructure. The means to charge electrically powered MASS is an initiative which will pay off as soon as this kind of vessel is part of the logistic chain of maritime transportation. The electric power provided to MASS may also be produced by renewable sources using solar panels, wind generators, wave energy convertors or geothermal systems. The benefits to port-adjacent areas will obviously be multiple, from cleaner air to better environmental conditions and, consequently, quality of life.

C.4 Links to main existing and ongoing projects

Further info for the main existing and ongoing projects may be obtained from the following links*:

The Maritime Unmanned Navigation through Intelligence in Networks (MUNIN)

http://www.unmanned-ship.org/munin/

The ReVolt

https://www.dnvgl.com/technology-innovation/revolt/index.html

The Advanced Autonomous Waterborne Applications Initiative (AAWA)

https://www.rolls-royce.com/~/media/Files/R/Rolls-Royce/documents/customers/marine/ship-intel/aawa-whitepaper-210616.pdf

The YARA Birkeland

https://www.yara.com/news-and-media/press-kits/yara-birkeland-press-kit/

The Cyber-enabled ship project

https://www.lr.org/en/cyber-safe-for-marine/

The Autonomous Marine Operations and Systems (AMOS)

https://www.ntnu.edu/amos

The Maritime Autonomous Surface Ship (MASS)

http://www.imo.org/en/MediaCentre/HotTopics/Pages/Autonomous-shipping.aspx

References

[1] Rosmorport Set to Test Unmanned Hopper Barge in 2019. [Online]. Available from: http://en.portnews.ru/news/272861/. [Accessed 20 March 2021].

[2] 119 Enova-Millioner Til ASKOs Autonome Fartøy. [Online]. Available from: http://presse.enova.no/news/119-enova-millioner-til-askos-autonome-fartoey-362196. [Accessed 20 March 2021].

[3] Releases news NYK Line, 2018. NYK to Participate in Demonstration Project to Remotely Operate a Ship. [Online]. Available from: https://www.nyk.com/english/news/2018/20180810_01.html. [Accessed 20 March 2021].

[4] International Network for Autonomous Ships (INAS). [Online]. Available from: http://www.autonomous-ship.org/. [Accessed 20 March 2021].

[5] MUNIN – Maritime Unmanned Navigation through Intelligence in Networks, 2016. [Online]. Available from: http://www.unmanned-ship.org/munin/. [Accessed 20 March 2021].

*All links accessed on March 2021.

[6] The ReVolt – A new inspirational ship concept. [Online]. Available from: http:// www.dnvgl.com/technology-innovation/revolt/index.html. [Accessed 20 March 2021].

[7] Rolls-Royce, 2016. Rolls-Royce publishes vision of the future of remote and autonomous shipping. [Online]. Available from: http://www.rolls-royce.com/media/press-releases/2016/21-06-2016-rr-publishes-vision-of-the-future-of-remote-and-autonomous-shipping.aspx. [Accessed 20 March 2021].

[8] Yara Birkeland press kit, 2020. Yara Birkeland status. [Online]. Available from: https://www.yara.com/news-and-media/press-kits/yara-birkeland-press-kit/. [Accessed 20 March 2021].

[9] Lloyd's Register (2016). "Cyber-Enabled Ships. Deploying Information and Communications Technology in Shipping" – Lloyd's Register's Approach to Assurance, 1st Edition.

[10] NTNU AMOS – Centre for Autonomous Marine Operations and Systems. [Online]. Available from: https://www.ntnu.edu/amos. [Accessed 20 March 2021].

[11] Autonomous Shipping. [Online]. Available from: www.imo.org/en/MediaCentre/HotTopics/Pages/Autonomous-shipping.aspx. [Accessed 20 March 20

Further reading

Baker J. and Baker J. How Should Ports Prepare for Autonomous Shipping? [Internet]. Ship Technology. 2019 [cited 28 September 2019]. Available from: https://www.ship-technology.com/features/ports-autonomous-shipping.

Blanke M., Henriques M., and Bang J. (2016). A Pre-analysis on Autonomous Ships. The Technical University of Denmark (DTU).

Bryant D. T. (1991). The Human Element in Shipping Casualties. Report prepared for the Dept. of Transport, Marine Directorate, United Kingdom.

Burmeistera H. and Bruhna W. Can Unmanned Ships Improve Navigational Safety [online]. 2014. Publications.lib.chalmers.se. Available from: http://publications.lib.chalmers.se/records/fulltext/198207/local_198207.pdf.

Chong J. C. Impact of maritime autonomous surface ships (MASS) on VTS operations. World Maritime University Dissertations. 647; 2018. Available from: https://commons.wmu.se/all_dissertations/647.

COLREG (1972). Convention on the International Regulations for Preventing Collisions at Sea. IMO.

Holder C., Khurana V., Harrison F., and Jacobs L. Robotics and Law: Key Legal and Regulatory Implications of the Robotics Age (Part I of II). Computer Law & Security Review. 2016;32:383–402.

Holder C., Khurana V., Hook J., Bacon G., and Day R. Robotics and Law: Key Legal and Regulatory Implications of the Robotics Age (Part II of II). Computer Law & Security Review. 2016;32:557–576.

Danish Maritime Authority Report (2017). Analysis of Regulatory Barriers to Autonomous Ships. Available from: https://www.dma.dk/Documents/Publikationer/Analysis of Regulatory Barriers to the Use of Autonomous Ships.pdf.

Deployable and Autonomous Mooring System (2004). Florida Atlantic University Libraries. Available from http://digitalcommons.fau.edu/patents/56.

Endsley M. R. and Jones D. G. (2012). Designing for Situation Awareness: An Approach to User Centered Design, 2nd Edition. London: Taylor and Francis.

FRAUNHOFER-CML (2019). Autonomous Vehicles' Impact on Port Infrastructure Requirements. Hamburg Port Authority AöR.

Futurenautics Autonomous Ships – White Paper [online]. 2016. Available from: http://www.futurenautics.com/2016/11/white-paper-autonomous-ships.

Jokioinen E. (2016). Autonomous Ships – The Next Steps. Available from: https://www.rolls-royce.com/~/media/Files/R/Rolls-Royce/documents/customers/marine/ship-intel/rr-ship-intel-aawa-8pg.pdf.

International Underwriting Association of London Limited (2016). Cyber Risks and Insurance. An Introduction to Cross Class Cyber Liabilities. London. Available from: https://www.maritimelondon.com/wp-content/uploads/2016/01/005_Cyber_Risks_Combined_110116.pdf.

IMO LEG 105/11/1 (2018). Proposal for a Regulatory Scoping Exercise and Gap Analysis With Respect to Maritime Autonomous Surface Ships (MASS).

IMO (1974). International Convention for the Safety of Life at Sea (SOLAS).

IMO (1997). International Safety Management Code (ISM Code) and Guidelines on the Implementation of the ISM Code. London: International Maritime Organization.

IMO (2003). ISPS Code: International Ship and Port Facility Security Code and SOLAS Amendments Adopted 12 December 2002. London.

IMO (2011). MARPOL Consolidated Edition 2011: Articles, Protocols, Annexes and Unified Interpretations of the International Convention for the Prevention of Pollution from Ships, 1973, as Modified by the 1978 and 1997 Protocols.

IMO (2016). E-navigation. Available from: http://www.imo.org/en/OurWork/safety/navigation/pages/enavigation.aspx.

Komianos A. The Autonomous Shipping Era. Operational, Regulatory, and Quality Challenges. TransNav, the International Journal on Marine Navigation and Safety of Sea Transportation. 2018;12(2):335–348.

Komianos A. (2019). Towards the Autonomous Shipp: Operational, Regulatory, Quality Challenges and Other Maritime Transportation Issues. London, UK: Austin Macauley Publishers Ltd.

Koopman P. and Wagner M. (2016). Challenges in Autonomous Vehicle Testing and Validation.

MacKinnon S. N., Man Y., Lundh M., and Porathe T. (2015). "Command and Control of Unmanned Vessels: Keeping Shore Based Operators in-the-Loop." ATENA Conferences System, NAV 2015 18th International Conference on Ships and Shipping Research, Milan, Italy, June 24–25.

Lee J. D. and See K. A. Trust in Automation: Designing for Appropriate Reliance. Human Factors: The Journal of the Human Factors and Ergonomics Society. 2004;46(1):50–80.

Man Y., Weber R., Cimbritz J., Lundh M., and MacKinnon SN. (2018). Human Factor Issues During Remote Ship Monitoring Tasks: An Ecological Lesson for System Design in a Distributed Context. International Journal of Industrial Ergonomics. 2018;68:231–244.

Man Y., Lundh M., and Porathe T. (2014). "Seeking Harmony in Shore-based Unmanned Ship Handling – From the Perspective of Human Factors, What Is the Difference We Need to Focus on from Being Onboard to Onshore?" Proceedings of the Fifth International Conference on Applied Human Factors and Ergonomics AHFE 2014, Kraków, Poland.

Man Y., Lundh M., Porathe T., and MacKinnon S. From Desk to Field – Human Factor Issues in Remote Monitoring and Controlling of Autonomous Unmanned Vessels. Procedia Manufacturing. 2015;3:2674–2681.

Marine Insurance (1906). Act. Available from: http://www.legislation.gov.uk/ukpga/Edw7/6/41/contents.

MASS UK Code of Practice Version 2 (2018). The Society of Maritime Industries on Behalf of Maritime UK.

McCallum M. C., Raby M., and Rothblum A. M. (1996). Procedures for Investigating and Reporting Human Factors and Fatigue Contributions to Marine Casualties. Washington, DC: U.S. Dept. of Transportation, U.S. Coast Guard Report No. CG-D-09-97. AD-A323392.

Mindel D. Our Robots Ourselves, Robotics, and the Myths of Autonomy [Internet]. Amazon.com. 2015 [cited 28 September 2019]. Available from: https://www.amazon.com/Our-Robots-Ourselves-Robotics-Autonomy-ebook/dp/B00SI02AWK.

MUNIN (2015c). D8.8: Final Report: Shore Control Centre. Available from: http://www.unmanned-ship.org/munin/wp-content/uploads/2015/09/MUNIN-D8-8-Final-Report-Shore-Control-Centre-CTH-final.pdf.

MUNIN (2016). MUNIN Report Summary, European Commission. Available from: http://cordis.europa.eu/result/rcn/181600_en.html.

Noma T. Existing conventions and unmanned ships – need for changes? [MSc]. World Maritime University; 2016.

Quinisio B. Development of a strategy for management of autonomous ships by coastal states. World Maritime University Dissertations. 602; 2018. Available from: https://commons.wmu.se/all_dissertations/602.

Parasuraman R. and Riley V. Humans and Automation: Use, Misuse, Disuse, Abuse. Human Factors: The Journal of the Human Factors and Ergonomics Society. 1997;39:230–253. Available from: http://www.ise.ncsu.edu/nsf_itr/794B/papers/Parasuraman_Riley_1997_HF.pdf.

Porathe T., Prison J., and Man Y. (2014). "Situation Awareness in Remote Control Centres for Unmanned Ships." Proceedings of Human Factors in Ship Design and Operation, The Royal Institution of Naval Architects, London, UK, February 26–27.

Reilly G. and Jorgensen J. (2016). "Classification Considerations for Cyber Safety and Security in the Smart Ship Era." Proceedings of the International Smart Ships Technology Conference, London, UK, January 26–27.

Rødseth H. and Brage Mo B. (2014). MUNIN Deliverable D 6.7: Maintenance Indicators and Maintenance Management Principles for Autonomous Engine Room.

Rodseth H. Risk Assessment for an Unmanned Merchant Ship. TransNav, The International Journal on Marine Navigation and Safety of Sea Transportation. 2015;9(3).

Rothblum A. Human Error and Marine Safety [Internet]. 2000. Available from: http://bowles-langley.com/wp content/files_mf/humanerrorandmarinesafety 26.pdf. [Accessed 6 September 2019].

Van Hooydonk E. The Law of Unmanned Merchant Shipping – An Exploration. The Journal of International Maritime Law. 2014;20(3):403–423. Available from: http://www.ericvanhooydonk.be/media/54f3185ce9304.pdf.

Young M. S., Brookhuis K. A., Wickens C. D., and Hancock P. A. State of Science: Mental Workload in Ergonomics. Ergonomics. 2015;58(1):1–17. DOI: 10.1080/00140139.2014.956151.

Index

www.ingramcontent.com/pod-product-compliance
Lightning Source LLC
LaVergne TN
LVHW011556200926
848484LV00001B/323